Systems & Control: Foundations & Applications

Series Editor

Christopher I. Byrnes, Washington University

Associate Editors

S.-I. Amari, University of Tokyo
B.D.O. Anderson, Australian National University, Canberra
Karl Johan Äström, Lund Institute of Technology, Sweden
Jean-Pierre Aubin, EDOMADE, Paris
H.T. Banks, North Carolina State University, Raleigh
John S. Baras, University of Maryland, College Park
A. Bensoussan, INRIA, Paris
John Burns, Virginia Polytechnic Institute, Blacksburg
Han-Fu Chen, Academia Sinica, Beijing
M.H.A. Davis, Imperial College of Science and Technology, London
Wendell Fleming, Brown University, Providence, Rhode Island
Michel Fliess, CNRS-ESE, Gif-sur-Yvette, France
Keith Glover, University of Cambridge, England
Diederich Hinrichsen, University of Bremen, Germany
Alberto Isidori, University of Rome
B. Jakubczyk, Polish Academy of Sciences, Warsaw
Hidenori Kimura, University of Osaka
Arthur J. Krener, University of California, Davis
H. Kunita, Kyushu University, Japan
Alexander Kurzhansky, Russian Academy of Sciences, Moscow
Harold J. Kushner, Brown University, Providence, Rhode Island
Anders Lindquist, Royal Institute of Technology, Stockholm
Andrzej Manitius, George Mason University, Fairfax, Virginia
Clyde F. Martin, Texas Tech University, Lubbock, Texas
Sanjoy K. Mitter, Massachusetts Institute of Technology, Cambridge
Giorgio Picci, University of Padova, Italy
Boris Pshenichnyj, Glushkov Institute of Cybernetics, Kiev
H.J. Sussman, Rutgers University, New Brunswick, New Jersey
T.J. Tarn, Washington University, St. Louis, Missouri
V.M. Tikhomirov, Institute for Problems in Mechanics, Moscow
Pravin P. Varaiya, University of California, Berkeley
Jan C. Willems, University of Grönigen, The Netherlands
W.M. Wonham, University of Toronto

Jerzy Zabczyk

Mathematical Control Theory:
An Introduction

Birkhäuser
Boston · Basel · Berlin

QA
402.3
.Z315
1992

Jerzy Zabczyk
Institute of Mathematics
Polish Academy of Sciences
00-950 Warsaw
Poland

Library of Congress Cataloging-in-Publication Data

Zabczyk, Jerzy.
    Mathematical control theory : an introduction / Jerzy Zabczyk.
       p.   cm. -- (Systems and control)
    Includes bibliographical references and index.
    ISBN 0-8176-3645-5 (alk. paper) . -- ISBN 3-7643-3645-5 (alk. paper)
    1. Control theory. I. Title. II. Series: Systems & control.
QA402.3.Z315 1992   92-12101
515' .64--dc20   CIP

Printed on acid-free paper

© Birkhäuser Boston 1992

Copyright is not claimed for works of U.S. Government employees.
All rights reserved. No part of this publication may be reproduced, stored in a retrieval system, or transmitted, in any form or by any means, electronic, mechanical, photocopying, recording, or otherwise, without prior permission of the copyright owner.

Permission to photocopy for internal or personal use of specific clients is granted by Birkhäuser Boston for libraries and other users registered with the Copyright Clearance Center (CCC), provided that the base fee of $5.00 per copy, plus $0.20 per page is paid directly to CCC, 21 Congress Street, Salem, MA 01970, U.S.A. Special requests should be addressed directly to Birkhäuser Boston, 675 Massachusetts Avenue, Cambridge, MA 02139, U.S.A.

ISBN 0-8176-3645-5
ISBN 3-7643-3645-5

Camera-ready copy prepared by the Authors in TeX.
Printed and bound by Quinn-Woodbine, Woodbine, NJ.
Printed in the U.S.A.

9 8 7 6 5 4 3 2 1

CONTENTS

Preface

Introduction 1

§ 0.1. Problems of mathematical control theory 1
§ 0.2. Specific models 3
    Bibliographical notes 8

PART I. Elements of classical control theory 10

Chapter 1. Controllability and observability 10

§ 1.1. Linear differential equations 10
§ 1.2. The controllability matrix 14
§ 1.3. Rank condition 17
§ 1.4. A classification of control systems 21
§ 1.5. Kalman decomposition 23
§ 1.6. Observability 25
    Bibliographical notes 27

Chapter 2. Stability and stabilizability 28

§ 2.1. Stable linear systems 28
§ 2.2. Stable polynomials 32
§ 2.3. The Routh theorem 34
§ 2.4. Stability, observability, and Liapunov equation 40
§ 2.5. Stabilizability and controllability 43
§ 2.6. Detectability and dynamical observers 46
    Bibliographical notes 49

Chapter 3. Realization theory 50

§ 3.1. Impulse response and transfer functions 50
§ 3.2. Realizations of the impulse response function 54
§ 3.3. The characterization of transfer functions 60
    Bibliographical notes 61

Chapter 4. Systems with constraints 62

§ 4.1. Bounded sets of parameters 62
§ 4.2. Positive systems 64
Bibliographical notes 72

PART II. Nonlinear control systems 73

Chapter 1. Controllability and observability of nonlinear systems . 73

§ 1.1. Nonlinear differential equations 73
§ 1.2. Controllability and linearization 77
§ 1.3. Lie brackets 81
§ 1.4. The openness of attainable sets 84
§ 1.5. Observability 88
Bibliographical notes 91

Chapter 2. Stability and stabilizability 92

§ 2.1. Differential inequalities 92
§ 2.2. The main stability test 95
§ 2.3. Linearization 100
§ 2.4. The Liapunov function method 103
§ 2.5. La Salle's theorem 105
§ 2.6. Topological stability criteria 108
§ 2.7. Exponential stabilizability and the robustness problem . . 112
§ 2.8. Necessary conditions for stabilizability 116
§ 2.9. Stabilization of the Euler equations 117
Bibliographical notes 120

Chapter 3. Realization theory 121

§ 3.1. Input-output maps 121
§ 3.2. Partial realizations 122
Bibliographical notes 126

PART III. Optimal control 127

Chapter 1. Dynamic programming 127

§ 1.1. Introductory comments 127
§ 1.2. Bellman's equation and the value function 128
§ 1.3. The linear regulator problem and the Riccati equation . . 133
§ 1.4. The linear regulator and stabilization 136
Bibliographical notes 141

Chapter 2. Dynamic programming for impulse control 142

§ 2.1. Impulse control problems 142
§ 2.2. An optimal stopping problem 144
§ 2.3. Iterations of convex mappings 145
§ 2.4. The proof of Theorem 2.1 146
   Bibliographical notes 151

Chapter 3. The maximum principle 152

§ 3.1. Control problems with fixed terminal time 152
§ 3.2. An application of the maximum principle 155
§ 3.3. The maximum principle for impulse control problems . . . 157
§ 3.4. Separation theorems 162
§ 3.5. Time-optimal problems 164
   Bibliographical notes 169

Chapter 4. The existence of optimal strategies 170

§ 4.1. A control problem without an optimal solution 170
§ 4.2. Fillipov's theorem 171
   Bibliographical notes 175

PART IV. Infinite dimensional linear systems 176

Chapter 1. Linear control systems 176

§ 1.1. Introduction 176
§ 1.2. Semigroups of operators 177
§ 1.3. The Hille–Yosida theorem 185
§ 1.4. Phillips' theorem 188
§ 1.5. Important classes of generators and Lions' theorem . . . 190
§ 1.6. Specific examples of generators 194
§ 1.7. The integral representation of linear systems 202
   Bibliographical notes 205

Chapter 2. Controllability 206

§ 2.1. Images and kernels of linear operators 206
§ 2.2. The controllability operator 209
§ 2.3. Various concepts of controllability 212
§ 2.4. Systems with self-adjoint generators 213
§ 2.5. Controllability of the wave equation 218
   Bibliographical notes 220

Chapter 3. Stability and stabilizability 221

§ 3.1. Various concepts of stability 221
§ 3.2. Liapunov's equation 226
§ 3.3. Stabilizability and controllability 227
        Bibliographical notes 231

Chapter 4. Linear regulators in Hilbert spaces 232

§ 4.1. Introduction 232
§ 4.2. The operator Riccati equation 234
§ 4.3. The finite horizon case 236
§ 4.4. The infinite horizon case: Stabilizability and detectability . 240
        Bibliographical notes 243

Appendix 244

§ A.1. Metric spaces 244
§ A.2. Banach spaces 245
§ A.3. Hilbert spaces 247
§ A.4. Bochner's integral 248
§ A.5. Spaces of continuous functions 250
§ A.6. Spaces of measurable functions 250

References 252

Notations 256

Index 257

PREFACE

Control theory originated around 150 years ago when the performance of mechanical governors started to be analysed in a mathematical way. Such governors act in a stable way if all the roots of some associated polynomials are contained in the left half of the complex plane. One of the most outstanding results of the early period of control theory was the Routh algorithm, which allowed one to check whether a given polynomial had this property. Questions of stability are present in control theory today, and, in addition, to technical applications, new ones of economical and biological nature have been added. Control theory has been strongly linked with mathematics since World War II. It has had considerable influence on the calculus of variations, the theory of differential equations and the theory of stochastic processes.

The aim of *Mathematical Control Theory* is to give a self-contained outline of mathematical control theory. The work consciously concentrates on typical and characteristic results, presented in four parts preceded by an introduction. The introduction surveys basic concepts and questions of the theory and describes typical, motivating examples.

Part I is devoted to structural properties of linear systems. It contains basic results on controllability, observability, stability and stabilizability. A separate chapter covers realization theory. Toward the end more special topics are treated: linear systems with bounded sets of control parameters and the so-called positive systems.

Structural properties of nonlinear systems are the content of Part II, which is similar in setting to Part I. It starts from an analysis of controllability and observability and then discusses in great detail stability and stabilizability. It also presents typical theorems on nonlinear realizations.

Part III concentrates on the question of how to find optimal controls. It discusses Bellman's optimality principle and its typical applications to the linear regulator problem and to impulse control. It gives a proof of Pontriagin's maximum principle for classical problems with fixed control intervals as well as for time-optimal and impulse control problems. Existence problems are considered in the final chapters, which also contain the basic Fillipov theorem.

Part IV is devoted to infinite dimensional systems. The course is limited to linear systems and to the so-called semigroup approach. The first chapter treats linear systems without control and is, in a sense, a concise presentation of the theory of semigroups of linear operators. The following two chapters concentrate on controllability, stability and stabilizability of linear systems and the final one on the linear regulator problem in Hilbert spaces.

Besides classical topics the book also discusses less traditional ones. In particular great attention is paid to realization theory and to geometrical methods of analysis of controllability, observability and stabilizability of linear and nonlinear systems. One can find here recent results on positive, impulsive and infinite dimensional systems. To preserve some uniformity of style discrete systems as well as stochastic ones have not been included. This was a conscious compromise. Each would be worthy of a separate book.

Control theory is today a separate branch of mathematics, and each of the topics covered in this book has an extensive literature. Therefore the book is only an introduction to control theory.

Knowledge of basic facts from linear algebra, differential equations and calculus is required. Only the final part of the book assumes familiarity with more advanced mathematics.

Several unclear passages and mistakes have been corrected due to remarks of Professor W. Szlenk. The presentation of realization theory owes much to discussions with Professor B. Jakubczyk. I thank both of them very much for their help.

Finally some comments about the arrangement of the material. Successive numbers of paragraphs, theorems, lemmas, formulae, examples, exercises are preceded by the number of the chapter. When referring to a paragraph from some other part of the book the Roman number of the part is added. Numbers of paragraphs, formulae and examples from the Introduction are preceded by 0 and those from the Appendix by the letter A.

Jerzy Zabczyk

INTRODUCTION

In the first part of the Introduction, in a schematic way, basic questions of control theory are formulated. The second part describes several specific models of control systems giving physical interpretation of the introduced parameters. The second part is not necessary for the understanding of the following considerations, although mathematical versions of the discussed models will often appear.

§0.1. Problems of mathematical control theory

A departure point of control theory is the differential equation

$$\dot{y} = f(y, u), \quad y(0) = x \in \mathbb{R}^n, \tag{0.1}$$

with the right hand side depending on a parameter u from a set $U \subset \mathbb{R}^m$. The set U is called *the set of control parameters*. Differential equations depending on a parameter have been objects of the theory of differential equations for a long time. In particular an important question of continuous dependence of the solutions on parameters has been asked and answered under appropriate conditions. Problems studied in mathematical control theory are, however, of different nature, and a basic role in their formulation is played by the concept of *control*. One distinguishes controls of two types: *open* and *closed loop*. An *open loop control* can be basically an arbitrary function $u(\cdot) \colon [0, +\infty) \longrightarrow U$, for which the equation

$$\dot{y}(t) = f(y(t)), u(t)), \quad t \geq 0, \ y(0) = x, \tag{0.2}$$

has a well defined solution.

A *closed loop control* can be identified with a mapping $k \colon \mathbb{R}^n \longrightarrow U$, which may depend on $t \geq 0$, such that the equation

$$\dot{y}(t) = f(y(t), k(y(t))), \quad t \geq 0, \ y(0) = x, \tag{0.3}$$

has a well defined solution. The mapping $k(\cdot)$ is called *feedback*. Controls are called also *strategies* or *inputs*, and the corresponding solutions of (0.2) or (0.3) are *outputs* of the system.

One of the main aims of control theory is to find a strategy such that the corresponding output has desired properties. Depending on the properties involved one gets more specific questions.

Controllability. One says that a state $z \in \mathbf{R}^n$ is *reachable* from x in time T, if there exists an open loop control $u(\,\cdot\,)$ such that, for the output $y(\,\cdot\,)$, $y(0) = x$, $y(T) = z$. If an arbitrary state z is reachable from an arbitrary state x in a time T, then the system (0.1) is said to be *controllable*. In several situations one requires a weaker property of transfering an arbitrary state into a given one, in particular into the origin. A formulation of effective characterizations of controllable systems is an important task of control theory only partially solved.

Stabilizability. An equally important issue is that of stabilizability. Assume that for some $\bar{x} \in \mathbf{R}^n$ and $\bar{u} \in U$, $f(\bar{x}, \bar{u}) = 0$. A function $k\colon \mathbf{R}^n \longrightarrow U$, such that $k(\bar{x}) = \bar{u}$, is called a *stabilizing feedback* if $\bar{x}$ is a stable equilibrium for the system

$$\dot{y}(t) = f(y(t), k(y(t))), \quad t \geq 0, \ y(0) = x. \tag{0.4}$$

In the theory of differential equations there exist several methods to determine whether a given equilibrium state is a stable one. The question of whether, in the class of all equations of the form (0.4), there exists one for which $\bar{x}$ is a stable equilibrium is of a new qualitative type.

Observability. In many situations of practical interest one observes not the state $y(t)$ but its function $h(y(t))$, $t \geq 0$. It is therefore often necessary to investigate the pair of equations

$$\dot{y} = f(y, u), \quad y(0) = x, \tag{0.5}$$
$$w = h(y). \tag{0.6}$$

Relation (0.6) is called an *observation equation*. The system (0.5)-(0.6) is said to be *observable* if, knowing a control $u(\,\cdot\,)$ and an observation $w(\,\cdot\,)$, on a given interval $[0, T]$, one can determine uniquely the initial condition x.

Stabilizability of partially observable systems. The constraint that one can use only a partial observation w complicates considerably the stabilizability problem. Stabilizing feedback should be a function of the observation only, and therefore it should be "factorized" by the function $h(\,\cdot\,)$. This way one is led to a closed loop system of the form

$$\dot{y} = f(y, k(h(y))), \quad y(0) = x. \tag{0.7}$$

There exists no satisfactory theory which allows one to determine when there exists a function $k(\,\cdot\,)$ such that a given $\bar{x}$ is a stable equilibrium for (0.7).

Realization. In connection with the full system (0.5)-(0.6) one poses the problem of realization.

For a given initial condition $x \in \mathbf{R}^n$, system (0.5)-(0.6) defines a mapping which transforms open loop controls $u(\cdot)$ onto outputs given by (0.6): $w(t) = h(y(t))$, $t \in [0,T]$. Denote this transformation by $\mathcal{R}$. What are its properties? What conditions should a transformation $\mathcal{R}$ satisfy to be given by a system of the type (0.5)-(0.6)? How, among all the possible "realizations" (0.5)-(0.6) of a transformation $\mathcal{R}$, do we find the simplest one? The transformation $\mathcal{R}$ is called an *input–output* map of the system (0.5)-(0.6).

Optimality. Besides the above problems of structural character, in control theory, with at least the same intensity, one asks optimality questions. In the so-called time-optimal problem one is looking for a control which not only transfers a state x onto z but does it in the minimal time T. In other situations the time $T > 0$ is fixed and one is looking for a control $u(\cdot)$ which minimizes the integral

$$\int_0^T g(y(t), u(t))\, dt + G(y(T)),$$

in which g and G are given functions. A related class of problems consists of optimal impulse control. They require however a modified concept of strategy.

Systems on manifolds. Difficulties of a different nature arise if the state space is not $\mathbf{R}^n$ or an open subset of $\mathbf{R}^n$ but a differential manifold. This is particularly so if one is interested in the global properties of a control system. The language and methods of differential geometry in control theory are starting to play a role similar to the one they used to play in classical mechanics.

Infinite dimensional systems. The problems mentioned above problems do not lose their meanings if, instead of ordinary differential equations, one takes, as a description of a model, a partial differential equation of parabolic or hyperbolic type. The methods of solutions, however become, much more complicated.

§ 0.2. Specific models

The aim of the examples introduced in this paragraph is to show that the models and problems discussed in control theory have an immediate real meaning.

Example 0.1. *Electrically heated oven.* Let us consider a simple model of an electrically heated oven, which consists of a jacket with a coil directly heating the jacket and of an interior part. Let T_0 denote the outside temperature. We make a simplifying assumption, that at an arbitrary moment

$t \geq 0$, temperatures in the jacket and in the interior part are uniformly distributed and equal to $T_1(t), T_2(t)$. We assume also that the flow of heat through a surface is proportional to the area of the surface and to the difference of temperature between the separated media. Let $u(t)$ be the intensity of the heat input produced by the coil at moment $t \geq 0$. Let moreover a_1, a_2 denote the area of exterior and interior surfaces of the jacket, c_1, c_2 denote heat capacities of the jacket and the interior of the oven and r_1, r_2 denote radiation coefficients of the exterior and interior surfaces of the jacket. An increase of heat in the jacket is equal to the amount of heat produced by the coil reduced by the amount of heat which entered the interior and exterior of the oven. Therefore, for the interval $[t, t + \Delta t]$, we have the following balance:

$$c_1(T_1(t+\Delta t)-T_1(t)) \approx u(t)\Delta t - (T_1(t)-T_2(t))a_1 r_1 \Delta t - (T_1(t)-T_0)a_2 r_2 \Delta t.$$

Similarly, an increase of heat in the interior of the oven is equal to the amount of heat radiated by the jacket:

$$c_2(T_2(t+\Delta t) - T_2(t)) = (T_1(t) - T_2(t))a_1 r_2 \Delta t.$$

Dividing the obtained identities by Δt and taking the limit, as $\Delta t \downarrow 0$, we obtain

$$c_1 \frac{dT_1}{dt} = u - (T_1 - T_2)a_1 r_1 - (T_1 - T_0)a_2 r_2,$$
$$c_2 \frac{dT_2}{dt} = (T_1 - T_2)a_1 r_1.$$

Let us remark that, according to the physical interpretation, $u(t) \geq 0$ for $t \geq 0$. Introducing new variables $x_1 = T_1 - T_0$ and $x_2 = T_2 - T_0$, we have

$$\frac{d}{dt}\begin{bmatrix} x_1 \\ x_2 \end{bmatrix} = \begin{bmatrix} -\frac{r_1 a_1 + r_2 a_2}{c_1} & \frac{r_1 a_1}{c_1} \\ \frac{r_1 a_1}{c_2} & -\frac{r_1 a_1}{c_2} \end{bmatrix} \begin{bmatrix} x_1 \\ x_2 \end{bmatrix} + \begin{bmatrix} c_1^{-1} \\ 0 \end{bmatrix} u.$$

It is natural to limit the considerations to the case when $x_1(0) \geq 0$ and $x_2(0) \geq 0$. It is physically obvious that if $u(t) \geq 0$ for $t \geq 0$, then also $x_1(t) \geq 0$, $x_2(t) \geq 0$, $t \geq 0$. One can prove this mathematically; see § I.4.2.

Let us assume that we want to obtain, in the interior part of the oven, a temperature T and keep it at this level infinitely long. Is this possible? Does the answer depend on initial temperatures $T_1 \geq T_0$, $T_2 \geq T_0$?

The obtained model is a typical example of a positive control system discussed in detail in § I.4.2.

Example 0.2. *Rigid body.* When studying the motion of a spacecraft it as convenient to treat it as a rigid body rotating around a fixed point O.

§ 0.2. Specific models

Let us regard this point as the origin of an inertial system. In the inertial system the motion is described by

$$\dot{H} = F,$$

where H denotes the total angular momentum and F the torque produced by $2r$ gas jets located symetrically with respect to O. The torque F is equal to $u_1 b_1 + \ldots + u_r b_r$, where $b_1, \ldots, b_r$ are vectors fixed to the spacecraft and $u_1, \ldots, u_r$ — are thrusts of the jets.

Let $\{e_1, e_2, e_3\}$ and $\{r_1, r_2, r_3\}$ be orthonormal bases of the inertial system and the rotating one. There exists exactly one matrix R such that $r_i = Re_i$, $i = 1, 2, 3$. It determines completely the position of the body. Let Ω be the angular velocity measured in the inertial system, given in the moving system by the formula $\omega = R\Omega$. It is not difficult to show (see [2]) that

$$\dot{R} = SR, \quad \text{where} \quad S = \begin{bmatrix} 0 & \omega_3 & -\omega_2 \\ -\omega_3 & 0 & \omega_1 \\ \omega_2 & -\omega_1 & 0 \end{bmatrix}.$$

Let J be the inertia matrix. Then the total angular momentum H is given by $H = R^{-1}J$. Inserting this expresion into the equation of motion we obtain

$$\frac{d}{dt}(R^{-1}J\omega) = R^{-1}\left(\sum_{i=1}^{r} u_i b_i\right).$$

After an elementary transformation we arrive at the Euler equation

$$J\dot{\omega} = SJ\omega + \sum_{i=1}^{r} u_i b_i.$$

This equation together with

$$\dot{R} = SR$$

characterizes completely the motion of the rotating object.

From a practical point of view the following questions are of interest:

Is it possible to transfer a given pair (R_0, ω_0) to a desirable pair $(R_1, 0)$, using some steering functions $u_1(\cdot), \ldots, u_r(\cdot)$?

Can one find feedback controls which slow down the rotational movement independently on the initial angular velocity?

While the former question was concerned with controllability, the latter one was about stabilizability.

Example 0.3. *Watt's governor.* Let J be the moment of inertia of the flywheel of a steam engine with the rotational speed $\omega(t)$, $t \geq 0$. Then

$$J\dot{\omega} = u - p,$$

where u and p are torques produced by the action of steam and the weight of the cage respectively. This is the basic movement equation, and Watt's governor is a technical implementation of a feedback whose aim is a stable action of the steam engine. It consists of two arms with weights m at their ends, fixed symetrically to a vertical rod, rotating with speed l, where l is a positive number. Let φ denote the angle between the arms and the rod, equal to the angle between the bars supporting the arms and attached to the sleeve moving along the rod and to the rod itself. The position of the sleeve determines the intensity of the steam input. If b is the coefficient of the frictional force in the hinge joints and g the gravitational acceleration, then (see [45])

$$m\ddot{\varphi} = ml^2\omega^2 \sin\varphi\cos\varphi - mg\sin\varphi - b\dot{\varphi}.$$

One is looking for a feedback of the form

$$u = \bar{u} + k(\cos\varphi - \cos\overline{\varphi}),$$

where $\bar{u}, \overline{\varphi}$ and k are fixed numbers. Introducing a new variable $\psi = \dot{\varphi}$, one obtains a system of three equations

$$\dot{\varphi} = \psi,$$
$$\dot{\psi} = l^2\omega^2 \sin\varphi\cos\varphi - g\sin\varphi - \frac{b}{m}\psi,$$
$$\dot{\omega} = k\cos\varphi + (\bar{u} - p - k\cos\overline{\varphi}).$$

Assume that p is constant and let $(\varphi_0, \psi_0, \omega_0)^*$ be an equilibrium state of the above system. For what parameters of the system is the equilibrium position stable for values p from a given interval? This question has already been asked by Maxwell. An answer is provided by stability theory.

Example 0.4. *Electrical filter.* An electrical filter consists of a capacitor with capacity C, a resistor with resistance R, two inductors with inductance L and a voltage source [45]. Let $U(t)$ be the voltage of the input and $I(t)$ the current in one of the inductors at a moment $t \in \mathbf{R}$. From Kirchoff's law

$$L^2C\frac{d^3I}{dt^3} + RLC\frac{d^2I}{dt^2} + 2L\frac{dI}{dt} + RI = U.$$

The basic issue here is a relation between the voltage $U(\,\cdot\,)$ and current $I(\,\cdot\,)$. They can be regarded as the control and the output of the system.

Let us assume that the voltage $U(\,\cdot\,)$ is a periodic function of a period ω. Is it true that the current $I(\,\cdot\,)$ is also periodic? Let $\alpha(\omega)$ be the amplitude of $I(\,\cdot\,)$. For what range of ω is the amplitude $\alpha(\omega)$ very large or conversely very close to zero? In the former case we deal with the amplification of the frequences ω and in the latter one the frequences ω are "filtered" out.

Example 0.5. *Soft landing.* Let us consider a spacecraft of total mass M moving vertically with the gas thruster directed toward the landing surface. Let h be the height of the spacecraft above the surface, u the thrust of its engine produced by the expulsion of gas from the jet. The gas is a product of the combustion of the fuel. The combustion decreases the total mass of the spacecraft, and the thrust u is proportional to the speed with which the mass decreases. Assuming that there is no atmosphere above the surface and that g is gravitational acceleration, one arrives at the following equations [26]:

$$M\ddot{h} = -gM + u, \qquad (0.8)$$
$$\dot{M} = -ku, \qquad (0.9)$$

with the initial conditions $M(0) = M_0$, $h(0) = h_0$, $\dot{h}(0) = h_1$; k a positive constant. One imposes additional constraints on the control parameter of the type $0 \leq u \leq \alpha$ and $M \geq m$, where m is the mass of the spacecraft without fuel. Let us fix $T > 0$. The soft landing problem consists of finding a control $u(\,\cdot\,)$ such that for the solutions $M(\,\cdot\,)$, $h(\,\cdot\,)$ of equation (0.8)

$$M(t) \geq m, \quad h(t) \geq 0, \quad t \in [0,T], \quad \text{and} \quad h(T) = \dot{h}(T) = 0.$$

The problem of the existence of such a control is equivalent to the controllability of the system (0.8)-(0.9).

A natural optimization question arises when the moment T is not fixed and one is minimizing the landing time. The latter problem can be formulated equivalently as the *minimum fuel problem*. In fact, let $v = \dot{h}$ denote the velocity of the spacecraft, and let $M(t) > 0$ for $t \in [0,T]$. Then

$$\frac{\dot{M}(t)}{M(t)} = -k\dot{v}(t) - gk, \quad t \in [0,T].$$

Therefore, after integration,

$$M(T) = e^{-v(T)k - gkT + v(0)k} M(0).$$

Thus a soft landing is taking place at a moment $T > 0$ ($v(T) = 0$) if and only if

$$M(T) = e^{-gkT} e^{v(0)k} M(0).$$

Consequently, the minimization of the landing time T is equivalent to the minimization of the amount of fuel $M(0) - M(T)$ needed for landing.

Example 0.6. *Optimal consumption.* The capital $y(t) \geq 0$ of an economy at any moment t is divided into two parts: $u(t)y(t)$ and $(1-u(t))y(t)$, where $u(t)$ is a number from the interval $[0,1]$. The first part goes for investments and contributes to the increase in capital according to the formula

$$\dot{y} = uy, \quad y(0) = x > 0.$$

The remaining part is for consumption evaluated by the *satisfaction*

$$J_T(x, u(\,\cdot\,)) = \int_0^T ((1-u(t))y(t))^\alpha \, dt + ay^\alpha(T). \qquad (0.10)$$

In definition (0.10), the number a is nonnegative and $\alpha \in (0,1)$. In the described situation one is trying to divide the capital to maximize the satisfaction.

Example 0.7. *Heating a rod.* To fix ideas let us denote by $y(t,\xi)$ the temperature of a rod at the moment $t \geq 0$ and at the point $\xi \in [0,L]$, heated with the intensity $u(t)b(\xi)$, where the value $u(t)$ can be arbitrarily chosen: $u(t) \in \mathbf{R}$. Assuming additionally that the ends of the rods are insulated: $u(t,0) = u(t,L)$, $t \geq 0$, and applying elementary laws of the heat conduction, one arrives at the following parabolic equation describing the evolution of the temperature in the rod:

$$\frac{\partial y}{\partial t}(t,\xi) = \sigma^2 \frac{\partial^2 y}{\partial \xi^2}(t,\xi) + u(t)b(\xi), \quad t > 0, \ \xi \in (0,L), \qquad (0.11)$$

$$y(t,0) = y(t,L) = 0, \quad t > 0, \qquad (0.12)$$

$$y(0,\xi) = x(\xi), \quad \xi \in [0,L]. \qquad (0.13)$$

The parameter σ in (0.11) is the heat capacity of the rod, and the function $x(\,\cdot\,)$ is the initial distribution of the temperature.

Let us assume that $\hat{x}(\xi)$, $\xi \in [0,L]$ is the required distribution of the temperature and $T > 0$ a fixed moment. The question, whether one can heat the rod in such a way to obtain $y(T,\xi) = \hat{x}(\xi)$, $\xi \in [0,L]$, is identical to asking whether, for system (0.11)-(0.12), the initial state $x(\,\cdot\,)$ can be transferred onto $\hat{x}(\,\cdot\,)$ in time T. A practical situation may impose additional constraints on the control parameter u of the type $u \in [0,\bar{u}]$. Under such a constraint, the problem of finding control transferring $x(\,\cdot\,)$ and $\hat{x}(\,\cdot\,)$ in minimal time is mathematically well posed and has a clear physical meaning.

Bibliographical notes

In the development of mathematical control theory the following works played an important rôle: J.C. Maxwell, *On governors* [39], N. Wiener, *Cybernetics or control and communication in the animal and the machine* [58], R. Bellman, *Dynamic Programming* [5], L.S. Pontriagin, W.G. Boltianski, R.W. Gamkrelidze and E.F. Miszczenko, *Matematičeskaja teorija optymal'nych processow* [45], R.E. Kalman, *On the general theory of control systems* [33], T. Ważewski, *Systèmes de commande et équations au contingent* [57], J.L. Lions, *Contrôle optimale de systèmes par des équations aux dérivées partielles* [38], W.M. Wonham, *Linear multivariable control: A geometric approach* [61].

The model of an electrically heated oven is borrowed from [4]. The dynamics of a rigid body is discussed in W.I. Arnold [2].

A detailed derivation of the equations describing Watt's governor can be found in [45]. The electrical filter is a typical example of an automatic system. The soft landing and optimal consumption models are extensively discussed in [27]. Interesting information about heating processes can be found in [50] and [12]. Recently, control applications to biological sciences have been intensively studied. The interested reader is referred to the conference proceedings [16].

PART I

ELEMENTS OF CLASSICAL CONTROL THEORY

Chapter 1
Controllability and observability

In this chapter basic information about linear differential equations are recalled and the main concepts of control theory, controllability and obervability, are studied. Specific formulae for controls transferring one state onto another as well as algebraic characterizations of controllable and observable systems are obtained. A complete classification of controllable systems with one dimensional input is given.

§1.1. Linear differential equations

The basic object of classical control theory is a linear system described by a differential equation

$$\frac{dy}{dt} = Ay(t) + Bu(t), \quad y(0) = x \in \mathbf{R}^n, \tag{1.1}$$

and an observation relation

$$w(t) = Cy(t), \quad t \geq 0. \tag{1.2}$$

Linear transformations $A\colon \mathbf{R}^n \longrightarrow \mathbf{R}^n$, $B\colon \mathbf{R}^m \longrightarrow \mathbf{R}^n$, $C\colon \mathbf{R}^m \longrightarrow \mathbf{R}^k$ in (1.1) and (1.2) will be identified with representing matrices and elements of $\mathbf{R}^n$, $\mathbf{R}^m$, $\mathbf{R}^k$ with one column matrices. The set of all matrices with n rows and m columns will be denoted by $\mathbf{M}(n,m)$ and the identity transformation as well as the identity matrix by I. The scalar product $\langle x, y \rangle$ and the norm $|x|$, of elements $x, y \in \mathbf{R}^n$ with coordinates $\xi_1, \ldots, \xi_n$ and $\eta_1, \ldots, \eta_n$, are defined by

$$\langle x, y \rangle = \sum_{j=1}^n \xi_j \eta_j, \quad |x| = \left(\sum_{j=1}^n \xi_j^2 \right)^{1/2}.$$

§ 1.1. Linear differential equations

The adjoint transformation of a linear transformation A as well as the transpose matrix of A are denoted by A^*. A matrix $A \in \mathbf{M}(n,n)$ is called *symmetric* if $A = A^*$. The set of all symmetric matrices is partially ordered by the relation $A_1 \geq A_2$ if $\langle A_1 x, x \rangle \geq \langle A_2 x, x \rangle$ for arbitrary $x \in \mathbf{R}^n$. If $A \geq 0$ then one says that matrix A is *nonnegative definite* and if, in addition, $\langle Ax, x \rangle > 0$ for $x \neq 0$ that A is *positive definite*. Treating $x \in \mathbf{R}^n$ as an element of $\mathbf{M}(n,1)$ we have $x^* \in \mathbf{M}(1,n)$. In particular we can write $\langle x, y \rangle = x^* y$ and $|x|^2 = x^* x$. The inverse transformation of A and the inverse matrix of A will be denoted by A^{-1}.

If $F(t) = [f_{ij}(t); i = 1,\ldots,n, j = 1,\ldots,m] \in \mathbf{M}(n,m)$, $t \in [0,T]$, then, by definition,

$$\int_0^T F(t)\,dt = \left[\int_0^T f_{ij}(t)\,dt,\ i = 1,\ldots,n;\ j = 1,\ldots,n\right], \tag{1.3}$$

under the condition that elements of $F(\cdot)$ are integrable.

Derivatives of the 1st and 2nd order of a function $y(t)$, $t \in \mathbf{R}$, are denoted by $\frac{dy}{dt}$, $\frac{d^2y}{dt^2}$ or by $\dot{y}$, $\ddot{y}$ and the nth order derivative, by $\frac{d^{(n)}y}{dt^{(n)}}$.

We will need some basic results on linear equations

$$\frac{dq}{dt} = A(t)q(t) + a(t), \quad q(t_0) = q_0 \in \mathbf{R}^n, \tag{1.4}$$

on a fixed interval $[0,T]$; $t_0 \in [0,T]$, where $A(t) \in \mathbf{M}(n,n)$, $A(t) = [a_{ij}(t); i = 1,\ldots,n, j = 1,\ldots,m]$, $a(t) \in \mathbf{R}^n$, $a(t) = (a_i(t); i = 1,\ldots,n)$, $t \in [0,T]$.

Theorem 1.1. *Assume that elements of the function $A(\cdot)$ are locally integrable. Then there exists exactly one function $S(t)$, $t \in [0,T]$ with values in $\mathbf{M}(n,n)$ and with absolutely continuous elements such that*

$$\frac{d}{dt}S(t) = A(t)S(t) \quad \text{for almost all } t \in [0,T], \tag{1.5}$$

$$S(0) = I. \tag{1.6}$$

In addition, a matrix $S(t)$ is invertible for an arbitrary $t \in [0,T]$, and the unique solution of the equation (1.4) is of the form

$$q(t) = S(t)S^{-1}(t_0)q_0 + \int_{t_0}^t S(t)S^{-1}(s)a(s)\,ds, \quad t \in [0,T]. \tag{1.7}$$

Because of its importance we will sketch a proof of the theorem.

Proof. Equation (1.4) is equivalent to the integral equation

$$q(t) = a_0 + \int_{t_0}^t A(s)q(s)\,ds + \int_{t_0}^t a(s)\,ds, \quad t \in [0,T].$$

The formula

$$\mathcal{L}y(t) = a_0 + \int_{t_0}^t a(s)\,ds + \int_{t_0}^t A(s)y(s)\,ds, \quad t \in [0,T],$$

defines a continuous transformation from the space of continuous functions $C[0,T;\mathbf{R}^n]$ into itself, such that for arbitrary $y(\,\cdot\,), \tilde{y}(\,\cdot\,) \in C[0,T;\mathbf{R}^n]$

$$\sup_{t\in[0,T]} |\mathcal{L}y(t) - \mathcal{L}\tilde{y}(t)| \leq \left(\int_0^T |A(s)|\,ds\right) \sup_{t\in[0,T]} |y(t) - \tilde{y}(t)|.$$

If $\int_0^T |A(s)|\,ds < 1$, then by Theorem A.1 (the contraction mapping principle) the equation $q = \mathcal{L}q$ has exactly one solution in $C[0,T;\mathbf{R}^n]$ which is the solution of the integral equation. The case $\int_0^T |A(s)|\,ds \geq 1$ can be reduced to the previous one by considering the equation on appropriately shorter intervals. In particular we obtain the existence and uniqueness of a matrix valued function satifying (1.5) and (1.6).

To prove the second part of the theorem let us denote by $\psi(t)$, $t \in [0,T]$, the matrix solution of

$$\frac{d}{dt}\psi(t) = -\psi(t)A(t), \quad \psi(0) = I,\, t \in [0,T].$$

Assume that, for some $t \in [0,T]$, $\det S(t) = 0$. Let $T_0 = \min\{t \in [0,T];\, \det S(t) = 0\}$. Then $T_0 > 0$, and for $t \in [0, T_0)$

$$0 = \frac{d}{dt}\left(S(t)S^{-1}(t)\right) = \left(\frac{d}{dt}S(t)\right)S^{-1}(t) + S(t)\frac{d}{dt}S^{-1}(t).$$

Thus

$$-A(t) = S(t)\frac{d}{dt}S^{-1}(t),$$

and consequently

$$\frac{d}{dt}S^{-1}(t) = -S^{-1}(t)A(t), \quad t \in [0, T_0),$$

so $S^{-1}(t) = \psi(t)$, $t \in [0, T_0)$.

Since the function $\det \psi(t)$, $t \in [0,T]$, is continuous and

$$\det \psi(t) = \frac{1}{\det S(t)}, \quad t \in [0, T_0),$$

therefore there exists a finite $\lim_{t\uparrow T_0} \det \psi(t)$. This way $\det S(T_0) = \lim_{t\uparrow T_0} S(t) \neq 0$, a contradiction. The validity of (1.6) follows now by elementary calculation. □

§ 1.1. Linear differential equations

The function $S(t)$, $t \in [0, T]$ will be called *the fundamental solution* of equation (1.4). It follows from the proof that the fundamental solution of the "adjoint" equation

$$\frac{dp}{dt} = -A^*(t)p(t), \quad t \in [0, T],$$

is $(S^*(t))^{-1}$, $t \in [0, T]$.

Exercise 1.1. Show that for $A \in \mathbf{M}(n, n)$ the series

$$\sum_{n=1}^{+\infty} \frac{A^n}{n!} t^n, \quad t \in \mathbf{R},$$

is uniformly convergent, with all derivatives, on an arbitrary finite interval.

The sum of the series from Exercise 1.1 is often denoted by $\exp(tA)$ or e^{tA}, $t \in \mathbf{R}$. We check easily that

$$e^{tA} e^{sA} = e^{(t+s)A}, \quad t, s \in \mathbf{R},$$

in particular

$$(e^{tA})^{-1} = e^{-tA}, \quad t \in \mathbf{R}.$$

Therefore the solution of (1.1) has the form

$$y(t) = e^{tA} x + \int_0^t e^{(t-s)A} Bu(s) \, ds \tag{1.8}$$

$$= S(t)x + \int_0^t S(t-s) Bu(s) \, ds, \quad t \in [0, T],$$

where $S(t) = \exp tA$, $t \geq 0$.

The majority of the concepts and results from Part I discussed for systems (1.1)-(1.2) can be extended to time dependent matrices $A(t) \in \mathbf{M}(n,n)$, $B(t) \in \mathbf{M}(n,n)$, $C(t) \in \mathbf{M}(k,n)$, $t \in [0,T]$, and therefore for systems

$$\frac{dy}{dt} = A(t)y(t) + B(t)u(t), \quad y(0) = x \in \mathbf{R}^n, \tag{1.9}$$

$$w(t) = C(t)y(t), \quad t \in [0, T]. \tag{1.10}$$

Some of these extensions will appear in the exercises.

Generalizations to arbitrary finite dimensional spaces of states and control parameters E and U are immediate.

§1.2. The controllability matrix

An arbitrary function $u(\cdot)$ defined on $[0, +\infty)$ locally integrable and with values in $\mathbf{R}^m$ will be called a *control, strategy* or *input* of the system (1.1)-(1.2). The corresponding solution of equation (1.1) will be denoted by $y^{x,u}(\cdot)$, to underline the dependence on the initial condition x and the input $u(\cdot)$. Relationship (1.2) can be written in the following way:

$$w(t) = Cy^{x,u}(t), \quad t \in [0,T].$$

The function $w(\cdot)$ is the *output* of the controlled system.

We will assume now that $C = I$ or equivalently that $w(t) = y^{x,u}(t)$, $t \geq 0$.

We say that a control u *transfers* a state a to a state b at the time $T > 0$ if

$$y^{a,u}(T) = b. \tag{1.11}$$

We then also say that the state a can be *steered* to b at time T or that the state b is *reachable* or *attainable* from a at time T.

The proposition below gives a formula for a control transferring a to b. In this formula the matrix Q_T, called the *controllability matrix* or *controllability Gramian*, appears:

$$Q_T = \int_0^T S(r) BB^* S^*(r) \, dr, \quad T > 0.$$

We check easily that Q_T is symmetric and nonnegative definite (see the beginning of §1.1).

Proposition 1.1. *Assume that for some $T > 0$ the matrix Q_T is nonsingular. Then*

(i) for arbitrary $a, b \in \mathbf{R}^n$ the control

$$\hat{u}(s) = -B^* S^*(T-s) Q_T^{-1}(S(T)a - b), \quad s \in [0,T], \tag{1.12}$$

transfers a to b at time T;

(ii) among all controls $u(\cdot)$ steering a to b at time T the control $\hat{u}$ minimizes the integral $\int_0^T |u(s)|^2 \, ds$. Moreover,

$$\int_0^T |\hat{u}(s)|^2 \, ds = \langle Q_T^{-1}(S(T)a - b), \, S(T)a - b \rangle. \tag{1.13}$$

Proof. It follows from (1.12) that the control $\hat{u}$ is smooth or even analytic. From (1.8) and (1.12) we obtain that

$$y^{a,\hat{u}}(T) = S(T)a - \left(\int_0^T S(T-s) BB^* S^*(T-s) \, ds \right) (Q_T^{-1}(S(T)a - b))$$

$$= S(T)a - Q_T(Q_T^{-1}(S(T)a - b)) = b.$$

§ 1.2. The controllability matrix

This shows (i). To prove (ii) let us remark that the formula (1.13) is a consequence of the following simple calculations:

$$\int_0^T |\hat{u}(s)|^2 \, ds = \int_0^T |B^* S^*(T-s) Q_T^{-1}(S(T)a - b)|^2 \, ds =$$

$$= \langle \int_0^T S(T-s) B B^* S^*(T-s) (Q_T^{-1}(S(T)a - b)) \, ds, \, Q_T^{-1}(S(T)a - b) \rangle$$

$$= \langle Q_T Q_T^{-1}(S(T)a - b), \, Q_T^{-1}(S(T)a - b) \rangle$$

$$= \langle Q_T^{-1}(S(T)a - b), \, S(T)a - b \rangle.$$

Now let $u(\cdot)$ be an arbitrary control transferring a to b at time T. We can assume that $u(\cdot)$ is square integrable on $[0, T]$. Then

$$\int_0^T \langle u(s), \hat{u}(s) \rangle \, ds = -\int_0^T \langle u(s), B^* S^*(T-s) Q_T^{-1}(S(T)b - a) \rangle \, ds$$

$$= -\langle \int_0^T S(T-s) B u(s) \, ds, \, Q_T^{-1}(S(T)a - b) \rangle$$

$$= \langle S(T)a - b, \, Q_T^{-1}(S(T)a - b) \rangle.$$

Hence

$$\int_0^T \langle u(s), \hat{u}(s) \rangle \, ds = \int_0^T \langle \hat{u}(s), \hat{u}(s) \rangle \, ds.$$

From this we obtain that

$$\int_0^T |u(s)|^2 \, ds = \int_0^T |\hat{u}(s)|^2 \, ds + \int_0^T |u(s) - \hat{u}(s)|^2 \, ds$$

and consequently the desired minimality property. □

Exercise 1.2. Write equation

$$\frac{d^2 y}{dt^2} = u, \quad y(0) = \xi_1, \quad \frac{dy}{dt}(0) = \xi_2, \quad \begin{bmatrix} \xi_1 \\ \xi_2 \end{bmatrix} \in \mathbf{R}^2,$$

as a first order system. Prove that for the new system, the matrix Q_T is nonsingular, $T > 0$. Find the control u transferring the state $\begin{bmatrix} \xi_1 \\ \xi_2 \end{bmatrix}$ to $\begin{bmatrix} 0 \\ 0 \end{bmatrix}$ at time $T > 0$ and minimizing the functional $\int_0^T |u(s)|^2 \, ds$. Determine the minimal value m of the functional. Consider $\xi_1 = 1$, $\xi_2 = 0$.

Answer. The required control is of the form

$$\hat{u}(s) = -\frac{12}{T^3} \left(\frac{\xi_1 T}{2} + \frac{\xi_2 T^2}{3} - \frac{s T \xi_2}{2} - s \xi_1 \right), \quad s \in [0\, T],$$

and the minimal value m of the fuctional is equal to

$$m = \frac{12}{T^3}\left((\xi_1)^2 + \xi_1\xi_2 T - \frac{2T^2}{3}(\xi_2)^2\right).$$

In particular, when $\xi_1 = 1$, $\xi_2 = 0$,

$$\hat{u}(s) = \frac{12}{T^3}(s - \frac{T}{2}), \quad s \in [0,T], \quad m = \frac{12}{T^3}.$$

We say that a state b is *attainable* or *reachable* from $a \in \mathbb{R}^n$ if it is attainable or reachable at some time $T > 0$.

System (1.1) is called *controllable* if an arbitrary state $b \in \mathbb{R}^n$ is attainable from any state $a \in \mathbb{R}^n$ at some time $T > 0$. Instead of saying that system (1.1) is controllable we will frequently say that the *pair (A, B) is controllable*.

If for arbitrary $a, b \in \mathbb{R}^n$ the attainablity takes place at a given time $T > 0$, we say that the system is *controllable at time T*. Proposition 1.1 gives a sufficient condition for the system (1.1) to be controllable. It turns out that this condition is also a necessary one.

The following result holds.

Proposition 1.2. *If an arbitrary state $b \in \mathbb{R}^n$ is attainable from 0, then the matrix Q_T is nonsingular for an arbitrary $T > 0$.*

Proof. Let, for a control u and $T > 0$,

$$\mathcal{L}_T u = \int_0^T S(r)Bu(T-r)\, dr. \tag{1.14}$$

The formula (1.14) defines a linear operator from $U_T = L^1[0,T; \mathbb{R}^m]$ into $\mathbb{R}^n$. Let us remark that

$$\mathcal{L}_T u = y^{0,u}(T). \tag{1.15}$$

Let $E_T = \mathcal{L}_T(U_T)$, $T > 0$. It follows from (1.14) that the family of the linear spaces E_T is nondecreasing in $T > 0$. Since $\bigcup_{T>0} E_T = \mathbb{R}^n$, taking into account the dimensions of E_T, we have that $E_{\widetilde{T}} = \mathbb{R}^n$ for some $\widetilde{T}$. Let us remark that, for arbitrary $T > 0$, $v \in \mathbb{R}^n$ and $u \in U_T$,

$$\langle Q_T v, v \rangle = \langle \left(\int_0^T S(r)BB^*S^*(r)\, dr\right)v, v\rangle \tag{1.16}$$

$$= \int_0^T |B^*S^*(r)v|^2\, dr,$$

$$\langle \mathcal{L}_T u, v \rangle = \int_0^T \langle u(r), B^*S^*(T-r)v \rangle\, dr. \tag{1.17}$$

§ 1.3. Rank condition

From identities (1.16) and (1.17) we obtain $Q_T v = 0$ for some $v \in \mathbf{R}^n$ if the space E_T is orthogonal to v or if the function $B^* S^*(\,\cdot\,)v$ is identically equal to zero on $[0,T]$. It follows from the analiticity of this function that it is equal to zero everywhere. Therefore if $Q_T v = 0$ for some $T > 0$ then $Q_T v = 0$ for all $T > 0$ and in particular $Q_{\widetilde{T}} v = 0$. Since $E_{\widetilde{T}} = \mathbf{R}^n$ we have that $v = 0$, and the nonsingularity of Q_T follows. □

A sufficient condition for controllability is that the rank of B is equal to n. This follows from the next exercise.

Exercise 1.3. Assume rank $B = n$ and let B^+ be a matrix such that $BB^+ = I$. Check that the control

$$u(s) = \frac{1}{T} B^+ e^{(s-T)A}(b - e^{TA}a), \quad s \in [0,T],$$

transfers a to b at time $T \geq 0$.

§ 1.3. Rank condition

We now formulate an algebraic condition equivalent to controllability. For matrices $A \in \mathbf{M}(n,n)$, $B \in \mathbf{M}(n,m)$ denote by $[A|B]$ the matrix $[B, AB, \ldots, A^{n-1}B] \in \mathbf{M}(n, nm)$ which consists of consecutively written columns of matrices $B, AB, \ldots, A^{n-1}B$.

Theorem 1.2. *The following conditions are equivalent.*
  (i) *An arbitrary state $b \in \mathbf{R}^n$ is attainable from 0.*
  (ii) *System (1.1) is controllable.*
  (iii) *System (1.1) is controllable at a given time $T > 0$.*
  (iv) *Matrix Q_T is nonsingular for some $T > 0$.*
  (v) *Matrix Q_T is nonsingular for an arbitrary $T > 0$.*
  (vi) *rank $[A|B] = n$.*

Condition (vi) is called the *Kalman rank condition*, or the rank condition for short.

The proof will use the Cayley–Hamilton theorem. Let us recall that a *characteristic polynomial* $p(\,\cdot\,)$ of a matrix $A \in \mathbf{M}(n,n)$ is defined by

$$p(\lambda) = \det(\lambda I - A), \quad \lambda \in \mathbf{C}. \tag{1.18}$$

Let

$$p(\lambda) = \lambda^n + a_1 \lambda^{n-1} + \ldots + a_n, \quad \lambda \in \mathbf{C}. \tag{1.19}$$

The Cayley–Hamilton theorem has the following formulation (see [3, 358-359]):

Theorem 1.3. *For arbitrary $A \in \mathbf{M}(n,n)$, with the characteristic polynomial (1.19),*

$$A^n + a_1 A^{n-1} + \ldots + a_n I = 0.$$

Symbolically, $p(A) = 0$.

Proof of Theorem 1.2. Equivalences (i)-(v) follow from the proofs of Propositions 1.1 and 1.2 and the identity

$$y^{a,u}(T) = \mathcal{L}_T u + S(T)a.$$

To show the equivalences to condition (vi) it is convenient to introduce a linear mapping l_n from the Cartesian product of n copies $\mathbf{R}^m$ into $\mathbf{R}^n$:

$$l_n(u_0, \ldots, u_{n-1}) = \sum_{j=0}^{n-1} A^j B u_j, \quad u_j \in \mathbf{R}^m, \ j = 0, \ldots, n-1.$$

We prove first the following lemma.

Lemma 1.1. *The transformation $\mathcal{L}_T$, $T > 0$, has the same image as l_n. In particular $\mathcal{L}_T$ is onto if and only if l_n is onto.*

Proof. For arbitrary $v \in \mathbf{R}^n$, $u \in L^1[0, T; \mathbf{R}^m]$, $u_j \in \mathbf{R}^m$, $j = 0, \ldots, n-1$:

$$\langle \mathcal{L}_T u, v \rangle = \int_0^T \langle u(s), B^* S^*(T-s)v \rangle \, ds,$$

$$\langle l_n(u_0, \ldots, u_{n-1}), v \rangle = \langle u_0, B^* v \rangle + \ldots + \langle u_{n-1}, B^*(A^*)^{n-1} v \rangle.$$

Suppose that $\langle l_n(u_0, \ldots, u_{n-1}), v \rangle = 0$ for arbitrary $u_0, \ldots, u_{n-1} \in \mathbf{R}^m$. Then $B^* v = 0, \ldots, B^*(A^*)^{n-1} v = 0$. From Theorem 1.3, applied to matrix A^*, it follows that for some constants $c_0, \ldots, c_{n-1}$

$$(A^*)^n = \sum_{k=0}^{n-1} c_k (A^*)^k.$$

Thus, by induction, for abitrary $l = 0, 1, \ldots$ there exist constants $c_{l,0}, \ldots, c_{l,n-1}$ such that

$$(A^*)^{n+l} = \sum_{k=0}^{n-1} c_{l,k} (A^*)^k.$$

Therefore $B^*(A^*)^k v = 0$ for $k = 0, 1, \ldots$. Taking into account that

$$B^* S^*(t) v = \sum_{k=0}^{+\infty} B^*(A^*)^k v \frac{t^k}{k!}, \quad t \geq 0,$$

we deduce that for arbitrary $T > 0$ and $t \in [0, T]$

$$B^* S^*(t) v = 0,$$

so $\langle \mathcal{L}_T u, v \rangle = 0$ for arbitrary $u \in L^1[0,T; \mathbf{R}^m]$.

Assume, conversely, that for arbitrary $u \in L^1[0,T; \mathbf{R}^n]$, $\langle \mathcal{L}_T u, v \rangle = 0$. Then $B^* S^*(t) v = 0$ for $t \in [0,T]$. Differentiating the identity

$$\sum_{k=0}^{+\infty} B^*(A^*)^k v \frac{t^k}{k!} = 0, \quad t \in [0,T],$$

$0, 1, \ldots, (n-1)$-times and inserting each time $t = 0$, we obtain that $B^*(A^*)^k v = 0$ for $k = 0, 1, \ldots, n-1$. And therefore

$$\langle l_n(u_0, \ldots, u_{n-1}), v \rangle = 0 \quad \text{for arbitrary } u_0, \ldots, u_{n-1} \in \mathbf{R}^m.$$

This implies the lemma. □

Assume that the system (1.1) is controllable. Then the transformation $\mathcal{L}_T$ is onto $\mathbf{R}^n$ for arbitrary $T > 0$ and, by the above lemma, the matrix $[A|B]$ has rank n. Conversely, if the rank of $[A|B]$ is n then the mapping l_n is onto $\mathbf{R}^n$ and also, therefore, the transformation $\mathcal{L}_T$ is onto $\mathbf{R}^n$ and the controllability of (1.1) follows. □

If the rank condition is satisfied then the control $\hat{u}(\,\cdot\,)$ given by (1.12) transfers a to b at time T. We now give a different, more explicit, formula for the transfer control involving the matrix $[A|B]$ instead of the controllability matrix Q_T.

Note that if rank $[A|B] = n$ then there exists a matrix $K \in \mathbf{M}(mn, n)$ such that $[A|B]K = I \in \mathbf{M}(n, n)$ or equivalently there exist matrices $K_1, K_2, \ldots, K_n \in \mathbf{M}(m, n)$ such that

$$BK_1 + ABK_2 + \ldots + A^{n-1} B K_n = I. \tag{1.20}$$

Let, in addition, φ be a function of class C^{n-1} from $[0,T]$ into R such that

$$\frac{d^j \varphi}{ds^j}(0) = \frac{d^j \varphi}{ds^j}(T) = 0, \quad j = 0, 1, \ldots, n-1, \tag{1.21}$$

$$\int_0^T \varphi(s) ds = 1 \tag{1.22}$$

Proposition 1.3. *Assume that rank $[A|B] = n$ and (1.20)-(1.22) hold. Then the control*

$$\tilde{u}(s) = K_1 \psi(s) + K_2 \frac{d\psi}{ds}(s) + \ldots + K_n \frac{d^{n-1}\psi}{ds^{n-1}}(s), \quad s \in [0,T]$$

where

$$\psi(s) = S(s-T)(b - S(T)a)\varphi(s), \quad s \in [0,T] \tag{1.23}$$

transfers a to b at time $T \geq 0$.

Proof. Taking into account (1.21) and integrating by parts $(j-1)$ times, we have

$$\int_0^T S(T-s)BK_j \frac{d^{j-1}}{ds^{j-1}}\psi(s)\,ds = \int_0^T e^{A(T-s)}BK_j \frac{d^{j-1}}{ds^{j-1}}\psi(s)\,ds$$

$$= \int_0^T e^{A(T-s)}A^{j-1}BK_j\psi(s)\,ds$$

$$= \int_0^T S(T-s)A^{j-1}BK_j\psi(s)\,ds,$$

$$j = 1, 2, \ldots, n.$$

Consequently

$$\int_0^T S(T-s)B\tilde{u}(s)\,ds = \int_0^T S(t-s)[A|B]K\psi(s)\,ds$$

$$= \int_0^T S(T-s)\psi(s)\,ds.$$

By the definition of ψ and by (1.22) we finally have

$$y^{a,\tilde{u}}(T) = S(T)a + \int_0^T S(T-s)(S(s-T)(b-S(T)a))\varphi(s)\,ds$$

$$= S(T)a + (b - S(T)a)\int_0^T \varphi(s)\,ds$$

$$= b.$$

$\square$

Remark. Note that Proposition 1.3 is a generalization of Exercise 1.3.

Exercise 1.4. Assuming that $U = \mathbf{R}$ prove that the system describing the electrically heated oven from Example 0.1 is controllable.

Exercise 1.5. Let L_0 be a linear subspace dense in $L^1[0,T;\mathbf{R}^m]$. If system (1.1) is controllable then for arbitrary $a, b \in \mathbf{R}^n$ there exists $u(\cdot) \in L_0$ transferring a to b at time T.

Hint. Use the fact that the image of the closure of a set under a linear continuous mapping is contained in the closure of the image of the set.

Exercise 1.6. If system (1.1) is controllable then for arbitrary $T > 0$ and arbitrary $a, b \in \mathbf{R}^n$ there exists a control $u(\cdot)$ of class C^∞ transferring a to b at time T and such that

$$\frac{d^{(j)}u}{dt^{(j)}}(0) = \frac{d^{(j)}u}{dt^{(j)}}(T) = 0 \quad \text{for } j = 0, 1, \ldots.$$

Exercise 1.7. Assuming that the pair (A, B) is controllable, show that the system
$$\dot{y} = Ay + Bv$$
$$\dot{v} = u,$$
with the state space $\mathbf{R}^{n+m}$ and the set of control parameters $\mathbf{R}^m$, is also controllable. Deduce that for arbitrary $a, b \in \mathbf{R}^n$, $u_0, u_1 \in \mathbf{R}^m$ and $T > 0$ there exists a control $u(\,\cdot\,)$ of class C^∞ transferring a to b at time T and such that $u(0) = u_0$, $u(T) = u_1$.

Hint. Use Exercise 1.6 and the Kalman rank condition.

Exercise 1.8. Suppose that $A \in \mathbf{M}(n, n)$, $B \in \mathbf{M}(n, m)$. Prove that the system
$$\frac{d^2 y}{dt^2} = Ay + Bu, \quad y(0) \in \mathbf{R}^n, \quad \frac{dy}{dt}(0) \in \mathbf{R}^n,$$
is controllable in $\mathbf{R}^{2n}$ if and only if the pair (A, B) is controllable.

Exercise 1.9. Consider system (1.9) on $[0, T]$ with integrable matrix-valued functions $A(t)$, $B(t)$, $t \in [0, T]$. Let $S(t)$, $t \in [0, T]$ be the fundamental solution of the equation $\dot{q} = Aq$. Assume that the matrix
$$Q_T = \int_0^T S(T) S^{-1}(s) B(s) B^*(s) (S^{-1}(s))^* S^*(T)\, ds$$
is positive definite. Show that the control
$$\hat{u}(s) = B^*(S^{-1}(s))^* S^*(T) Q_T^{-1}(b - S(T)a), \quad s \in [0, T],$$
transfers a to b at time T minimizing the functional $u \longrightarrow \int_0^T |u(s)|^2\, ds$.

§1.4. A classification of control systems

Let $y(t)$, $t \geq 0$, be a solution of the equation (1.1) corresponding to a control $u(t)$, $t \geq 0$, and let $P \in \mathbf{M}(n, n)$ and $S \in \mathbf{M}(m, m)$ be nonsingular matrices. Define
$$\tilde{y}(t) = Py(t), \quad \tilde{u}(t) = Su(t), \quad t \geq 0.$$
Then
$$\frac{d}{dt}\tilde{y}(t) = P\frac{d}{dt}y(t) = PAy(t) + PBu(t)$$
$$= PAP^{-1}\tilde{y}(t) + PBS^{-1}\tilde{u}(t)$$
$$= \tilde{A}\tilde{y}(t) + \tilde{B}\tilde{u}(t), \quad t \geq 0,$$
where
$$\tilde{A} = PAP^{-1}, \quad \tilde{B} = PBS^{-1}. \tag{1.24}$$

The control systems described by (A, B) and $(\widetilde{A}, \widetilde{B})$ are called *equivalent* if there exist nonsingular matrices $P \in \mathbf{M}(n,n)$, $S \in \mathbf{M}(m,m)$, such that (1.24) holds. Let us remark that P^{-1} and S^{-1} can be regarded as transition matrices from old to new bases in $\mathbf{R}^n$ and $\mathbf{R}^m$ respectively. The introduced concept is an equivalence relation. It is clear that a pair (A, B) is controllable if and only if $(\widetilde{A}, \widetilde{B})$ is controllable.

We now give a complete description of equivalent classes of the introduced relation in the case when $m = 1$.

Let us first consider a system

$$\frac{d^{(n)}}{dt^{(n)}}z + a_1 \frac{d^{(n-1)}}{dt^{(n-1)}}z + \ldots + a_n z = u, \tag{1.25}$$

with initial conditions

$$z(0) = \xi_1, \quad \frac{dz}{dt}(0) = \xi_2, \quad \ldots, \quad \frac{d^{(n-1)}z}{dt^{(n-1)}}(0) = \xi_n. \tag{1.26}$$

Let $z(t), \frac{dz}{dt}(t), \ldots, \frac{d^{(n-1)}z}{dt^{(n-1)}}(t)$, $t \geq 0$, be coordinates of a function $y(t), t \geq 0$, and $\xi_1, \ldots, \xi_n$ coordinates of a vector x. Then

$$\dot{y} = \widetilde{A}y + \widetilde{B}u, \quad y(0) = x \in \mathbf{R}^n, \tag{1.27}$$

where matrices $\widetilde{A}$ and $\widetilde{B}$ are of the form

$$\widetilde{A} = \begin{bmatrix} 0 & 1 & \ldots & 0 & 0 \\ 0 & 0 & \ldots & 0 & 0 \\ \vdots & \vdots & \ddots & \vdots & \vdots \\ 0 & 0 & \ldots & 0 & 1 \\ -a_n & -a_{n-1} & \ldots & -a_2 & -a_1 \end{bmatrix}, \quad \widetilde{B} = \begin{bmatrix} 0 \\ \vdots \\ 0 \\ 1 \end{bmatrix}. \tag{1.28}$$

We easily check that on the main diagonal of the matrix $[\widetilde{A}|\widetilde{B}]$ there are only ones and above the diagonal only zeros. Therefore rank $[\widetilde{A}|\widetilde{B}] = n$ and, by Theorem 1.2, the pair $(\widetilde{A}, \widetilde{B})$ is controllable. Interpreting this result in terms of the initial system (1.21)-(1.22) we can say that for two arbitrary sequences of n numbers $\xi_1, \ldots, \xi_n$ and $\eta_1, \ldots, \eta_n$ and for an arbitrary positive number T there exists an analytic function $u(t)$, $t \in [0, T]$, such that for the corresponding solution $z(t)$, $t \in [0, T]$, of the equation (1.25)-(1.26)

$$z(T) = \eta_1, \quad \frac{dz}{dt}(T) = \eta_2, \quad \ldots, \quad \frac{d^{(n-1)}z}{dt^{(n-1)}}(T) = \eta_n.$$

Theorem 1.4 states that an arbitrary controllable system with the one dimensional space of control parameters is equivalent to a system of the form (1.25)-(1.26).

Theorem 1.4. *If $A \in M(n,n)$, $b \in M(n,1)$ and the system*
$$\dot{y} = Ay + bu, \quad y(0) = x \in \mathbf{R}^n \tag{1.29}$$
is controllable then it is equivalent to exactly one system of the form (1.28). Moreover the numbers $a_1, \ldots, a_n$ in the representation (1.24) are identical to the coefficients of the characteristic polynomial of the matrix A:
$$p(\lambda) = \det[\lambda I - A] = \lambda^n + a_1 \lambda^{n-1} + \ldots + a_n, \quad \lambda \in \mathbf{C}. \tag{1.30}$$

Proof. By the Cayley–Hamilton theorem, $A^n + a_1 A^{n-1} + \ldots + a_n I = 0$. In particular
$$A^n b = -a_1 A^{n-1} b - \ldots - a_n b.$$
Since rank $[A|b] = n$, therefore vectors $e_1 = A^{n-1}b, \ldots, e_n = b$ are linearly independent and form a basis in $\mathbf{R}^n$. Let $\xi_1(t), \ldots, \xi_n(t)$ be coordinates of the vector $y(t)$ in this basis, $t \geq 0$. Then

$$\frac{d\xi}{dt} = \begin{bmatrix} -a_1 & 1 & 0 & \cdots & 0 & 0 \\ -a_2 & 0 & 1 & \cdots & 0 & 0 \\ \vdots & \vdots & \vdots & \ddots & \vdots & \vdots \\ -a_{n-1} & 0 & 0 & \cdots & 0 & 1 \\ -a_n & 0 & 0 & \cdots & 0 & 0 \end{bmatrix} \xi + \begin{bmatrix} 0 \\ 0 \\ \vdots \\ 0 \\ 1 \end{bmatrix} u. \tag{1.31}$$

Therefore an arbitrary controllable system (1.29) is equivalent to (1.31) and the numbers $a_1, \ldots, a_n$ are the coefficients of the characteristic polynomial of A. On the other hand, direct calculation of the determinant of $[\lambda I - \widetilde{A}]$ gives
$$\det(\lambda I - \widetilde{A}) = \lambda^n + a_1 \lambda^{n-1} + \ldots + a_n = p(\lambda), \quad \lambda \in \mathbf{C}.$$
Therefore the pair $(\widetilde{A}, \widetilde{B})$ is equivalent to the system (1.31) and consequently also to the pair (A, b). $\square$

Remark. The problem of an exact description of the equivalence classes in the case of arbitrary m is much more complicated; see [39] and [67].

§1.5. Kalman decomposition

Theorem 1.2 gives several characterizations of controllable systems. Here we deal with uncontrollable ones.

Theorem 1.5. *Assume that*
$$\text{rank}\,[A|B] = l < n.$$
There exists a nonsingular matrix $P \in M(n,n)$ such that
$$PAP^{-1} = \begin{bmatrix} A_{11} & A_{12} \\ 0 & A_{22} \end{bmatrix}, \quad PB = \begin{bmatrix} B_1 \\ 0 \end{bmatrix},$$

where $A_{11} \in \mathbf{M}(l,l)$, $A_{22} \in \mathbf{M}(n-l, n-l)$, $B_1 \in \mathbf{M}(l,m)$. *In addition the pair*

$$(A_{11}, B_1)$$

is controllable.

The theorem states that there exists a basis in $\mathbf{R}^n$ such that system (1.1) written with respect to that basis has a representation

$$\dot{\xi}_1 = A_{11}\xi_1 + A_{12}\xi_2 + B_1 u, \quad \xi_1(0) \in \mathbf{R}^l,$$
$$\dot{\xi}_2 = A_{22}\xi_2, \quad \xi_2(0) \in \mathbf{R}^{n-l},$$

in which (A_{11}, B_1) is a controllable pair. The first equation describes the so-called *controllable part* and the second the *completely uncontrollable* part of the system.

Proof. It follows from Lemma 1.1 that the subspace $E_0 = \mathcal{L}_T(L^1[0,T; \mathbf{R}^m])$ is identical with the image of the transformation l_n. Therefore it consists of all elements of the form $Bu_1 + ABu_1 + \ldots + A^{n-1}Bu_n$, $u_1, \ldots, u_n \in \mathbf{R}^m$ and is of dimension l. In addition it contains the image of B and by the Cayley-Hamilton theorem, it is invariant with respect to the transformation A. Let E_1 be any linear subspace of $\mathbf{R}^n$ complementing E_0 and let $e_1, \ldots, e_l$ and $e_{l+1}, \ldots, e_n$ be bases in E_0 and E_1 and P the transition matrix from the new to the old basis. Let $\widetilde{A} = PAP^{-1}$, $\widetilde{B} = PB$,

$$\widetilde{A}\begin{bmatrix}\xi_1 \\ \xi_2\end{bmatrix} = \begin{bmatrix}A_{11}\xi_1 + A_{12}\xi_2 \\ A_{21}\xi_1 + A_{22}\xi_2\end{bmatrix}, \quad \widetilde{B} = \begin{bmatrix}B_1 u \\ B_2 u\end{bmatrix},$$

$\xi_1 \in \mathbf{R}^l$, $\xi_2 \in \mathbf{R}^{n-l}$, $u \in \mathbf{R}^m$. Since the space E_0 is invariant with respect to A, therefore

$$\widetilde{A}\begin{bmatrix}\xi_1 \\ 0\end{bmatrix} = \begin{bmatrix}A_{11}\xi_1 \\ 0\end{bmatrix}, \quad \xi_1 \in \mathbf{R}^l.$$

Taking into account that $B(\mathbf{R}^m) \subset E_0$,

$$B_2 u = 0 \quad \text{dla } u \in \mathbf{R}^m.$$

Consequently the elements of the matrices A_{22} and B_2 are zero. This finishes the proof of the first part of the theorem. To prove the final part, let us remark that for the nonsingular matrix P

$$\text{rank}[A|B] = \text{rank}\,(P[A|B]) = \text{rank}\,[\widetilde{A}|\widetilde{B}].$$

Since

$$[\widetilde{A}|\widetilde{B}] = \begin{bmatrix}B_1 & A_{11}B_1 & \ldots & A_{11}^{n-1}B_1 \\ 0 & 0 & \ldots & 0\end{bmatrix},$$

so

$$l = \text{rank}[\widetilde{A}|\widetilde{B}] = \text{rank}\,[A_{11}|B_1].$$

Taking into account that $A_{11} \in \mathbf{M}(l,l)$, one gets the required property. $\square$

§ 1.6. Observability

Remark. Note that the subspace E_0 consists of all points attainable from 0. It follows from the proof of Theorem 1.5 that E_0 is the smallest subspace of R^n invariant with respect to A and containing the image of B, and it is identical to the image of the transformation represented by $[A|B]$.

Exercise 1.10. Give a complete classification of controllable systems when $m = 1$ and the dimension of E_0 is $l < n$.

§ 1.6. Observability

Assume that $B = 0$. Then system (1.1) is identical with the linear equation
$$\dot{z} = Az, \quad z(0) = x. \tag{1.32}$$
The observation relation (1.2) we leave unchanged:
$$w = Cz. \tag{1.33}$$
The solution to (1.32) will be denoted by $z^x(t)$, $t \geq 0$. Obviously
$$z^x(t) = S(t)x, \quad x \in \mathbf{R}^n.$$
The system (1.32)-(1.33), or the pair (A, C), is said to be *observable* if for arbitrary $x \in \mathbf{R}^n$, $x \neq 0$, there exists a $t > 0$ such that
$$w(t) = Cz^x(t) \neq 0.$$
If for a given $T > 0$ and for arbitrary $x \neq 0$ there exists $t \in [0, T]$ with the above property, then the system (1.32)-(1.33) or the pair (A, C) are said to be *observable at time T*. Let us introduce the so-called *observability matrix*:
$$R_T = \int_0^T S^*(r) C^* C S(r)\, dr.$$
The following theorem, dual to Theorem 1.2, holds.

Theorem 1.6. *The following conditions are equivalent.*
  (i) *System (1.32)-(1.33) is observable.*
  (ii) *System (1.32)-(1.33) is observable at a given time $T > 0$.*
  (iii) *The matrix R_T is nonsingular for some $T > 0$.*
  (iv) *The matrix R_T is nonsingular for arbitrary $T > 0$.*
  (v) $\mathrm{rz}\,[A^*|C^*] = n.$

Proof. Analysis of the function $w(\,\cdot\,)$ implies the equivalence of (i) and (ii). Besides,
$$\int_0^T |w(r)|^2\, dr = \int_0^T |Cz^x(r)|^2\, dr$$
$$= \int_0^T \langle S^*(r) C^* C S(r) x, x \rangle\, dr$$
$$= \langle R_T x, x \rangle.$$

Therefore observability at time $T \geq 0$ is equivalent to $\langle R_T x, x \rangle \neq 0$ for arbitrary $x \neq 0$ and consequently to nonsingularity of the nonnegative, symmetric matrix R_T. The remaining equivalences are consequences of Theorem 1.2 and the observation that the controllability matrix corresponding to (A^*, C^*) is exactly R_T. □

Example 1.1. Let us consider the equation
$$\frac{d^{(n)}z}{dt^{(n)}} + a_1 \frac{d^{(n-1)}z}{dt^{(n-1)}} + \ldots + a_n z = 0 \tag{1.34}$$

and assume that
$$w(t) = z(t), \quad t \geq 0. \tag{1.35}$$

Matrices A and C corresponding to (1.34)-(1.35) are of the form

$$A = \begin{bmatrix} 0 & 1 & \ldots & 0 & 0 \\ 0 & 0 & \ldots & 0 & 0 \\ \vdots & \vdots & \ddots & \vdots & \vdots \\ 0 & 0 & \ldots & 0 & 1 \\ -a_n & -a_{n-1} & \ldots & -a_2 & -a_1 \end{bmatrix}, \quad C = [1, 0, \ldots, 0].$$

We check directly that rank $[A^*|C^*] = n$ and thus the pair (A, C) is observable.

The next theorem is analogous to Theorem 1.5 and gives a decomposition of system (1.32)-(1.33) into observable and completely unobservable parts.

Theorem 1.7. *Assume that rank $[A^*|C^*] = l < n$. Then there exists a nonsingular matrix $P \in \mathbf{M}(n, n)$ such that*

$$PAP^{-1} = \begin{bmatrix} A_{11} & 0 \\ A_{21} & A_{22} \end{bmatrix}, \quad CP^{-1} = [C_1, 0],$$

where $A_{11} \in \mathbf{M}(l, l)$, $A_{22} \in \mathbf{M}(n - l, n - l)$ and $C_1 \in \mathbf{M}(k, l)$ and the pair (A_{11}, C_1) is observable.

Proof. The theorem follows directly from Theorem 1.5 and from the observation that a pair (A, C) is observable if and only if the pair (A^*, C^*) is controllable. □

Remark. It follows from the above theorem that there exists a basis in $\mathbf{R}^n$ such that the system (1.1)-(1.2) has representation

$$\dot{\xi}_1 = A_{11}\xi_1 + B_1 u,$$
$$\dot{\xi}_2 = A_{21}\xi_1 + A_{22}\xi_2 + B_2 u,$$
$$\eta = C_1 \xi_1,$$

and the pair (A_{11}, C_1) is observable.

Bibliographical notes

Basic concepts of the chapter are due to R. Kalman [33]. He is also the author of Theorems 1.2, 1.5, 1.6 and 1.7. Exercise 1.3 as well as Proposition 1.3 are due to R. Triggiani [56]. A generalisation of Theorem 1.4 to arbitrary m leads to the so-called controllability indices discussed in [59] and [61].

Chapter 2
Stability and stabilizability

In this chapter stable linear systems are characterized in terms of associated characteristic polynomials and Liapunov equations. A proof of the Routh theorem on stable polynomials is given as well as a complete description of completely stabilizable systems. Luenberger's observer is introduced and used to illustrate the concept of detectability.

§2.1. Stable linear systems

Let $A \in \mathbf{M}(n,n)$ and consider linear systems

$$\dot{z} = Az, \quad z(0) = x \in \mathbf{R}^n. \tag{2.1}$$

Solutions of equation (2.1) will be denoted by $z^x(t)$, $t \geq 0$. In accordance with earlier notations we have that

$$z^x(t) = S(t)x = (\exp tA)x, \quad t \geq 0.$$

The system (2.1) is called *stable* if for arbitrary $x \in \mathbf{R}^n$

$$z^x(t) \longrightarrow 0, \quad \text{as } t \uparrow +\infty.$$

Instead of saying that (2.1) is stable we will often say that the matrix A is stable. Let us remark that the concept of stability does not depend on the choice of the basis in $\mathbf{R}^n$. Therefore if P is a nonsingular matrix and A is a stable one, then matrix PAP^{-1} is stable.

In what follows we will need the Jordan theorem [4] on canonical representation of matrices. Denote by $\mathbf{M}(n,m;\mathbf{C})$ the set of all matrices with n rows and m columns and with complex elements. Let us recall that a number $\lambda \in \mathbf{C}$ is called an *eigenvalue* of a matrix $A \in \mathbf{M}(n,n;\mathbf{C})$ if there exists a vector $a \in \mathbf{C}^n$, $a \neq 0$, such that $Aa = \lambda a$. The set of all eigenvalues of a matrix A will be denoted by $\sigma(A)$. Since $\lambda \in \sigma(A)$ if and only if the matrix $\lambda I - A$ is singular, therefore $\lambda \in \sigma(A)$ if and only if $p(\lambda) = 0$, where p is a *characteristic polynomial* of A: $p(\lambda) = \det[\lambda I - A]$, $\lambda \in \mathbf{C}$. The set $\sigma(A)$ consists of at most n elements and is nonempty.

§ 2.1. Stable linear systems

Theorem 2.1. *For an arbitrary matrix $A \in \mathbf{M}(n,n; \mathbf{C})$ there exists a nonsingular matrix $P \in \mathbf{M}(n,n; \mathbf{C})$ such that*

$$PAP^{-1} = \begin{bmatrix} J_1 & 0 & \cdots & 0 & 0 \\ 0 & J_2 & \cdots & 0 & 0 \\ \vdots & \vdots & \ddots & \vdots & \vdots \\ 0 & 0 & \cdots & J_{r-1} & 0 \\ 0 & 0 & \cdots & 0 & J_r \end{bmatrix} = \tilde{A}, \tag{2.2}$$

where $J_1, J_2, \ldots, J_r$ are the so-called Jordan blocks

$$J_k = \begin{bmatrix} \lambda_k & \gamma_k & \cdots & 0 & 0 \\ 0 & \lambda_k & \cdots & 0 & 0 \\ \vdots & \vdots & \ddots & \vdots & \vdots \\ 0 & 0 & \cdots & \lambda_k & \gamma_k \\ 0 & 0 & \cdots & 0 & \lambda_k \end{bmatrix}, \quad \gamma_k \neq 0 \text{ or } J_k = [\lambda_k], \ k = 1, \ldots, r.$$

In the representation (2.2) *at least one Jordan block corresponds to an eigenvalue $\lambda_k \in \sigma(A)$. Selecting matrix P properly one can obtain a representation with numbers $\gamma_k \neq 0$ given in advance.*

For matrices with real elements the representation theorem has the following form:

Theorem 2.2. *For an arbitrary matrix $A \in \mathbf{M}(n,n)$ there exists a nonsingular matrix $P \in \mathbf{M}(n,n)$ such that* (2.2) *holds with "real" blocks I_k. Blocks I_k, $k = 1, \ldots, r$, corresponding to real eigenvalues $\lambda_k = \alpha_k \in \mathbf{R}$ are of the form*

$$[\alpha_k] \quad \text{or} \quad \begin{bmatrix} \alpha_k & \gamma_k & \cdots & 0 & 0 \\ 0 & \alpha_k & \cdots & 0 & 0 \\ \vdots & \vdots & \ddots & \vdots & \vdots \\ 0 & 0 & \cdots & \alpha_k & \gamma_k \\ 0 & 0 & \cdots & 0 & \alpha_k \end{bmatrix}, \quad \gamma_k \neq 0, \ \gamma_k \in \mathbf{R},$$

and corresponding to complex eingenvalues $\lambda_k = \alpha_k + i\beta_k$, $\beta_k \neq 0$, $\alpha_k, \beta_k \in \mathbf{R}$,

$$\begin{bmatrix} K_k & L_k & \cdots & 0 & 0 \\ 0 & K_k & \cdots & 0 & 0 \\ \vdots & \vdots & \ddots & \vdots & \vdots \\ 0 & 0 & \cdots & K_k & L_k \\ 0 & 0 & \cdots & 0 & K_k \end{bmatrix} \quad \text{where } K_k = \begin{bmatrix} \alpha_k & \beta_k \\ -\beta_k & \alpha_k \end{bmatrix}, \ L_k = \begin{bmatrix} \gamma_k & 0 \\ 0 & \gamma_k \end{bmatrix},$$

compare [4].

We now prove the following theorem.

Theorem 2.3. *Assume that $A \in \mathbf{M}(n,n)$. The following conditions are equivalent:*

(i) $z^x(t) \longrightarrow 0$ *as* $t \uparrow +\infty$, *for arbitrary* $x \in \mathbf{R}^n$.

(ii) $z^x(t) \longrightarrow 0$ *exponentially as* $t \uparrow +\infty$, *for arbitrary* $x \in \mathbf{R}^n$.

(iii) $\omega(A) = \sup\{\operatorname{Re}\lambda; \lambda \in \sigma(A)\} < 0$.

(iv) $\int_0^{+\infty} |z^x(t)|^2\, dt < +\infty$ *for arbitrary* $x \in \mathbf{R}^n$.

For the proof we will need the following lemma.

Lemma 2.1. *Let $\omega > \omega(A)$. For arbitrary norm $\|\cdot\|$ on $\mathbf{R}^n$ there exist constants M such that*
$$\|z^x(t)\| \leq Me^{\omega t}\|x\| \quad \text{for } t \geq 0 \text{ and } x \in \mathbf{R}^n.$$

Proof. Let us consider equation (2.1) with the matrix A in the Jordan form (2.2)
$$\dot{x} = \widetilde{A}w, \quad w(0) = x \in \mathbf{C}^n.$$
For $a = a_1 + ia_2$, where $a_1, a_2 \in \mathbf{R}^n$ set $\|a\| = \|a_1\| + \|a_2\|$. Let us decompose vector $w(t)$, $t \geq 0$ and the initial state x into sequences of vectors $w_1(t), \ldots, w_r(t)$, $t > 0$ and $x_1, \ldots, x_r$ according to the decomposition (2.2). Then
$$\dot{w}_k = J_k w_k, \quad w_k(0) = x_k, \quad k = 1, \ldots, r.$$
Let $j_1, \ldots, j_r$ denote the dimensions of the matrices $J_1, \ldots, J_r$, $j_1 + j_2 + \ldots + j_r = n$.

If $j_k = 1$ then
$$w_k(t) = e^{\lambda_k t} x_k, \quad t \geq 0.$$
So $\|w_k(t)\| = e^{(\operatorname{Re}\lambda_k)t}\|x_k\|$, $t \geq 0$.

If $j_k > 1$, then
$$w_k(t) = e^{\lambda_k t} \sum_{l=0}^{j_k-1} \begin{bmatrix} 0 & \gamma_k & \cdots & 0 & 0 \\ 0 & 0 & \cdots & 0 & 0 \\ \vdots & \vdots & \ddots & \vdots & \vdots \\ 0 & 0 & \cdots & 0 & \gamma_k \\ 0 & 0 & \cdots & 0 & 0 \end{bmatrix}^l x_k \frac{t^l}{l!}.$$

So
$$\|w_k(t)\| \leq e^{(\operatorname{Re}\lambda_k)t}\|x_k\| \sum_{l=0}^{j_k-1} (M_k)^l \frac{t^l}{l!}, \quad t \geq 0,$$
where M_k is the norm of the transformation represented by
$$\begin{bmatrix} 0 & \gamma_k & \cdots & 0 & 0 \\ 0 & 0 & \cdots & 0 & 0 \\ \vdots & \vdots & \ddots & \vdots & \vdots \\ 0 & 0 & \cdots & 0 & \gamma_k \\ 0 & 0 & \cdots & 0 & 0 \end{bmatrix}.$$

§ 2.1. Stable linear systems

Setting $\omega_0 = \omega(A)$ we get

$$\sum_{k=1}^{r} \|w_k(t)\| \leq e^{\omega_0 t} q(t) \sum_{k=1}^{r} \|x_k\|, \quad t \geq 0,$$

where q is a polynomial of order at most $\max(j_k - 1)$, $k = 1, \ldots, r$. If $\omega > \omega_0$ and

$$M_0 = \sup \left\{ q(t) e^{(\omega_0 - \omega)t}, \ t \geq 0 \right\},$$

then $M_0 < +\infty$ and

$$\sum_{k=1}^{r} \|w_k(t)\| \leq M_0 e^{\omega t} \sum_{k=1}^{r} \|x_k\|, \quad t \geq 0.$$

Therefore for a new constant M_1

$$\|w(t)\| \leq M_1 e^{\omega t} \|x\|, \quad t \geq 0.$$

Finally

$$\|z^x(t)\| = \|Pw(t)P^{-1}\| \leq M_1 e^{\omega t} \|P\| \|P^{-1}\| \|x\|, \quad t \geq 0,$$

and this is enough to define $M = M_1 \|P\| \|P^{-1}\|$. □

Proof of the theorem. Assume $\omega_0 \geq 0$. There exist $\lambda = \alpha + i\beta$, $\operatorname{Re} \lambda = \alpha \geq 0$ and a vector $a \neq 0$, $a = a_1 + ia_2$, $a_1, a_2 \in \mathbf{R}^n$ such that

$$A(a_1 + ia_2) = (\alpha + i\beta)(a_1 + ia_2).$$

The function

$$z(t) = z_1(t) + iz_2(t) = e^{(\alpha + i\beta)t} a, \quad t \geq 0,$$

as well as its real and imaginary parts, is a solution of (2.1). Since $a \neq 0$, either $a_1 \neq 0$ or $a_2 \neq 0$. Let us assume, for instance, that $a_1 \neq 0$ and $\beta \neq 0$. Then

$$z_1(t) = e^{\alpha t}(\cos \beta t)a_1 - (\sin \beta t)a_2, \quad t \geq 0.$$

Inserting $t = 2\pi k/\beta$, we have

$$|z_1(t)| = e^{\alpha t}|a_1|$$

and, taking $k \uparrow +\infty$, we obtain $z_1(t) \not\to 0$.

Now let $\omega_0 < 0$ and $\alpha \in (0, -\omega_0)$. Then by the lemma

$$|z^x(t)| \le Me^{-\alpha t}|x| \quad \text{for } t \ge 0 \text{ and } x \in \mathbf{R}^n.$$

This implies (ii) and therefore also (i).

It remains to consider (iv). It is clear that it follows from (ii) and thus also from (iii). Let us assume that condition (iv) holds and $\omega_0 \ge 0$. Then $|z_1(t)| = e^{\alpha t}|a_1|$, $t \ge 0$, and therefore

$$\int_0^{+\infty} |z_1(t)|^2\, dt = +\infty,$$

a contradiction. The proof is complete. □

Exercise 2.1. The matrix

$$A = \begin{bmatrix} 0 & 1 \\ -2 & -2 \end{bmatrix}$$

corresponds to the equation $\ddot{z} + 2\dot{z} + 2z = 0$. Calculate $\omega(A)$. For $\omega > \omega(A)$ find the smallest constant $M = M(\omega)$ such that

$$|S(t)| \le Me^{\omega t}, \quad t \ge 0.$$

Hint. Prove that $|S(t)| = \varphi(t)e^{-t}$, where

$$\varphi(t) = \frac{1}{2}\left(2 + 5\sin^2 t + (20\sin^2 t + 25\sin^4 t)^{1/2}\right)^{1/2}, \quad t \ge 0.$$

§ 2.2. Stable polynomials

Theorem 2.3 reduces the problem of determining whether a matrix A is stable to the question of finding out whether all roots of the characteristic polynomial of A have negative real parts. Polynomials with this property will be called *stable*. Because of its importance, several efforts have been made to find necessary and sufficient conditions for the stability of an arbitrary polynomial

$$p(\lambda) = \lambda^n + a_1\lambda^{n-1} + \ldots + a_n, \quad \lambda \in \mathbf{C}, \tag{2.3}$$

with real coefficients, in term of the coefficients $a_1, \ldots, a_n$. Since there is no general formula for roots of polynomials of order greater than 4, the existence of such conditions is not obvious. Therefore their formulation in the nineteenth century by Routh was a kind of a sensation. Before formulating and proving a version of the Routh theorem we will characterize

§2.2. Stable polynomials

stable polynomials of degree smaller than or equal to 4 using only the fundamental theorem of algebra. We deduce also a useful necessary condition for stability.

Theorem 2.4. (1) *Polynomials with real coefficients*:

  (i) $\lambda + a$,
  (ii) $\lambda^2 + a\lambda + b$,
  (iii) $\lambda^3 + a\lambda^2 + b\lambda + c$,
  (iv) $\lambda^4 + a\lambda^3 + b\lambda^2 + c\lambda + d$

are stable if and only if, respectively

  (i)* $a > 0$,
  (ii)* $a > 0$, $b > 0$,
  (iii)* $a > 0$, $b > 0$, $c > 0$ *and* $ab > c$,
  (iv)* $a > 0$, $b > 0$, $c > 0$, $d > 0$ *and* $abc > c^2 + a^2 d$.

(2) *If polynomial (2.3) is stable then all its coefficients* $a_1, \ldots, a_n$ *are positive*.

Proof. (1) Equivalence (i)$\Longleftrightarrow$(i)* is obvious.

To prove (ii)$\Longleftrightarrow$(ii)* assume that the roots of the polynomial are of the form $\lambda_1 = -\alpha + i\beta$, $\lambda_2 = -\alpha - i\beta$, $\beta \neq 0$. Then $p(\lambda) = \lambda^2 + 2\alpha\lambda + \beta^2$, $\lambda \in \mathbb{C}$ and therefore the stability conditions are $a > 0$ and $b > 0$. If the roots λ_1, λ_2 of the polynomial p are real then $a = -(\lambda_1 + \lambda_2)$, $b = \lambda_1 \lambda_2$. Therefore they are negative if only if $a > 0$, $b > 0$.

To show that (iii) $\Longleftrightarrow$ (iii)* let us remark that the fundamental theorem of algebra implies the following decomposition of the polynomial, with real coefficients α, β, γ:

$$p(\lambda) = \lambda^3 + a\lambda^2 + b\lambda + c = (\lambda + \alpha)(\lambda^2 + \beta\lambda + \gamma), \quad \lambda \in \mathbb{C}.$$

It therefore follows from (i) and (ii) that the polynomial p is stable if only if $\alpha > 0$, $\beta > 0$ and $\gamma > 0$. Comparing the coefficients gives

$$a = \alpha + \beta, \quad b = \gamma + \alpha\beta, \quad c = \alpha\gamma,$$

and therefore $ab - c = \beta(\alpha^2 + \gamma + \alpha\beta) = \beta(\alpha^2 + b)$.

Assume that $a > 0$, $b > 0$, $c > 0$ and $ab - c > 0$. It follows from $b > 0$ and $ab - c > 0$ that $\beta > 0$. Since $c = \alpha\gamma$, α and γ are either positive or negative. They cannot, however, be negative because then $b = \gamma + \alpha\beta < 0$. Thus $\alpha > 0$ and $\gamma > 0$ and consequently $\alpha > 0$, $\beta > 0$, $\gamma > 0$. It is clear from the above formulae that the positivity of α, β, γ implies inequalities (iii)*. To prove (iv)$\Longleftrightarrow$(iv)* we again apply the fundamental theorem of algebra to obtain the representation

$$\lambda^4 + a\lambda^3 + b\lambda^2 + c\lambda + d = (\lambda^2 + \alpha\lambda + \beta)(\lambda^2 + \gamma\lambda + \delta)$$

and the stability condition $\alpha > 0$, $\beta > 0$, $\gamma > 0$, $\delta > 0$.

From the decomposition

$$a = \alpha + \gamma, \quad b = \alpha\gamma + \beta + \delta, \quad c = \alpha\delta + \beta\gamma, \quad d = \beta\delta,$$

we check directly that

$$abc - c^2 - a^2 d = \alpha\gamma\left((\beta - \delta)^2 + ac\right).$$

It is therefore clear that $\alpha > 0$, $\beta > 0$, $\gamma > 0$ and $\delta > 0$, and then (iv)* holds. Assume now that the inequalities (iv)* are true. Then $\alpha\gamma > 0$, and, since $a = \alpha + \gamma > 0$, therefore $\alpha > 0$ and $\delta > 0$. Since, in addition, $d = \beta\delta > 0$ and $c = \alpha\delta + \beta\gamma > 0$, so $\beta > 0$, $\delta > 0$. Finally $\alpha > 0$, $\beta > 0$, $\gamma > 0$, $\delta > 0$, and the polynomial p is stable.

(2) By the fundamental theorem of algebra, the polynomial p is a product of polynomials of degrees at most 2 which, by (1), have positive coefficients. This implies the result. □

Exercise 2.2. Find necessary and sufficient conditions for the polynomial

$$\lambda^2 + a\lambda + b$$

with complex coefficients a and b to have both roots with negative real parts.

Hint. Consider the polynomial $(\lambda^2 + a\lambda + b)(\lambda^2 + \bar{a}\lambda + \bar{b})$ and apply Theorem 2.4.

Exercise 2.3. Equation

$$L^2 C \dddot{z} + RLC\ddot{z} + 2L\dot{z} + Rz = 0, \quad R > 0, \ L > 0, \ C > 0,$$

describes the action of the electrical filter from Example 0.4. Check that the associated characteristic polynomial is stable.

§2.3. The Routh theorem

We now give the theorem, mentioned earlier, which allows us to check, in a finite number of steps, that a given polynomial $p(\lambda) = \lambda^n + a_1\lambda^{n-1} + \ldots + a_n$, $\lambda \in \mathbb{C}$, with real coefficients is stable. As we already know, a stable polynomial has all coefficients positive, but this condition is not sufficient for stability if $n > 3$. Let U and V be polynomials with real coefficients given by

$$U(x) + iV(x) = p(ix), \quad x \in \mathbb{R}.$$

Let us remark that $\deg U = n$, $\deg V = n - 1$ if n is an even number and $\deg U = n - 1$, $\deg V = n$, if n is an odd number. Denote $f_1 = U$, $f_2 = V$ if

§ 2.3. The Routh theorem

$\deg U = n$, $\deg V = n-1$ and $f_1 = V$, $f_2 = U$ if $\deg V = n$, $\deg U = n-1$. Let $f_3, f_4, \ldots, f_m$ be polynomials obtained from f_1, f_2 by an application of the Euclid algorithm. Thus $\deg f_{k+1} < \deg f_k$, $k = 2, \ldots, m-1$ and there exist polynomials $\kappa_1, \ldots, \kappa_m$ such that

$$f_{k-1} = \kappa_k f_k - f_{k+1}, \quad f_{m-1} = \kappa_m f_m.$$

Moreover the polynomial f_m is equal to the largest commun divisor of f_1, f_2 multiplied by a constant.

The following theorem is due to F.J. Routh [51].

Theorem 2.5. *A polynomial p is stable if and only if $m = n+1$ and the signs of the leading coefficients of the polynomials $f_1, \ldots, f_{n+1}$ alternate.*

Proof. Let $\Gamma(r)$, $r > 0$ be an counterclockwise oriented curve composed of the segment $I(r)$ with ends ir and $-ir$ and the semicircle $S(r)$, $S(r, \theta) = re^{i\theta}$, $-\frac{1}{2}\pi \le \theta \le \frac{1}{2}\pi$.

We will use the following result from elementary complex analysis:

Lemma 2.2. *If a polynomial p has no roots on the curve $\Gamma(r)$ and D_r is the number of all roots inside of $\Gamma(r)$, taking into account their multiplicity, then*

$$\frac{1}{i}\int_{\Gamma(r)} \frac{p'(\lambda)}{p(\lambda)} d\lambda = 2\pi D_r.$$

Let us assume that p is a stable polynomial. Then $D_r = 0$ for arbitrary $r > 0$. Let us remark that

$$\frac{p'(\lambda)}{p(\lambda)} = \frac{n}{\lambda}(1 - q(\lambda)),$$

where

$$q(\lambda) = \frac{\lambda^{n-2} + b_1 \lambda^{n-3} + \ldots + b_{n-2}}{\lambda^{n-1} + c_1 \lambda^{n-2} + \ldots + c_{n-1}}, \quad \lambda \in \mathbb{C},$$

for some constants $b_1, \ldots, b_{n-2}, c_1, \ldots, c_{n-2}$. There exist numbers $M > 0$ and $r_0 > 0$ such that

$$\sup_{\lambda \in S(r)} |q(\lambda)| \le \frac{M}{r}, \quad r > r_0.$$

Therefore

$$\frac{1}{i}\int_{S(r)} \frac{p'(\lambda)}{p(\lambda)} d\lambda = \frac{n}{i}\int_{S(r)} \frac{1}{\lambda} d\lambda - \frac{n}{i}\int_{S(r)} \frac{q(\lambda)}{\lambda} d\lambda,$$

$$\lim_{r \uparrow +\infty} \frac{1}{i}\int_{I(r)} \frac{p'(\lambda)}{p(\lambda)} d\lambda = n\pi. \tag{2.4}$$

Let $\gamma(r)$ be the oriented image of $I(r)$ under the transformation $\lambda \longrightarrow p(\lambda)$. Then

$$\frac{1}{i}\int_{I(r)} \frac{p'(\lambda)}{p(\lambda)} d\lambda = \frac{1}{i}\int_{\gamma(r)} \frac{1}{\lambda} d\lambda.$$

Let us assume that the degree of p is an even number $n = 2m$. Considerations for n odd are analogical, with the imaginary axis replaced by the real one, and therefore will be omitted. Let $x_1 < x_2 < \ldots < x_l$ be real roots of the polynomial $f_1 = U$. Then $0 \le l \le n$, and points $p_1 = p(ix_1), \ldots, p_l = p(ix_l)$ are exactly intersection points of the arc $\gamma(r)$ and of the imaginary axis. The oriented arc $\gamma(r)$ consists of oriented subarcs $\gamma_1(r), \gamma_2(r), \ldots, \gamma_l(r), \gamma_{l+1}(r)$ with respective endpoints $p_0(r) = p(-ri)$, $p_1, \ldots, p_l$, $p(ri) = p_{l+1}(r)$, and $r > 0$ is a positive number such that $-r < x_1$, $x_l < r$.

Denote by $\mathbb{C}_+$ and $\mathbb{C}_-$ respectively the right and the left closed half-planes of the complex plane $\mathbb{C}$. If γ is an oriented smooth curve, not passing through 0, contained in $\mathbb{C}_+$, with initial and final points a and b, then

$$\frac{1}{i}\int_\gamma \frac{1}{\lambda} d\lambda = \frac{1}{i} \ln\left|\frac{b}{a}\right| + \operatorname{Arg} b - \operatorname{Arg} a,$$

where $\operatorname{Arg} a$, $\operatorname{Arg} b$ denote the arguments of a and b included in $[-\frac{1}{2}\pi, \frac{1}{2}\pi]$. Similarly, for curves γ contained completely in $\mathbb{C}_-$

$$\frac{1}{i}\int_\gamma \frac{1}{\lambda} d\lambda = \frac{1}{i} \ln\left|\frac{b}{a}\right| + \arg b - \arg a,$$

where $\arg a$, $\arg b \in [\frac{1}{2}\pi, \frac{3}{2}\pi]$. It follows from the above formulae that

Lemma 2.3. *If the oriented smooth curve γ is contained in $\mathbb{C}_+$ and $b = i\beta$, $\beta \ne 0$, $\beta \in \mathbb{R}$, then*

$$\frac{1}{i}\int_\gamma \frac{1}{z} dz = \frac{1}{i} \ln\left|\frac{\beta}{a}\right| + \frac{\pi}{2} \operatorname{sgn} \beta - \operatorname{Arg} a.$$

If, in addition, $a = i\alpha$, $\alpha \ne 0$, $\alpha \in \mathbb{R}$, then

$$\frac{1}{i}\int_\gamma \frac{1}{z} dz = \frac{1}{i} \ln\frac{|\beta|}{|\alpha|} + \varepsilon(\alpha,\beta)\pi,$$

where

$$\varepsilon(\alpha,\beta) = \begin{cases} 1 & \text{for } \alpha < 0,\ \beta > 0, \\ 0 & \text{for } \alpha\beta > 0, \\ -1 & \text{for } \alpha > 0,\ \beta < 0. \end{cases}$$

§2.3. The Routh theorem

If the curve γ is in $\mathbb{C}_-$ then the above formulae remain true with $\operatorname{Arg} a$ replaced by $\arg a$, $\tfrac{1}{2}\operatorname{sgn}\beta$ replaced by $\pi - \tfrac{1}{2}\pi\operatorname{sgn}\beta$ and ε replaced by $-\varepsilon$.

From Lemma 2.3

$$\frac{1}{i}\int_{\gamma(r)} \frac{1}{\lambda}d\lambda = \sum_{j=1}^{l+1} \frac{1}{i}\int_{\gamma_j(r)} \frac{1}{\lambda}d\lambda \tag{2.5}$$

$$= \frac{1}{i}\int_{\gamma_1(r)} \frac{1}{\lambda}d\lambda + \sum_{j=2}^{l} \left(\varepsilon_j + \ln\frac{|p_j|}{|p_{j-1}|}\right) + \frac{1}{i}\int_{\gamma_{l+1}(r)} \frac{1}{\lambda}d\lambda,$$

where $\varepsilon_j = 1, -1$ or 0 for $j = 2, \ldots, l$.

Let us assume additionally that $n = 2k$, where k is an even number. Since

$$p_0(r) = p(-ir) = r^n(1 + c_0(r)), \tag{2.6}$$

$$p_{l+1}(r) = p(ir) = r^n(1 + c_1(r)), \tag{2.7}$$

where $\lim_{r\uparrow+\infty} c_0(r) = \lim_{r\uparrow+\infty} c_1(r) = 0$, therefore $p_0(r) \in \mathbb{C}_+$, $p_{l+1}(r) \in \mathbb{C}_+$ for sufficiently large $r > 0$ and

$$\lim_{r\uparrow+\infty} \operatorname{Arg} p_0(r) = \lim_{r\uparrow+\infty} \operatorname{Arg} p_{l+1}(r) = 0. \tag{2.8}$$

From (2.5) and by Lemma 2.3

$$\frac{1}{i}\int_{\gamma(r)} \frac{1}{\lambda}d\lambda = \ln\frac{|p_1|}{|p_0(r)|} + \frac{\pi}{2}\varepsilon_1 - \operatorname{Arg} p_0(r) + \sum_{j=2}^{l}\left(\varepsilon_j\pi + \ln\frac{|p_j|}{|p_{j-1}|}\right)$$

$$+ \ln\frac{|p_{l+1}(r)|}{|p_l|} + \frac{\pi}{2}\varepsilon_{l+1} + \operatorname{Arg} p_{l+1}(r),$$

$$= \ln\frac{|p_{l+1}(r)|}{|p_0(r)|} + \varepsilon_0\frac{\pi}{2} + \sum_{j=2}^{l}\varepsilon_j\pi + \varepsilon_{l+1}\frac{\pi}{2} + \operatorname{Arg} p_{l+1}(r), \quad r > 0,$$

with $\varepsilon_1, \ldots, \varepsilon_{l+1} = 0, 1$ or -1. Using (2.6), (2.7) and (2.8) we have

$$\lim_{r\uparrow+\infty} \frac{1}{i}\int_{\gamma(r)} \frac{1}{\lambda}d\lambda = \varepsilon_1\frac{\pi}{2} + \varepsilon_{l+1}\frac{\pi}{2} + \sum_{j=2}^{l}\varepsilon_j\pi.$$

Taking into account (2.4)

$$\varepsilon_1 \cdot \frac{1}{2}\pi + \varepsilon_{l+1} \cdot \frac{1}{2}\pi + \sum_{j=2}^{l}\varepsilon_j\pi = n\pi.$$

This way $l = n$ and $\varepsilon_1 = \ldots = \varepsilon_{n+1} = 1$. Similarly, we obtain the identical result for an odd number k.

The following considerations are analogous to those in the proof of the classical Sturm theorem or the number of roots of a polynomial contained in a given interval.

If $\varepsilon_1 = \ldots = \varepsilon_{n+1} = 1$ then for sufficiently large $r > 0$ the curve $\gamma(r)$ crosses the imaginary axis passing either from the first to the second quadrant of the complex plane or from the third to the fourth quadrant. Therefore for any number x close to but smaller than x_k, the signs of $U(x)$, $V(x)$ are identical and for any number x close to but greater than x_k the signs of $U(x)$, $V(x)$ are opposite. Denote by $Z(x)$ the number of sign changes in the sequence obtained from $f_1(x), \ldots, f_m(x)$ by deleting all zeros (two neighbouring numbers, say f_k, f_{k+1}, form a sign change if $f_k f_{k+1} < 0$). Let us remark that the function $Z(x)$, $x \in \mathbf{R}$, can change its value only when passing through a root of some polynomial $f_1, \ldots, f_m$. If, however, for some $k = 2, \ldots, m-1$ and $\tilde{x} \in \mathbf{R}$, $f_k(\tilde{x}) = 0$, then

$$f_{k-1}(\tilde{x}) = -f_{k+1}(\tilde{x}).$$

Since polynomials f_1 and f_2 have no common divisors the same is true for f_{k-1}, f_{k+1} and, in particular, $f_{k-1}(\tilde{x}) \neq 0$, $f_{k+1}(\tilde{x}) \neq 0$. Consequently, $f_{k-1}(\tilde{x})f_{k+1}(\tilde{x}) < 0$, and for $x \neq \tilde{x}$ but close to $\tilde{x}$ the sequence of the signs of the numbers $f_{k-1}(x)$, $f_k(x)$, $f_{k+1}(x)$ can be one of the following types: $+, +, -; +, -, -; -, +, +; -, -, +$. Therefore, the function $Z(z)$, $x \in \mathbf{R}$ does not change its value when x passes through $\tilde{x}$. Since the polynomial f_m is of degree zero, Z may change its value only at the roots of f_1. In particular, it takes constant values $Z(-\infty)$, $Z(+\infty)$ for $x < x_1$ and $x > x_n$ respectively. But for x close to but smaller than a root x_k the signs of $f_1(x)$ and $f_2(x)$ are the same, and for x close to but greater than x_k, they are opposite. Therefore the function Z increases at x_k by 1. So $Z(+\infty) - Z(-\infty) = n$. Since for arbitrary x, $Z(x) \geq 0$, $Z(x) \leq m-1 \leq n$, hence $Z(+\infty) = n + Z(-\infty) \geq n$ and $m = n+1$, $Z(+\infty) = n$, $Z(-\infty) = 0$. It is easy to see that $Z(+\infty)$ is equal to the number of sign changes in the sequence of the leading coefficients of the polynomials $f_1, \ldots, f_{n+1}$, and since $Z(+\infty) = n$, these signs alternate. This shows the theorem in one direction.

To prove the opposite implication let us remark that the equality $m = n+1$ implies that U and V have no common divisors. In particular the polynomial p has no purely imaginary roots, and Lemma 2.2 is applicable. From the first part of the proof

$$\lim_{r \uparrow +\infty} \frac{1}{i} \int_{S(r)} \frac{p'(\lambda)}{p(\lambda)} d\lambda = n\pi,$$

so

$$\frac{1}{i} \int_{I(r)} \frac{p'(\lambda)}{p(\lambda)} d\lambda \longrightarrow n\pi - D_r.$$

§ 2.3. The Routh theorem

On the other hand, for x sufficiently large, $Z(x) = Z(+\infty) = n$, and, consequently, $Z(-\infty) = 0$. Again, from the first part of the proof, the function Z can change its value only at the roots of the polynomial $f_1 = U$, and the change in fact takes place only when the signs of the pair f_1, f_2 alternate. Each such change implies that the curve $\gamma(r)$ passes either from the first to the second or from the third to the fourth quadrants of the complex plane (we use here that n is an even number). The number of such passages must be equal exactly to n, and the arguments of the initial and final points of the curve $\gamma(r)$ tend either to 0 or π, as $r \uparrow +\infty$, so

$$\frac{1}{i}\int_{I(r)} \frac{p'(\lambda)}{p(\lambda)} d\lambda = \frac{1}{i}\int_{\gamma(r)} \frac{1}{z} dz \longrightarrow n\pi, \quad r \uparrow +\infty.$$

Hence finally $D_r = 0$ for $r > 0$. □

Let us apply the above theorem to polynomials of degree 4,

$$p(\lambda) = \lambda^4 + a\lambda^3 + b\lambda^2 + c\lambda + d, \quad \lambda \in \mathbf{C}.$$

In this case

$$U(x) = x^4 - bx^2 + d = f_1(x),$$
$$V(x) = -ax^3 + cx = f_2(x), \quad x \in \mathbf{R}.$$

Performing appropriate divisions we obtain

$$f_3(x) = \left(b - \frac{c}{a}\right)x^2 - d,$$
$$f_4(x) = -\left(c - ad\left(b - \frac{c}{a}\right)^{-1}\right)x,$$
$$f_5(x) = d.$$

The leading coefficients of the polynomials $f_1, f_2, \ldots, f_5$ are

$$1, \quad -a, \quad \left(b - \frac{c}{a}\right), \quad -\left(c - ad\left(b - \frac{c}{a}\right)^{-1}\right), \quad d.$$

We obtain therefore the following necessary and sufficient conditions for the stability of the polynomial p:

$$a > 0, \quad b - \frac{c}{a} > 0, \quad c - ad\left(b - \frac{c}{a}\right) > 0, \quad d > 0,$$

equivalent to those stated in Theorem 2.4.

We leave as an exercise the proof that the Routh theorem leads to an explicit stability algorithm. To formulate it we have to define the so-called *Routh array*.

For arbitrary sequences (α_k), (β_k), the *Routh sequence* (γ_k) is defined by
$$\gamma_k = -\frac{1}{\beta_1} \det \begin{bmatrix} \alpha_1 & \alpha_{k+1} \\ \beta_1 & \beta_{k+1} \end{bmatrix}, \quad k = 1, 2, \ldots$$

If $a_1, \ldots, a_n$ are coefficients of a polynomial p, we set additionaly $a_k = 0$ for $k > n = \deg p$. The *Routh array* is a matrix with infinite rows obtained from the first two rows:
$$1, a_2, a_4, a_6, \ldots,$$
$$a_1, a_3, a_5, a_7, \ldots,$$
by consequtive calculations of the Routh sequences from the two preceding rows. The calculations stop if 0 appears in the first column. The Routh algorithm can be now stated as the theorem

Theorem 2.6. *A polynomial p of degree n is stable if and only if the $n+1$ first elements of the first columns of the Routh array are positive.*

Exercise 2.4. Show that, for an arbitrary polynomial $p(\lambda) = \lambda^n + a_1 \lambda^{n-1} + \ldots + a_n$, $\lambda \in \mathbb{C}$, with complex coefficients $a_1, \ldots, a_n$, the polynomial $(\lambda^n + a_1\lambda^{n-1} + \ldots + a_n)(\lambda^n + \bar{a}_1\lambda^{n-1} + \ldots + \bar{a}_n)$ has real coefficients. Formulate necessary and sufficient conditions for the polynomial p to have all roots with negative real parts.

§2.4. Stability, observability, and the Liapunov equation

An important role in stability considerations is played by the so-called *matrix Liapunov equation* of the form
$$A^*Q + QA = -R, \tag{2.9}$$
in which are given a symmetric matrix $R \in \mathbf{M}(n,n)$ and a matrix $A \in \mathbf{M}(n,n)$ and unknown is a symmetric matrix $Q \in \mathbf{M}(n,n)$. A connection between the equation (2.9) and the stability question is illustrated by the following theorem:

Theorem 2.7. (1) *Assume that the pair (A,C) is observable and $R = C^*C$. If there exists a nonnegative definite matrix Q satisfying (2.9) then the matrix A is stable.*

(2) *If the matrix A is stable then for an arbitrary symmetric matrix R the equation (2.9) has exactly one symmetric solution Q. This solution is positive (nonnegative) definite if the matrix R is positive (nonnegative) definite.*

Proof. (1) It follows from (2.9) that
$$\frac{d}{dt}S^*(t)QS(t) = -S^*(t)RS(t), \quad t \geq 0. \tag{2.10}$$

§ 2.4. Stability, observability, and the Liapunov equation

Let (see § 1.6)

$$R_T = \int_0^T S^*(r) R S(r)\, dr = \int_0^T S^*(r) C^* C S(r)\, dr, \quad T \geq 0.$$

Integrating equation (2.10) from 0 to T we obtain that

$$Q = R_T + S^*(T) Q S(T).$$

It follows from the observability of (A, C) that the matrix R_T is positive definite and therefore the matrix Q is also positive definite. Let $x \in \mathbf{R}^n$ be a fixed vector different from 0. Define

$$v(t) = \langle Q z^x(t), z^x(t) \rangle, \quad t > 0.$$

It is enough to show that $v(t) \longrightarrow 0$ as $t \uparrow +\infty$.

Let us remark that

$$\begin{aligned} \frac{dv}{dt}(t) &= -\langle R z^x(t),\ z^x(t) \rangle \\ &= -|C z^x(t)|^2, \quad t \geq 0, \\ &\leq 0. \end{aligned}$$

The function v is nondecreasing on $[0, +\infty)$, and therefore the trajectory $z^x(t)$, $t \geq 0$, is bounded and $\sup(|z^x(t)|;\ t \geq 0) < +\infty$ for arbitrary $x \in \mathbf{R}^n$. This implies easily that for arbitrary $\lambda \in \sigma(A)$, $\operatorname{Re} \lambda \leq 0$. Assume that for an $\omega \in \mathbf{R}$, $i\omega \in \sigma(A)$. If $\omega = 0$ then, for a vector $a \neq 0$ with real elements, $Aa = 0$. Then $0 = \langle (A^* Q + Q A) a, a \rangle = -|Ca|^2$, a contradiction with Theorem 1.6(v). If $\omega \neq 0$ and $a_1, a_2 \in \mathbf{R}^n$ are vectors such that $a_1 + i a_2 \neq 0$ and

$$i\omega(a_1 + i a_2) = A a_1 + i A a_2,$$

then the function

$$z^{a_1}(t) = a_1 \cos \omega t - a_2 \sin \omega t, \quad t \geq 0, \tag{2.11}$$

is a periodic solution of the equation

$$\dot{z} = A z.$$

Periodicity and the formula

$$\frac{d}{dt} \langle Q z^{a_1}(t), z^{a_1}(t) \rangle = -|C z^{a_1}(t)|^2 \leq 0, \quad t \geq 0,$$

imply that for a constant $\gamma \geq 0$

$$\langle Qz^{a_1}(t), z^{a_1}(t)\rangle = \gamma, \quad t \geq 0.$$

Hence $|Cz^{a_1}(t)| = 0$ for $t \geq 0$. However, by the observability of (A, C) and (2.11), $a_1 = a_2 = 0$, a contradiction. So $\operatorname{Re}\lambda < 0$ for all $\lambda \in \sigma(A)$ and the matrix A is stable.

(2) Assume that the matrix A is stable. Then the matrix

$$Q = \int_0^{+\infty} S^*(r)RS(r)\,dr = \lim_{T\uparrow+\infty} R_T$$

is well defined. Moreover

$$\begin{aligned}
A^*Q + QA &= \int_0^{+\infty} (A^*S^*(r)RS(r) + S^*(r)RS(r)A)\,dr \\
&= \int_0^{+\infty} \frac{d}{dr}S^*(r)RS(r)\,dr \\
&= \lim_{T\uparrow+\infty} (S^*(T)RS(T) - R) \\
&= -R,
\end{aligned}$$

and therefore Q satisfies the Liapunov equation. It is clear that if the matrix R is positive (nonnegative) definite then also the matrix Q is positive (nonnegative) definite.

It remains to show the uniqueness. Assume that $\tilde{Q}$ is also a symmetric solution of the equation (2.19). Then

$$\frac{d}{dt}(S^*(t)(\tilde{Q} - Q)S(t)) = 0 \quad \text{for } t \geq 0.$$

Consequently

$$S^*(t)(\tilde{Q} - Q)S(t) = \tilde{Q} - Q \quad \text{for } t \geq 0.$$

Since $S(t) \longrightarrow 0$, $S^*(t) \longrightarrow 0$ as $t \uparrow +\infty$, $0 = \tilde{Q} - Q$. The proof of the theorem is complete. □

Remark. Note that the pair (A, I) is always observable, so if there exists a nonnegative solution Q of the equation

$$A^*Q + QA = -I,$$

then A is a stable matrix.

As a corollary we deduce the following result:

Theorem 2.8. *If a pair (A, C) is observable and for arbitrary $x \in \mathbf{R}^n$*

$$w(t) = Cz^x(t) \longrightarrow 0, \quad \text{as } t \uparrow +\infty,$$

then the matrix A is stable and consequently $z^x(t) \longrightarrow 0$ as $t \uparrow +\infty$ for arbitrary $x \in \mathbf{R}^n$.

Proof. For arbitrary $\beta > 0$ a nonnegative matrix

$$Q_\beta = \int_0^{+\infty} e^{-2\beta t} S^*(t) C^* C S(T) \, dt$$

is well defined and satisfies the equation

$$(A - \beta)^* Q + Q(A - \beta) = -C^* C.$$

Since, for sufficiently small $\beta > 0$, the pair $((A - \beta), C)$ is also observable, therefore for arbitrary $\beta > 0$, $\sup\{\operatorname{Re} \lambda; \lambda \in \sigma(A)\} \leq \beta$. One has only to show that the matrix A has no eigenvalues on the imaginary axis. For this purpose it is enough to repeat the reasoning from the final part of the proof of Theorem 2.7(1). □

Exercise 2.5. Deduce from Theorem 2.8 the following result:

If for arbitrary initial conditions

$$z(0) = \xi_1, \quad \frac{dz}{dt}(0) = \xi_2, \quad \ldots, \quad \frac{d^{(n-1)}z}{dt^{(n-1)}}(0) = \xi_n$$

solutions $z(t)$, $t > 0$ of the equation (1.30) tend to 0 as $t \longrightarrow +\infty$ then for arbitrary $k = 1, 2, \ldots, n-1$

$$\frac{d^{(k)}z}{dt^{(k)}} \longrightarrow 0, \quad \text{as } t \uparrow +\infty.$$

§ 2.5. Stabilizability and controllability

We say that the system

$$\dot{y} = Ay + Bu, \quad y(0) = x \in \mathbf{R}^n, \tag{2.12}$$

is *stabilizable* or that the pair (A, B) is *stabilizable* if there exists a matrix $K \in \mathbf{M}(m, n)$ such that the matrix $A + BK$ is stable. So if the pair (A, B) is stabilizable and a control $u(\cdot)$ is given in the *feedback* form

$$u(t) = Ky(t), \quad t \geq 0,$$

then all solutions of the equation

$$\dot{y}(t) = Ay(t) + BKy(t) = (A + BK)y(t), \quad y(0) = x, \ t \geq 0, \quad (2.13)$$

tend to zero as $t \uparrow +\infty$.

We say that system (2.12) is *completely stabilizable* if and only if for arbitrary $\omega > 0$ there exist a matrix K and a constant $M > 0$ such that for an arbitrary solution $y^x(t)$, $t \geq 0$, of (2.13)

$$|y^x(t)| \leq Me^{-\omega t}|x|, \quad t \geq 0. \quad (2.14)$$

By p_K we will denote the characteristic polynomial of the matrix $A + BK$. One of the most important results in the linear control theory is given by

Theorem 2.9. *The following conditions are equivalent:*
  (i) *System (2.12) is completely stabilizable.*
  (ii) *System (2.12) is controllable.*
  (iii) *For arbitrary polynomial $p(\lambda) = \lambda^n + \alpha_1 \lambda^{n-1} + \ldots + \alpha_n$, $\lambda \in \mathbb{C}$, with real coefficients, there exists a matrix K such that*

$$p(\lambda) = p_K(\lambda) \quad \text{for } \lambda \in \mathbb{C}.$$

Proof. We start with the implication (ii)$\Longrightarrow$(iii) and prove it in three steps.

Step 1. The dimension of the space of control parameters $m = 1$. It follows from §1.4 that we can limit our considerations to systems of the form

$$\frac{d^{(n)}z}{dt^{(n)}}(t) + a_1 \frac{d^{(n-1)}z}{dt^{(n-1)}}(t) + \ldots + a_n z(t) = u(t), \quad t \geq 0.$$

In this case, however, (iii) is obvious: It is enough to define the control u in the feedback form,

$$u(t) = (a_1 - \alpha_1) \frac{d^{(n-1)}z}{dt^{(n-1)}}(t) + \ldots + (a_n - \alpha_n)z(t), \quad t \geq 0,$$

and use the result (see §1.4) that the characteristic polynomial of the equation

$$\frac{d^{(n)}z}{dt^{(n)}} + \alpha_1 \frac{d^{(n-1)}z}{dt^{(n-1)}} + \ldots + \alpha_n z = 0,$$

or, equivalently, of the matrix

$$\begin{bmatrix} 0 & 1 & \cdots & 0 & 0 \\ 0 & 0 & \cdots & 0 & 0 \\ \vdots & \vdots & \ddots & \vdots & \vdots \\ 0 & 0 & \cdots & 0 & 1 \\ -\alpha_n & -\alpha_{n-1} & \cdots & -\alpha_2 & -\alpha_1 \end{bmatrix},$$

§2.5. Stabilizability and controllability

is exactly
$$p(\lambda) = \lambda^n + \alpha_1 \lambda^{n-1} + \ldots + \alpha_n \lambda, \quad \lambda \in \mathbf{C}.$$

Step 2. The following lemma allows us to reduce the general case to $m = 1$. Note that in its formulation and proof its vectors from $\mathbf{R}^n$ are treated as one-column matrices.

Lemma 2.4. *If a pair (A, B) is controllable then there exist a matrix $L \in \mathbf{M}(m, n)$ and a vector $v \in \mathbf{R}^m$ such that the pair $(A + BL, Bv)$ is controllable.*

Proof of the lemma. It follows from the controllability of (A, B) that there exists $v \in \mathbf{R}^m$ such that $Bv \neq 0$. We show first that there exist vectors $u_1, \ldots, u_{n-1}$ in $\mathbf{R}^m$ such that the sequence $e_1, \ldots, e_n$ defined inductively

$$e_1 = Bv, \; e_{l+1} = Ae_l + Bu_l \quad \text{for } l = 1, 2, \ldots, n-1 \qquad (2.15)$$

is a basis in $\mathbf{R}^n$. Assume that such a sequence does not exist. Then for some $k \geq 0$ vectors $e_1, \ldots, e_k$, corresponding to some $u_1, \ldots, u_k$ are linearly independent, and for arbitrary $u \in \mathbf{R}^m$ the vector $Ae_k + Bu$ belongs to the linear space E_0 spanned by $e_1, \ldots, e_k$. Taking $u = 0$ we obtain $Ae_k \in E_0$. Thus $Bu \in E_0$ for arbitrary $u \in \mathbf{R}^m$ and consequently $Ae_j \in E_0$ for $j = 1, \ldots, k$. This way we see that the space E_0 is invariant for A and contains the image of B. Controllability of (A, B) implies now that $E_0 = \mathbf{R}^n$, and compare the remark following Theorem 1.5. Consequently $k = n$ and the required sequences $e_1, \ldots, e_n$ and $u_1, \ldots, u_{n-1}$ exist. Let u_n be an arbitrary vector from $\mathbf{R}^m$.

We define the linear transformation L setting $Le_l = u_l$, for $l = 1, \ldots, n$. We have from (2.15)

$$\begin{aligned} e_{l+1} &= Ae_l + BLe_l = (A + BL)e_l \\ &= (A + BL)^l e_1 \\ &= (A + BL)^l Bv, \quad l = 0, 1, \ldots, n-1. \end{aligned}$$

Since
$$[A + BL | Bv] = [e_1, e_2, \ldots, e_n],$$
the pair $(A + BL, Bv)$ is controllable. $\square$

Step 3. Let a polynomial p be given and let L and v be the matrix and vector constructed in the Step 2. The system

$$\dot{y} = (A + BL)y + (Bv)u,$$

in which $u(\cdot)$ is a scalar control function, is controllable. It follows from Step 1 that there exists $k \in \mathbf{R}^n$ such that the characteristic polynomial of $(A + BL) + (Bv)k^* = A + B(L + vk^*)$ is identical with p.

The required feedback K can be defined as

$$K = L + vk^*.$$

We proceed to the proofs of the remaining implications. To show that (iii) $\Longrightarrow$ (ii) assume that (A, B) is not controllable, that rank $[A|B] = l < n$ and that K is a linear feedback. Let $P \in \mathbf{M}(n,n)$ be a nonsingular matrix from Theorem 1.5. Then

$$\begin{aligned} p_K(\lambda) &= \det[\lambda I - (A + BK)] \\ &= \det[\lambda I - (PAP^{-1} + PBKP^{-1})] \\ &= \det \begin{bmatrix} (\lambda I - (A_{11} + B_1 K_1)) & -A_{12} \\ 0 & (\lambda I - A_{22}) \end{bmatrix} \\ &= \det[\lambda I - (A_{11} + B_1 K_1)] \det[\lambda I - A_{22}], \quad \lambda \in \mathbf{C}, \end{aligned}$$

where $K_1 \in \mathbf{M}(m,n)$. Therefore for arbitrary $K \in \mathbf{M}(m,n)$ the polynomial p_K has a nonconstant divisor, equal to the characteristic polynomial of A_{22}, and therefore p_K can not be arbitrary. This way the implication (iii) $\Longrightarrow$ (ii) holds true.

Assume now that condition (i) holds but that the system is not controllable. By the above argument we have for arbitrary $K \in \mathbf{M}(m,n)$ that $\sigma(A_{22}) \subset \sigma(A + BK)$. So if for some $M > 0$, $\omega > 0$ condition (2.14) holds then

$$\omega \leq -\sup\{\operatorname{Re} \lambda; \ \lambda \in \sigma(A_{22})\},$$

which contradicts complete stabilizability. Hence (i) $\Longrightarrow$ (ii). Assume now that (ii) and therefore (iii) hold. Let $p(\lambda) = \lambda^n + a_1 \lambda^{n-1} + \ldots + a_n$, $\lambda \in \mathbf{C}$ be a polynomial with all roots having real parts smaller than $-\omega$ (e.g., $p(\lambda) = (\lambda + \omega + \varepsilon)^n, \varepsilon > 0$). We have from (iii) that there exists a matrix K such that $p_K(\cdot) = p(\cdot)$. Consequently all eigenvalues of $A + BK$ have real parts smaller than $-\omega$. By Theorem 2.3, condition (i) holds. The proof of Theorem 2.9 is complete. □

§2.6. Detectability and dynamical observers

We say that a pair (A, C) is *detectable* if there exists a matrix $L \in \mathbf{M}(n,k)$ such that the matrix $A + LC$ is stable. It follows from Theorem 2.9 that controllability implies stabilizability. Similarly, observability implies detectability. For if the pair (A, C) is observable then (A^*, C^*) is controllable and there exists a matrix K such that $A^* + C^* K$ is a stable matrix. Therefore the matrix $A + K^* C$ is stable and it is enough to set $L = K^*$.

We illustrate the importance of the concept of detectability by discussing the *dynamical observer* introduced by Luenberger.

§ 2.6. Detectability and dynamical observers

Let us consider system (1.1)-(1.2). Since the observation $w = Cy$, it is natural to define an *output stabilizing feedback* as $K \in \mathbf{M}(m,k)$ such that the matrix $A + BKC$ is stable. It turns out, however, that even imposing conditions like controllability of (A, B) and observability of (A, C) it is not possible, in general, to construct the desired K. In short, the *output stabilizability* is only very rarely possible.

Consider for instance the system

$$\frac{d^{(n)}z}{dt^{(n)}}(t) + \frac{d^{(n-1)}z}{dt^{(n-1)}}(t) + \ldots + a_n z = u \qquad (2.16)$$

with the observation relation

$$w(t) = z(t), \quad t \geq 0. \qquad (2.17)$$

We know from earlier considerations that if system (2.16)-(2.17) is written in the form (1.1)-(1.2) then pairs (A, B), (A, C) are respectively controllable and observable. Let a feedback strategy u be of the form $u = kw$ where k is a constant. Then the system (2.16) becomes

$$\frac{d^{(n)}z}{dt^{(n)}} + a_1 \frac{d^{(n-1)}z}{dt^{(n-1)}} + \ldots + a_n z = -kz,$$

and the corresponding characteristic polynomial is of the form

$$p(\lambda) = \lambda^n + a_1 \lambda^{n-1} + \ldots + a_{n-1}\lambda + a_n - k, \quad \lambda \in \mathbf{C}.$$

So if some of the coefficients $a_1, \ldots, a_{n-1}$ are negative then there exists no k such that the obtained system is stable.

Therefore it was necessary to extend the concept of stabilizability allowing feedback based on dynamical observers which we define as follows: We say that matrices $F \in \mathbf{M}(r,r)$, $H \in \mathbf{M}(r,n)$, $D \in \mathbf{M}(r,m)$ define an *r-dimensional dynamical observer* if for arbitrary control $u(\cdot)$, solution $z(\cdot)$ of the equation

$$\dot{z}(t) = Fz(t) + Hw(t) + Du(t), \quad t \geq 0, \qquad (2.18)$$

and the *estimator*

$$\hat{y}(t) = Vw(t) + Wz(t), \quad t \geq 0, \qquad (2.19)$$

one has

$$y(t) - \hat{y}(t) \longrightarrow 0, \quad \text{as } t \uparrow +\infty. \qquad (2.20)$$

The equation (2.18) and the number r are called respectively the *equation* and the *dimension* of the *observer*.

The following theorem gives an example of an n-dimensional observer and shows its applicability to stabilizability.

Theorem 2.10. (1) *Assume that the pair (A, C) is detectable and let L be a matrix such that the matrix $A + LC$ is stable. Then equations*

$$\dot{z} = (A + LC)z - Lw + Bu,$$
$$\hat{y} = z$$

define a dynamical observer of order n.

(2) *If, in addition, the pair (A, B) is stabilizable, K is a matrix such that $A + BK$ is stable and a control u is given by*

(2.21) $$u = K\hat{y},$$

then $y(t) \longrightarrow 0$ as $t \uparrow +\infty$.

Proof. (1) From the definition of the observer

$$\frac{d}{dt}(\hat{y} - y) = (A + LC)\hat{y} + Bu - LCy - Ay - Bu$$
$$= (A + LC)(\hat{y} - y).$$

Since the matrix $A + LC$ is stable, therefore $\hat{y}(t) - y(t) \longrightarrow 0$ exponentially as $t \uparrow +\infty$.

(2) If the control u is of the form (2.21), then

$$\dot{\hat{y}} = (A + BK)\hat{y} + LC(\hat{y} - y).$$

Therefore

$$\hat{y}(t) = S_K(t)y(0) + \int_0^t S_K(t-s)\varphi(s)\,ds, \quad t \geq 0,$$

where

$$S_K(t) = \exp t(A + BK),$$
$$\varphi(t) = LC(\hat{y}(t) - y(t)), \quad t \geq 0.$$

There exist constants $M > 0$, $\omega > 0$, such that

$$|S_K(t)| \leq Me^{-\omega t}, \quad t \geq 0.$$

Consequently for arbitrary $T > 0$ and $t \geq T$

$$\left|\int_0^t S_K(t-s)\varphi(s)\,ds\right| \leq Me^{-\omega t}\int_0^T e^{\omega s}|\varphi(s)|\,ds + \sup_{T \leq s \leq t}|\varphi(s)|\frac{M}{\omega}.$$

So we see that $\hat{y}(t) \longrightarrow 0$ as $t \uparrow +\infty$ and by (1), $y(t) \longrightarrow 0$ as $t \uparrow +\infty$. □

Remark. The obtained result can be summarised as follows. If the pair (A, C) is detectable and the pair (A, B) is stabilizable then the system (1.1)-(1.2) is output stabilizable with respect to the modified output $z(t)$, $t \geq 0$, based on the original observation $w(t)$, $t \geq 0$.

Bibliographical notes

In the proof of the Routh theorem we basically follow Gantmacher [28]. There exist proofs which do not use analytic function theory. In particular, in [43] the proof is based on Theorem 2.7 from § 2.4. For numerous modifications of the Routh algorithm we refer to [61]. The proof of Theorem 2.9 is due to M. Wonham [60]. The main aim of § 2.6 was to illustrate the concept of detectability.

Chapter 3
Realization theory

This chapter is devoted to the input–output map generated by a linear control system. The input–output map is characterized in terms of the impulse response function and the transfer function.

§3.1. Impulse response and transfer functions

The connection between a control $u(\,\cdot\,)$ and the observation $w(\,\cdot\,)$ of the system (1.1)-(1.2) is given by

$$w(t) = CS(t)x + \int_0^t Ce^{A(t-s)}Bu(s)\,ds \qquad (3.1)$$
$$= CS(t)x + \mathcal{R}u(t), \quad t \geq 0.$$

The operator $\mathcal{R}$ defined above is called the *input–output map* of (1.1)-(1.2). Thus

$$\mathcal{R}u(t) = \int_0^t \Psi(t-s)u(s)\,ds, \qquad (3.2)$$

where

$$\Psi(t) = Ce^{At}B, \quad t \geq 0. \qquad (3.3)$$

The function Ψ is called the *impulse response function* of the system (1.1)-(1.2). Obviously there exists a one-to-one correspondence between input–output mappings and impulse response functions. However, different triplets (A, B, C) may define the same impulse response functions. Each of them is called a *realization* of Ψ or, equivalently, a *realization* of $\mathcal{R}$.

The realization theory of linear systems is concerned with the structural properties of input–output mappings $\mathcal{R}$ or impulse response functions Ψ as well as with constructions of their realizations. The theory has practical motivations which we will discuss very briefly.

In various applications, matrices A, B and C defining system (1.1)-(1.2) can be obtained from a consideration of physical laws. It happens, however, that the actual system is too complicated to get its description in such a way. Then the system is tested by applying special inputs and observing the corresponding outputs, and the matrices A, B, C are chosen to match the experimental results in the best way.

Observed outputs corresponding to the *impulse inputs* lead directly to the impulse response function defined above. This follows from

§ 3.1. Impulse response and transfer functions

Proposition 3.1. *Let v be a fixed element of $\mathbf{R}^m$, $u_k(\cdot)$, $k = 1, 2, \ldots$, controls of the form*

$$u_k(t) = \begin{cases} kv & \text{for } t \in [0, \tfrac{1}{k}), \\ 0 & \text{for } t \geq \tfrac{1}{k}, \end{cases}$$

and $y_k(\cdot)$ the corresponding solutions of the equations

$$\frac{d}{dt} y_k = A y_k + B u_k, \quad y_k(0) = x.$$

Then the sequence (y_k) converges almost uniformly on $(0, +\infty)$ to the function

$$y(t) = S(t)x + S(t)Bv, \quad t \geq 0. \tag{3.4}$$

Proof. Let us remark that

$$y_k(t) = \begin{cases} S(t)x + k \int_0^t S(r) Bv \, dr & \text{for } t \in [0, \tfrac{1}{k}], \\ S(t)x + S\left(t - \tfrac{1}{k}\right) k \int_0^{1/k} S(r) Bv \, dr & \text{for } t \geq \tfrac{1}{k}. \end{cases}$$

Therefore, for $t \geq \tfrac{1}{k}$,

$$|y_k(t) - y(t)| = \left| S(t - \tfrac{1}{k}) \left[k \int_0^{1/k} S(r) Bv \, dr - S(\tfrac{1}{k}) Bv \right] \right|.$$

Moreover

$$\left| k \int_0^{1/k} S(r) Bv \, dr - S(\tfrac{1}{k}) Bv \right| \longrightarrow 0, \quad \text{as } k \uparrow +\infty,$$

and the lemma easily follows. □

The function

$$w(t) = C S(t) x + C e^{At} Bv, \quad t \geq 0,$$

can be regarded as an output corresponding to the impulse input $u(\cdot) = v \delta_{\{0\}}(\cdot)$, where $\delta_{\{0\}}(\cdot)$ is the Dirac distribution. If $x = 0$ then $w(t) = \Psi(t) v$, $t \geq 0$, and this is the required relation between the observation $w(\cdot)$ and the impulse response function Ψ.

It is convenient also to introduce the *transfer function* of (1.1)-(1.2) defined by

$$\Phi(\lambda) = C(\lambda I - A)^{-1} B, \quad \lambda \in \mathbf{C} \setminus \sigma(A). \tag{3.5}$$

Let us notice that if $\operatorname{Re} \lambda > \sup\{\operatorname{Re} \mu; \ \mu \in \sigma(A)\}$, then

$$\Phi(\lambda) = C \left[\int_0^{+\infty} e^{-\lambda t} S(t)\, dt \right] B = \int_0^{+\infty} e^{-\lambda t} \Psi(t)\, dt.$$

Hence the transfer function is the Laplace transform of the impulse response function, and they are in a one to one correspondence.

The transfer function can be constructed from the observations of outputs resulting from *periodic*, called also *harmonic*, inputs $u(\cdot)$:

$$u(t) = e^{it\omega} v, \quad v \in \mathbf{R}^m,\ \omega \in \mathbf{R},\ t \geq 0. \tag{3.6}$$

For them

$$w(t) = CS(t)x + e^{i\omega t} \int_0^t e^{-i\omega r} CS(r) Bv\, dr, \quad t \geq 0. \tag{3.7}$$

Practical methods are based on the following result.

Proposition 3.2. *Assume that $A \in \mathbf{M}(n,n)$ is a stable matrix. Let $u\colon \mathbf{R} \longrightarrow \mathbf{R}^m$ be a bounded, periodic function with a period γ and $y^x(t)$, $t \geq 0$, the solution of the equation*

$$\dot{y}(t) = Ay(t) + Bu(t), \quad t \geq 0,\ y(0) = x.$$

There exists exactly one periodic function $\hat{y}\colon \mathbf{R} \longrightarrow \mathbf{R}^n$ such that

$$|y^x(t) - \hat{y}(t)| \longrightarrow 0, \quad \text{as } t \uparrow +\infty.$$

Moreover, the period of $\hat{y}$ is also γ and

$$\hat{y}(t) = \int_0^{+\infty} S(r) Bu(t-r)\, dr, \quad t \in \mathbf{R}. \tag{3.8}$$

The function $\hat{y}(\cdot)$ is called sometimes a *periodic component* of the solution y.

Proof. The function $\hat{y}(\cdot)$ given by (3.8) is well defined, periodic with period γ. Since

$$y^x(t) = S(t)x + \int_0^t S(t-s) Bu(s)\, ds = S(t)x + \int_0^t S(r) Bu(t-r)\, dr$$

$$= S(t)x + \int_0^{+\infty} S(r) Bu(t-r)\, dr - \int_t^{+\infty} S(r) Bu(t-r)\, dr$$

$$= S(t)x + \hat{y}(t) - S(t) \int_0^{+\infty} S(r) Bu(-r)\, dr$$

$$= S(t)x + \hat{y}(t) - S(t)\hat{y}(0), \quad t \geq 0,$$

§ 3.1. Impulse response and transfer functions

therefore

$$|y^x(t) - \hat{y}(t)| \le |S(t)| \, |x - \hat{y}(0)| \longrightarrow 0, \quad \text{as } t \uparrow +\infty.$$

If $\hat{y}_1$ is an arbitrary function with the properties specified in the proposition, then

$$|\hat{y}(t) - \hat{y}_1(t)| \le |\hat{y}(t) - y^x(t)| + |y^x(t) - \hat{y}_1(t)|, \quad t \ge 0.$$

Consequently $|\hat{y}(\,\cdot\,) - \hat{y}_1(\,\cdot\,)| \longrightarrow 0$ as $t \uparrow +\infty$. By the periodicity of $\hat{y}$, $\hat{y}(t) - \hat{y}_1(t) = 0$ for all $t \ge 0$. □

It follows from Proposition 3.2 that the observed outputs of a stable system, corresponding to harmonic controls $e^{i\omega t}v$, $t \ge 0$, are approximately equal, with an error tending to zero, to

$$\hat{w}(t) = C\left[\int_0^{+\infty} S(r)e^{i\omega(t-r)}\,dr\right]Bv$$
$$= e^{i\omega t}C\left[\int_0^{+\infty} e^{-i\omega r}S(r)\,dr\right]Bv, \quad t \ge 0.$$

Taking into account that

$$\int_0^{+\infty} e^{-i\omega r}S(r)\,dr = (i\omega I - A)^{-1}, \quad \omega \in \mathbf{R},$$

we see that the periodic component of the output $w(\,\cdot\,)$ corresponding to the periodic input (3.6) is of the form

$$\hat{w}(t) = e^{i\omega t}\Phi(i\omega)v, \quad t \ge 0. \tag{3.9}$$

Assume that $m = k = 1$. Then the transfer function is meromorphic. Denote it by φ and consider for arbitrary $\omega \in \mathbf{R}$ the trigonometric representation

$$\varphi(i\omega) = |\varphi(i\omega)|e^{i\delta(\omega)}, \quad \delta(\omega) \in [0, 2\pi).$$

If $v = 1$ then the output $\hat{w}(\,\cdot\,)$ given by (3.9) is exactly

$$\hat{w}(t) = |\varphi(i\omega))|e^{i(\delta(\omega)+\omega t)}, \quad t \ge 0.$$

Thus, if $|\varphi(i\omega)|$ is large then system (1.1)-(1.2) amplifies the periodic oscillation of the period ω and, if $|\varphi(i\omega)|$ is small, filters them off.

Exercise 3.1. Find the transfer function for system (2.16)-(2.17).

Answer:

$$\varphi(\lambda) = p^{-1}(\lambda), \quad \text{gdy } p(\lambda) = \lambda^n + a_1\lambda^{n-1} + \ldots + a_n \neq 0.$$

Exercise 3.2. Consider the equation of an electrical filter, Example 0.4,

$$L^2C\frac{d^3z}{dt^3} + RLC\frac{d^2z}{dt^2} + 2L\frac{dz}{dt} + Rz = u,$$

where C, L, R are positive numbers. Find $\varphi(i\omega)$, $\omega \in \mathbf{R}$. Show that

$$\lim_{\omega \downarrow 0} |\varphi(i\omega)| = 1, \quad \lim_{\omega \uparrow +\infty} |\varphi(i\omega)|\frac{CL^2\omega^3}{R} = 1.$$

§3.2. Realizations of the impulse response function

A not arbitrary smooth function $\Psi(\cdot)$ with values in $\mathbf{M}(k,m)$, defined on $[0, +\infty)$, is the impulse response function of a linear system (1.1)-(1.2). Similarly, a not arbitrary operator $\mathcal{R}$ of the convolution type (3.2) is an input–output map of a linear system. In this paragraph we give two structural descriptions of the impulse response function and we show that among all triplets (A, B, C) defining the same function Ψ there exists, in a sense, a "best" one. Moreover, from Theorem 3.1 below, the following result will be obtained:

An operator $\mathcal{R}$ of the convolution type (3.2) is an input–output map of a linear system (1.1)-(1.2) if and only if it is of the form

$$\mathcal{R}u(t) = H(t) \int_0^t G(s)u(s)\,ds, \quad t \geq 0, \qquad (3.10)$$

where $H(\cdot)$ and $G(\cdot)$ are functions with values in $\mathbf{M}(k,n)$ and $\mathbf{M}(n,m)$, for some n.

Let us remark that operator (3.10) is the product of an integration and a multiplication operator.

Before formulating the next theorem we propose first to solve the following exercise.

Exercise 3.3. Continuous, real functions g, h and r defined on $[0, +\infty)$ satisfy the functional equation

$$g(t-s) = h(t)r(s), \quad t \geq s \geq 0,$$

if and only if either g and one of the functions h or r are identically zero, or if there exist numbers α, $\beta \neq 0$, $\gamma \neq 0$ such that

$$g(t) = \gamma e^{\alpha t}, \quad h(t) = \beta e^{\alpha t}, \quad r(t) = \frac{\gamma}{\beta}e^{-\alpha t}, \quad t \geq 0.$$

§ 3.2. Realizations of the impulse response function

Theorem 3.1. *A function Ψ of class C^1 defined on $[0,+\infty)$ and with values in $\mathbf{M}(k,m)$ is the impulse response function of a system (A,B,C) if and only if there exist a natural number n and functions $G\colon [0,+\infty) \longrightarrow \mathbf{M}(m,n)$, $H\colon [0,+\infty) \longrightarrow \mathbf{M}(n,k)$ of class C^1 such that*

$$\Psi(t-s) = H(t)G(s) \quad \text{for } t \geq s \geq 0. \tag{3.11}$$

Proof. If $\Psi(t) = Ce^{At}B$, $t \geq 0$, then it is sufficient to define

$$H(t) = Ce^{At}, \quad t \geq 0, \quad \text{and} \quad G(s) = e^{-As}B, \quad s \geq 0.$$

The converse implication requires a construction of a number n and matrices A, B, C in terms of the functions G, H. We show first the validity of the following lemma:

Lemma 3.1. *Assume that for some functions Ψ, G, H of class C^1, with values respectively in $\mathbf{M}(k,m)$, $\mathbf{M}(n,m)$, $\mathbf{M}(k,n)$, relation (3.11) holds. Then for arbitrary $T > 0$ there exist a natural number $\tilde{n}$ and functions $\tilde{G}(t)$, $\tilde{H}(t)$, $t \geq 0$ of class C^1 with values in $\mathbf{M}(\tilde{n},m)$ and $\mathbf{M}(k,\tilde{n})$ such that*

$$\Psi(t-s) = \tilde{H}(t)\tilde{G}(s), \quad s \leq t,\ 0 \leq s \leq T,$$

and the matrix

$$\int_0^T \tilde{G}(s)\tilde{G}^*(s)\,ds$$

is nonsingular and thus positive definite.

Proof of the lemma. Let $\tilde{n}$ be the rank of the matrix

$$W = \int_0^T G(s)G^*(s)\,ds.$$

If W is nonsingular then we define $\tilde{n} = n$, and the lemma is true. Assume therefore that $\tilde{n} < n$. Then there exists an orthogonal matrix P such that the diagonal of the matrix $PWP^{-1} = PWP^*$ is composed of eigenvalues of the matrix W. Therefore there exists a diagonal nonsingular matrix D such that

$$V = \begin{bmatrix} \tilde{I} & 0 \\ 0 & 0 \end{bmatrix} = (DP)W(DP)^* = \tilde{P}W\tilde{P}^*,$$

where $\tilde{I} \in \mathbf{M}(\tilde{n},\tilde{n})$ is the identity matrix and $\tilde{P}$ is a nonsingular one. Consequently, for $Q = (DP)^{-1}$,

$$QVQ^* = W.$$

This and the definition of W imply that

$$\int_0^T (QVQ^{-1}G(s) - G(s))(QVQ^{-1}G(s) - G(s))^*\,ds = 0.$$

Therefore $QVQ^{-1}G(s) = G(s)$, for $s \in [0,T]$.

Define $\widetilde{G}(t) = Q^{-1}G(t)$, $\widetilde{H}(t) = H(t)Q$, $t \geq 0$. Then

$$\Psi(t-s) = \widetilde{H}(t)V\widetilde{G}(s) \quad \text{for } t \geq s \text{ and } s \in [0,T].$$

It follows from the representation

$$\widetilde{H}(t) = \left[\widetilde{H}_1(t), \widetilde{H}_2(t)\right], \quad \widetilde{G}(s) = \begin{bmatrix} \widetilde{G}_1(s) \\ \widetilde{G}_2(s) \end{bmatrix},$$

where $\widetilde{H}_1(t) \in \mathbf{M}(k,\tilde{n})$, $\widetilde{H}_2(t) \in \mathbf{M}(k, n-\tilde{n})$, $G_1(s) \in \mathbf{M}(\tilde{n}, m)$, $\widetilde{G}_2(s) \in \mathbf{M}(n-\tilde{n}, m)$, $t,s \geq 0$, and from the identity $V^2 = V$ that

$$\Psi(t-s) = (\widetilde{H}(t)V)(V\widetilde{G}(s)) = \left[\widetilde{H}_1(t), 0\right] \begin{bmatrix} \widetilde{G}_1(s) \\ 0 \end{bmatrix}$$
$$= \widetilde{H}_1(t)\widetilde{G}_1(s), \quad t \geq s, \ s \in [0,T].$$

Hence

$$V = \int_0^T (Q^{-1}G(s))(Q^{-1}G(s))^* \, ds = \int_0^T \widetilde{G}(s)\widetilde{G}(s)^* \, ds,$$

and thus

$$\widetilde{I} = \int_0^T \widetilde{G}_1(s)\widetilde{G}_1^*(s) \, ds$$

is a nonsingular matrix of the dimension $\tilde{n}$. The proof of the lemma is complete. □

Continuation of the proof of Theorem 3.1. We can assume that functions H and G have the properties specified in the lemma. For arbitrary $t \geq s$, $0 \leq s \leq T$

$$0 = \frac{\partial \Psi}{\partial t}(t-s) + \frac{\partial \Psi}{\partial s}(t-s) = \dot{H}(t)G(s) + H(t)\dot{G}(s).$$

In particular

$$\dot{H}(t)G(s)G^*(s) = -H(t)\dot{G}(s)G^*(S), \quad t \geq s, \ s \in [0,T],$$

and thus for $t \geq T$ we obtain that

$$\dot{H}(t)\int_0^T G(s)G^*(s) \, ds = -H(t)\int_0^T \dot{G}(s)G^*(s) \, ds.$$

Define

$$A = -\left(\int_0^T \dot{G}(s)G^*(s) \, ds\right)\left(\int_0^T G(s)G^*(s) \, ds\right)^{-1}.$$

§ 3.2. Realizations of the impulse response function

Then $\dot{H}(t) = H(t)A$, $t \geq T$ and therefore $H(t) = H(T)S(t - T)$, $t \geq T$, where $S(r) = e^{rA}$, $r \geq 0$. However, for arbitrary $r \geq 0$.

$$\Psi(r) = \Psi((r+T) - T) = H(r+T)G(T) = H(T)S(r)G(T), \quad r \geq 0,$$

Consequently, defining $B = G(T)$ and $C = H(T)$, we obtain

$$\Psi(r) = Be^{rA}C, \quad r \geq 0,$$

the required representation. □

We now give a description of impulse response functions in terms of their Taylor expensions.

Theorem 3.2. *A function $\Psi: [0, +\infty) \longrightarrow \mathbf{M}(k, m)$ given in the form of an absolutely convergent series*

$$\Psi(t) = \sum_{j=0}^{+\infty} W_j \frac{t^j}{j!}, \quad t \geq 0,$$

is the impulse response function of a certain system (A, B, C) if and only if there exist numbers $a_1, a_2, \ldots, a_r$ such that

$$W_{j+r} = a_1 W_{j+r-1} + \ldots + a_r W_j, \quad j = 0, 1, \ldots. \tag{3.12}$$

Proof. (Necessity.) If $\Psi(t) = C(\exp tA)B$, $t \geq 0$, then

$$\Psi(t) = \sum_{j=0}^{+\infty} CA^j B \frac{t^j}{j!}, \quad t \geq 0,$$

and therefore $W_j = CA^j B$, $j = 0, 1, \ldots$.

If $p(\lambda) = \lambda^r - a_1 \lambda^{r-1} - \ldots - a_1$, $\lambda \in \mathbf{C}$, is the characteristic polynomial of A then, by the Cayley–Hamilton theorem,

$$A^r = a_1 A^{r-1} + \ldots + a_r I \tag{3.13}$$

and

$$A^{r+j} = a_1 A^{r+j-1} + \ldots + a_r A^j.$$

Therefore

$$CA^{r+j}B = a_1 CA^{r+j-1}B + \ldots + a_r CA^j B, \quad j = 0, \ldots,$$

the required relationship.

(Sufficiency.) Define

$$A = \begin{bmatrix} 0 & I & \cdots & 0 & 0 \\ 0 & 0 & \cdots & 0 & 0 \\ \vdots & \vdots & \ddots & \vdots & \vdots \\ 0 & 0 & \cdots & 0 & I \\ a_r I & a_{r-1} I & \cdots & a_2 I & a_1 I \end{bmatrix}, \quad B = \begin{bmatrix} W_0 \\ \vdots \\ W_{r-1} \end{bmatrix}, \quad C = [I, 0, \ldots, 0],$$

where I is the identity matrix of dimension k. Permuting rows and columns of the matrix A properly we can assume that it is of the form

$$\begin{bmatrix} \widetilde{A} & 0 & \cdots & 0 & 0 \\ 0 & \widetilde{A} & \cdots & 0 & 0 \\ \vdots & \vdots & \ddots & \vdots & \vdots \\ 0 & 0 & \cdots & \widetilde{A} & 0 \\ 0 & 0 & \cdots & 0 & \widetilde{A} \end{bmatrix} \quad \text{where } \widetilde{A} = \begin{bmatrix} 0 & 1 & \cdots & 0 & 0 \\ 0 & 0 & \cdots & 0 & 0 \\ \vdots & \vdots & \ddots & \vdots & \vdots \\ 0 & 0 & \cdots & 0 & 1 \\ a_1 & a_2 & \cdots & a_{r-1} & a_r \end{bmatrix}. \quad (3.14)$$

Since the characteristic polynomial of $\widetilde{A}$ is $\lambda^r - a_1 \lambda^{r-1} - \ldots - a_r$, $\lambda \in \mathbb{C}$, therefore $\widetilde{A}^r = a_1 \widetilde{A}^{r-1} + \ldots + a_r I$. The same equation holds for matrix (3.14) and consequently for A. We also check that

$$CA^j B = W_j \quad \text{for } j = 0, 1, \ldots, r - 1.$$

By (3.12),

$$CA^r B = a_1 CA^{r-1} B + \ldots + a_r CB = a_1 W_{r-1} + \ldots + a_r W_r = W_r.$$

Similarly, $CA^j B = W_j$ for $j \geq r$. $\square$

The next theorem shows that there exists a realization of an impulse response function with additional regularity properties.

Theorem 3.3. *For arbitrary matrices $A \in \mathbf{M}(n,n)$, $B \in \mathbf{M}(n,m)$, $C \in \mathbf{M}(k,n)$ there exist a natural number $\tilde{n}$ and matrices $\widetilde{A} \in \mathbf{M}(\tilde{n}, \tilde{n})$, $\widetilde{B} \in \mathbf{M}(\tilde{n}, m)$, $\widetilde{C} \in \mathbf{M}(k, \tilde{n})$ such that the pairs $(\widetilde{A}, \widetilde{B})$ and $(\widetilde{A}, \widetilde{C})$ are respectively controllable and observable and*

$$CS(t)B = \widetilde{C}\widetilde{S}(t)\widetilde{B} \quad \text{for } t \geq 0,$$

where $S(t) = \exp tA$, $\widetilde{S}(t) = \exp t\widetilde{A}$, $t \geq 0$.

Proof. Assume that the pair (A, B) is not controllable and let $l = \text{rank}[A|B]$. Let P, A_{11}, A_{12}, A_{22} and B_1 be matrices given by Theorem 1.5 and let $S(t) = \exp tA$, $S_{11}(t) = \exp(A_{11} t)$, $S_{22}(t) = \exp(A_{22} t)$, $t \geq 0$. There exists, in addition, a function $S_{12}(t)$, $t \geq 0$, with values in $\mathbf{M}(l, n-l)$, such that

$$PS(t)P^{-1} = \begin{bmatrix} S_{11}(t) & S_{12}(t) \\ 0 & S_{22}(t) \end{bmatrix}, \quad t \geq 0.$$

§3.2. Realizations of the impulse response function

Therefore,

$$CS(t)B = CP^{-1} \begin{bmatrix} S_{11}(t) & S_{12}(t) \\ 0 & S_{22}(t) \end{bmatrix} PB$$
$$= CP^{-1} \begin{bmatrix} S_{11} & S_{12}(t) \\ 0 & S_{22}(t) \end{bmatrix} \begin{bmatrix} B_1 \\ 0 \end{bmatrix}$$
$$= CP^{-1} \begin{bmatrix} S_{11}(t)B_1 \\ 0 \end{bmatrix} = C_1 S_{11}(t) B_1, \quad t \geq 0,$$

for some matrix C_1.

So we can assume, without any loss of generality, that the pair (A, B) is controllable. If the pair (A, C) is not observable and $\text{rank}[A|C] = l < n$ then, by Theorem 1.7, for a nonsingular matrix P, an observable pair $(A_{11}, C_1) \in \mathbf{M}(l, l) \times \mathbf{M}(k, l)$ and matrices $B_1 \in \mathbf{M}(l, n)$, $B_2 \in \mathbf{M}(n-l, m)$,

$$PAP^{-1} = \begin{bmatrix} A_{11} & 0 \\ A_{21} & A_{22} \end{bmatrix}, \quad CP^{-1} = [C_1, 0], \quad PB = \begin{bmatrix} B_1 \\ B_2 \end{bmatrix}.$$

It follows from the controllability of (A, B) that the pair (A_{11}, B_1) is controllable as well.

If $S(t) = \exp tA$, $S_{11}(t) = \exp tA_{11}$, $S_{22}(t) = \exp tA_{22}$, $t \geq 0$, then for an $\mathbf{M}(n-l, l)$ valued function $S_{21}(t)$, $t \geq 0$,

$$CS(t)B = [C_1, 0] PS(t)P^{-1} PB = [C_1, 0] \begin{bmatrix} S_{11}(t) & 0 \\ S_{21}(t) & S_{22}(t) \end{bmatrix} \begin{bmatrix} B_1 \\ B_2 \end{bmatrix}$$
$$= C_1 S_{11}(t) B_1, \quad t \geq 0.$$

This finishes the proof of the theorem. □

The following basic result is now an easy consequence of Theorem 3.3.

Theorem 3.4. *For any arbitrary impulse response function there exists a realization (A, B, C) such that the pairs (A, B), (A, C) are respectively controllable and observable. The dimension of the matrix A is the same for all controllable and observable realizations. Moreover, for an arbitrary realization $(\tilde{A}, \tilde{B}, \tilde{C})$, $\dim \tilde{A} \geq \dim A$.*

Proof. The first part of the theorem is a direct consequence of Theorem 3.3. Assume that (A, B, C) and $(\tilde{A}, \tilde{B}, \tilde{C})$ are two controllable and observable realizations of the same impulse response function and that $\dim \tilde{A} = \tilde{n}$, $\dim A = n$. Since

$$\Psi(t) = \sum_{j=0}^{+\infty} C A^j B \frac{t^j}{j!} = \sum_{j=0}^{+\infty} \tilde{C} \tilde{A}^j \tilde{B} \frac{t^j}{j!}, \quad t \geq 0,$$

so
$$\tilde{C}\tilde{A}^j\tilde{B} = CA^j B \quad \text{for } j = 0, 1, \ldots. \tag{3.15}$$

Moreover, the pair (A, B) is controllable and (A, C) is observable, and therefore the ranks of the matrices

$$\begin{bmatrix} C \\ CA \\ \vdots \\ CA^{r-1} \end{bmatrix}, \quad [B, AB, \ldots, A^{r-1}B]$$

are n for arbitrary $r \geq n$. Consequently the ranks of the matrices

$$\begin{bmatrix} C \\ CA \\ \vdots \\ CA^{r-1} \end{bmatrix} [B, AB, \ldots, A^{r-1}B] = \begin{bmatrix} CB & CAB & \ldots & CA^{r-1}B \\ CAB & CA^2B & \ldots & CA^r B \\ \vdots & \vdots & \ddots & \vdots \\ CA^{r-1}B & CA^r B & \ldots & CA^{2r-2}B \end{bmatrix} \tag{3.16}$$

are, for $r \geq n$, also equal to n. On the other hand, by (3.15), (3.16),

$$\begin{bmatrix} \tilde{C} \\ \tilde{C}\tilde{A} \\ \vdots \\ \tilde{C}\tilde{A}^{\tilde{n}-1} \end{bmatrix} [\tilde{B}, \tilde{A}\tilde{B}, \ldots, \tilde{A}^{\tilde{n}-1}\tilde{B}] = \begin{bmatrix} C \\ CA \\ \vdots \\ CA^{\tilde{n}-1} \end{bmatrix} [B, AB, \ldots, A^{\tilde{n}-1}B],$$

and therefore $\tilde{n} \leq n$. By the same argument $n \leq \tilde{n}$, so $n = \tilde{n}$. The proof that for arbitrary realization $(\tilde{A}, \tilde{B}, \tilde{C})$, $\tilde{n} \geq n$ is analogical. □

§3.3. The characterization of transfer functions

The transfer function of a given triplet (A, B, C) was defined by the formula
$$\Phi(\lambda) = C(\lambda I - A)^{-1}B, \quad \lambda \in \mathbb{C} \setminus \sigma(A).$$

Let $\Phi(\lambda) = [\varphi_{qr}(\lambda), q = 1, \ldots, k, r = 1, 2, \ldots, m]$, $\lambda \in \mathbb{C} \setminus \sigma(A)$. It follows, from the formulae for the inverse matrix $(\lambda I - A)^{-1}$, that functions φ_{qr} are rational with real coefficients for which the degree of the numerator is greater than the degree of the denominator. So they vanish as $|\lambda| \longrightarrow +\infty$. Moreover,

$$\Phi(\lambda) = \sum_{j=0}^{+\infty} \frac{1}{\lambda^{j+1}} CA^j B \quad \text{for } |\lambda| > |A|, \tag{3.17}$$

and the series in (3.17) is convergent with its all derivatives.

Theorem 3.5. *If $\Phi(\cdot)$ is an $\mathbf{M}(k,m;\mathbf{C})$ valued function, defined outside of a finite subset of $\mathbf{C}$, with rational elements having real coefficients vanishing as $|\lambda| \longrightarrow +\infty$, then the function $\Phi(\cdot)$ is of the form (3.9). Consequently, $\Phi(\cdot)$ is the transfer function of a system (A,B,C).*

Proof. There exists a polynomial $p(\lambda) = \lambda^r + a_1\lambda^{r-1} + \ldots + a_r$, $\lambda \in \mathbf{C}$, with real coefficients, such that for arbitrary elements φ_{qs}, $q = 1,\ldots,k$, $s = 1,\ldots,m$, of the matrix Φ, functions $p\varphi_{qs}$ are polynomials. Since $\Phi(\cdot)$ is rational for some $\gamma > 0$ and arbitrary λ, $|\lambda| > \gamma$,

$$\Phi(\lambda) = \sum_{j=0}^{+\infty} \frac{1}{\lambda^{j+1}} W_j.$$

The coefficients by λ^{-1}, λ^{-2}, ... of the expansion of $p(\lambda)\Phi(\lambda)$, $\lambda \in \mathbf{C}$, are zero, therefore

$$a_r W_0 + a_{r-1} W_1 + \ldots + a_1 W_{r-1} + W_r = 0,$$
$$a_r W_1 + a_{r-1} W_2 + \ldots + a_1 W_r + W_{r+1} = 0,$$
$$\vdots$$
$$a_r W_j + a_{r-1} W_{j+1} + \ldots + a_1 W_{r+j-1} + W_{r+j} = 0.$$

Hence the sequence $W_0, W_1, \ldots$ satisfies (3.12). Arguing as in the proof of Theorem 3.2, we obtain that there exist matrices A, B, C such that

$$W_j = CA^j B, \quad j = 0, 1, \ldots.$$

The proof is complete. □

Exercise 3.4. Assume that $k = m = 1$ and let

$$\varphi(\lambda) = \frac{1}{\lambda - a} + \frac{1}{(\lambda - b)^2}, \quad \lambda \in \mathbf{C} \setminus \{a, b\},$$

where a and b are some real numbers. Find the impulse response function $\Psi(\cdot)$ corresponding to φ. Construct its realization.
Answer. $\Psi(t) = e^{at} + te^{bt}$, $t \geq 0$.

Bibliographical notes

Realization theory was created by R. Kalman. He is the author of Theorems 3.2, 3.3 and 3.4. The proof of Theorem 3.1 is borrowed from R. Brockett [10].

Chapter 4
Systems with constraints

This chapter illustrates complications which arise in control theory when control parameters have to satisfy various constraints. Controllable systems with the set of controls being either a bounded neighbourhood of $0 \in \mathbf{R}^m$ or the positive cone $\mathbf{R}^m_+$ are characterized.

§4.1. Bounded sets of parameters

In the preceeding chapters sets of states and of control parameters were identical with $\mathbf{R}^n$ and $\mathbf{R}^m$. In several situations, however, it is necessary to consider controls taking values in subsets of $\mathbf{R}^m$ or to require that the system evolves in a given subset of $\mathbf{R}^n$. We will discuss two types of constraints for linear systems.
$$\dot{y} = Ay + Bu, \quad y(0) = x. \tag{4.1}$$
We start from the null controllability.

Proposition 4.1. *Assume that $0 \in \mathbf{R}^m$ is an interior point of a control set $U \subset \mathbf{R}^m$. Then there exists a neighbourhood V of $0 \in \mathbf{R}^n$ such that all its elements can be transferred to 0 by controls taking values in U.*

Proof. Let $T > 0$ be a given number. By Proposition 1.1 the control
$$\hat{u}(s) = -B^* S^*(T-s) Q_T^{-1} S(T) x, \quad s \in [0, T],$$
transfers the state x to 0 at the time T. Since the function $t \longrightarrow S^*(t)$ is continuous, therefore, for a constant $M > 0$,
$$|\hat{u}(s)| \leq M|x| \quad \text{for all } s \in [0, T], \ x \in \mathbf{R}^n.$$
So the result follows. $\square$

A possibility of transferring to 0 all points from $\mathbf{R}^n$ is discussed in the following theorem.

Theorem 4.1. *Assume that U is a bounded set.*

(i) If $0 \in \mathbf{R}^m$ is an interior point of U, (A, B) is a controllable pair and the matrix A is stable, then arbitrary $x \in \mathbf{R}^n$ can be transferred to 0 by controls with values in U only.

(ii) If the matrix A has at least one eingenvalue with a positive real part then there are states in $\mathbf{R}^n$ which cannot be transferred to 0 by controls with values in U only.

§4.1. Bounded sets of parameters

Proof. (i) If the matrix A is stable then the control $u(t) = 0$, for all $t \geq 0$, transfers an arbitrary state, in finite time, to a given in advance neighbourhood U of $0 \in \mathbf{R}^n$. It follows from Proposition 4.1 that if the neighbourhood V is sufficiently small then all its elements can be transferred to 0 by controls with values in U. Hence (i) holds.

(ii) Let $\lambda = \alpha + i\beta$ be an eigenvalue of A^* such that $\alpha > 0$. Assume that $\beta \neq 0$. Then there exist vectors $e_1, e_2 \in \mathbf{R}^n$, $|e_1| + |e_2| \neq 0$, such that

$$A^* e_1 + iA^* e_2 = (\alpha + i\beta)(e_1 + ie_2)$$

or equivalently

$$A^* e_1 = \alpha e_1 - \beta e_2, \quad A^* e_2 = \beta e_1 + \alpha e_2.$$

Let $u(\,\cdot\,)$ be a control taking values in U and $y(\,\cdot\,)$ the corresponding output:

$$\dot{y} = Ay + Bu, \quad y(0) = x.$$

Let v be the absolutely continuous function given by the formula

$$v(t) = \langle y(t), e_1\rangle^2 + \langle y(t), e_2\rangle^2, \quad t \geq 0.$$

Then, for almost all $t \geq 0$,

$$\begin{aligned}\dot{v}(t) &= 2(\langle \dot{y}(t), e_1\rangle\langle y(t), e_1\rangle + \langle \dot{y}(t), e_2\rangle\langle y(t), e_2\rangle) \\ &= 2(\langle y(t), A^* e_1\rangle\langle y(t), e_1\rangle + \langle y(t), A^* e_2\rangle\langle y(t), e_2\rangle) \\ &\quad + 2(\langle Bu(t), e_1\rangle\langle y(t), e_1\rangle + \langle Bu(t), e_2\rangle\langle y(t), e_2\rangle) \\ &= I_1(t) + I_2(t).\end{aligned}$$

From the definition of e_1 and e_2, $I_1(t) = 2\alpha v(t)$. Since the set U is bounded, therefore, for a constant $K > 0$, independent of $u(\,\cdot\,)$ and x,

$$|I_2(t)| \leq K\sqrt{v(t)}.$$

Consequently, for almost all $t \geq 0$,

$$\dot{v}(t) \geq \Psi(v(t)),$$

where $\Psi(\xi) = 2\alpha\xi - K\sqrt{\xi}$, $\xi \geq 0$. Since $\Psi(\xi) > 0$ for $\xi > \xi_0 = \left(\frac{K}{2\alpha}\right)^2$ therefore if $v(0) = \langle x, e_1\rangle^2 + \langle x, e_2\rangle^2 > \xi_0$ the function v is increasing and $v(t) \not\to 0$ as $t \uparrow +\infty$. Hence (ii) holds if $\beta \neq 0$. The proof of the result if $\beta = 0$ is similar. $\square$

Exercise 4.1. System
$$\frac{d^{(n)}}{dt^{(n)}}z + a_1\frac{d^{(n-1)}}{dt^{(n-1)}}z + \ldots + a_n z = u,$$
studied in §1.4, is controllable if $U = \mathbf{R}$. It follows from Theorem 4.1 that it is controllable to 0 also when $U = [-1, 1]$ and the polynomial $p(\lambda) = \lambda^n + a_1 \lambda^{n-1} + \ldots + a_n$, $\lambda \in \mathbf{C}$, is stable. In particular, the system
$$\frac{d^2 z}{dt^2} + \frac{dz}{dt} + z = u, \quad |u| \le 1,$$
is controllable to $0 \in \mathbf{R}^2$. Note however that the system
$$\frac{d^2 z}{dt^2} = u, \quad |u| \le 1,$$
is also controllable to $0 \in \mathbf{R}^2$, see Example III.3.2, although the corresponding polynomial p is not stable.

§4.2. Positive systems

Denote by E_+ and U_+ the sets of all vectors with nonnegative coordinates from $\mathbf{R}^n$ and $\mathbf{R}^m$ respectively. System (4.1) is said to be *positive* if for an arbitrary, locally integrable control $u(\,\cdot\,)$, taking values in U_+ and for an arbitrary initial condition $x \in E_+$, the output $y^{u,x}$ takes values in E_+.

Positive systems are of practical interest and are often seen in applications to heat processes. In particular, see the system modelling the electrically heated oven of Example 0.1.

We have the following:

Theorem 4.2. *System (4.1) is positive if and only if all elements of the matrix B and all elements of the matrix A outside of the main diagonal are nonnegative.*

Proof. Assume that system (4.1) is positive. Then for arbitrary vector $\bar{u} \in U_+$ and $t \ge 0$
$$\frac{1}{t}\int_0^t S(s) B \bar{u}\, ds \in E_+, \quad \text{where } S(s) = e^{As}, \ s \ge 0. \tag{4.2}$$

Letting t tend to 0 in (4.2) to the limit as $t \downarrow 0$ we get $B\bar{u} \in E_+$. Since $\bar{u}$ was an arbitrary element of U_+, all elements of B have to be nonnegative. Positivity of (4.1) also implies that for arbitrary $\bar{x} \in E_+$ and $t \ge 0$, $S(t)\bar{x} \in E_+$. Consequently all elements of the matrices $S(t)$, $t \ge 0$, are nonnegative. Assume that for some $i \ne j$ the element a_{ij} of A is negative. Then, for

§ 4.2. Positive systems

sufficiently small $t \geq 0$, the element in the i-th row and j-th column of the matrix

$$\frac{1}{t}S(t) = \left(\frac{1}{t}I + A\right) + t\left(\sum_{k=2}^{+\infty} \frac{A^k}{k!} t^{k-1}\right)$$

is also negative. This is a contradiction with the fact that all elements of $S(t)$, $t > 0$, are nonnegative.

Assume conversely that all elements of the matrix B and all elements of the matrix A outside of the main diagonal are nonnegative. Let $x \in E_+$ and let $u(\,\cdot\,)$ be a locally integrable function with values in U_+. There exists a number $\lambda > 0$ such that the matrix $\lambda I + A$ has only nonnegative elements and therefore also the matrix $e^{(\lambda I + A)t}$ has all nonnegative elements. But

$$e^{(\lambda I + A)t} = e^{\lambda t} S(t), \quad t \geq 0,$$

and we see that all elements of $S(t)$, $t \geq 0$, are nonnegative as well. Since

$$y^{x,u}(t) = S(t)x + \int_0^t S(t-s)Bu(s)\,ds, \quad t \geq 0,$$

vectors $y^{x,u}(t)$, $t \geq 0$, belong to E_+. This finishes the proof of the theorem. $\square$

To proceed further we introduce the following sets of attainable points:

$$O_t^+ = \{x \in E_+; x = y^{0,u}(s) \text{ for some } s \in [0,t] \text{ and } u \in L^1[0,s; U^+]\}, t \geq 0$$
$$O^+ = \bigcup_{t \geq 0} O_t^+.$$

System (4.1) is called *positively controllable at time $t > 0$* if the set O_t^+ is dense in E^+. System (4.1) is called *positively controllable* if the set O^+ is dense in E^+.

Theorem 4.3. *Let $e_1, \ldots, e_n$ and $\tilde{e}_1, \ldots, \tilde{e}_m$ be standard bases in $\mathbf{R}^n$ and $\mathbf{R}^m$ respectively.*

(i) *A positive system (4.1) is positively controllable at time $t > 0$ if and only if for arbitrary $i = 1, 2, \ldots, n$ there exists $j = 1, 2, \ldots, m$ and a constant $\mu > 0$ such that*

$$e_i = \mu B \tilde{e}_j.$$

(ii) *A positive system (4.1) is positively controllable if and only if for arbitrary $i = 1, 2, \ldots, n$ there exists $j = 1, 2, \ldots, m$ such that either*

$$e_i = \mu B \tilde{e}_j \quad \text{for some } \mu > 0$$

or

$$e_i = \lim_{k\uparrow +\infty} \frac{S(t_k)B\tilde{e}_j}{|S(t_k)B\tilde{e}_j|} \quad \textit{for some } t_k \uparrow +\infty.$$

We will need several lemmas.

Lemma 4.1. *A positive system* (4.1) *is positively controllable, respectively positively controllable at time $t > 0$ if the cone generated by*

$$S(s)B\tilde{e}_j, \quad s \geq 0, \ j = 1, 2, \ldots, m,$$

respectively generated by

$$S(s)B\tilde{e}_j, \quad s \in [0, t], \ j = 1, 2, \ldots, m,$$

are dense in E_+.

Proof of the lemma. Fix $s > 0$. Then

$$S(s)B\tilde{e}_j = \lim_{\delta \downarrow 0} y^{0,u_\delta}(s),$$

where

$$u_\delta(r) = \begin{cases} \delta B\tilde{e}_j & \text{for } r \in [s-\delta, s], \\ 0 & \text{for } r \in [0, s-\delta]. \end{cases}$$

On the other hand, piecewise constant functions are dense in $L^1[0, s; U_+]$, and vectors $y^{0,u}(s)$, where $u(\,\cdot\,) \in L^1[0, s; U_+]$, are limits of finite sums of the form

$$\sum_k S(k\delta) \int_0^\delta S(r) B u_{\delta k}\, dr,$$

where $u_{\delta k} \in U_+$, for $\delta > 0$, $k = 1, 2, \ldots$. Consequently, $y^{0,u}(s)$ is a limit of finite sums

$$\sum_{j=1}^m \sum_k \gamma_{\delta k} S(k\delta) B\tilde{e}_j,$$

with nonnegative numbers $\gamma_{\delta k}$. Hence the conclusion of the lemma follows. □

For arbitrary $x \in E_+$ define

$$s(x) = \{z \in E_+;\ \langle x, z\rangle = 0\}.$$

The set $s(x)$ will be called the *side* of U_+ determined by x.

Lemma 4.2. *Let K be a compact subset of E_+ and let $x \in E_+$, $x \neq 0$. If $s(x)$ contains a nonzero element of the convex cone $C(K)$, spanned by K, then $s(x)$ contains at least one element of K.*

§ 4.2. Positive systems

Proof of the lemma. We can assume that $0 \notin K$. Suppose that

$$z \in (C(K)) \cap (s(x)), \quad z \neq 0.$$

Then $\lim_l \sum_{j=1}^{k(l)} \alpha_{jl} x_{jl} = z$ for some $\alpha_{jl} > 0$, $x_{jl} \in K$, $j = 1, \ldots, k(l)$, $l = 1, 2, \ldots$. Setting

$$\beta_{jl} = \alpha_{jl}|x_{jl}| > 0, \quad y_{jl} = x_{jl}|x_{jl}|^{-1},$$

we obtain

$$0 = \langle x, z \rangle = \lim_l \sum_{j=1}^{k(l)} \alpha_{jl} \langle x, x_{jl} \rangle$$

$$= \lim_l \sum_{j=1}^{k(l)} \beta_{jl} \langle x, y_{jl} \rangle.$$

Moreover

$$\sum_{j=1}^{k(l)} \beta_{jl} = \sum_{j=1}^{k(l)} \beta_{jl}|y_{jl}| \geq \left| \sum_{j=1}^{k(l)} \beta_{jl} y_{jl} \right| \longrightarrow |z|, \quad l \uparrow +\infty,$$

and we can assume that for $l = 1, 2, \ldots$

$$\left[\sum_{j=1}^{k(l)} \beta_{jl} \langle x, y_{jl} \rangle \right] < \frac{|f|}{2l}, \quad \sum_{j=1}^{k(l)} \beta_{jl} \geq \frac{|f|}{2}.$$

Therefore for arbitrary $l = 1, 2, \ldots$ there is $j(l)$, such that $\langle x, y_{j(l),l} \rangle < 1/l$. Since the set K is compact, there exists a subsequence $(x_{j(l),l})$ convergent to $x_0 \in K$. Consequently there exists a subsequence (y_l) of the sequence $(y_{j(l),l})$ convergent to $x_0|x_0|^{-1}$. But

$$\langle x, x_0|x_0|^{-1} \rangle \lim_l \langle x, y_l \rangle = 0,$$

so finally $x_0 \in s(x)$ as required. □

Proof of the theorem. Sufficiency follows from Lemma 4.1. To show necessity we can assume, without any loss of generality, that $B\tilde{e}_j \neq 0$ for $j = 1, \ldots, m$.

(i) The set $\{S(s)B\tilde{e}_j; \ s \in [0,t], \ j = 1, \ldots, m\}$ does not contain 0 and is compact, therefore the cone $C_1 = C\{S(s)B\tilde{e}_j; \ s \in [0,t], \ j = 1, \ldots, m\}$ is closed. So if system (4.1) is positively controllable at time $t > 0$ then the cone C_1 must be equal to E_+. Taking into account Lemma 4.2 we get in particular that for arbitrary $i = 1, 2, \ldots, n$ there exist $s \in [0, t]$, $j = 1, 2, \ldots, m$ and $\mu > 0$ such that $e_i = \mu S(s) B \tilde{e}_j$.

Let $\lambda > 0$ be a number such that the matrix $\lambda I + A$ has all elements nonnegative. Then

$$e^{\lambda s} S(s) B \tilde{e}_j = e^{(\lambda I + A)s} B \tilde{e}_j = B \tilde{e}_j + \sum_{k=1}^{+\infty} \frac{1}{k!} (\lambda I + A)^k B \tilde{e}_j \geq B \tilde{e}_j.$$

Consequently for a positive $\nu > 0$

$$S(s) B \tilde{e}_j = \nu B \tilde{e}_j,$$

or equivalently
$$e_i = \mu \nu B \tilde{e}_j.$$

(ii) Assume that the system (4.1) is positive by controllable and denote by K the closure of the set $\{S(s)Be_j/|S(s)Be_j|;\ s \geq 0,\ j = 1, 2, \ldots\}$. It follows from the positive controllability of the system that $C(K) = E_+$. Therefore, by Lemma 4.2, for arbitrary $i = 1, 2, \ldots, n$ there exists $j = 1, 2, \ldots, m$ such that either

$$e_i = \mu S(s) B \tilde{e}_j \quad \text{for some } \mu > 0,\ s \geq 0$$

or

$$e_i = \lim_k \frac{S(t_k) B \tilde{e}_j}{|S(t_k) B e_j|} \quad \text{for some } t_k \uparrow +\infty.$$

Repeating now the arguments from the proof of (i) we obtain that the conditions of the theorem are necessary. □

We propose now to solve the following exercise.

Exercise 4.2. Assume that $n = 2$, $\beta, \gamma \geq 0$ and that $A = \begin{bmatrix} -\alpha & \beta \\ \gamma & -\delta \end{bmatrix}$, see Example 0.1. Let $S(t) = e^{At}$, $t \geq 0$. Prove that for arbitrary $x \in E_+$, $x \neq 0$, there exists

$$\lim_{t \uparrow +\infty} \frac{S(t)x}{|S(t)x|} = e$$

and e is an eigenvector of A.

Hint. Show that the eigenvalues λ_1, λ_2 of A are real. Consider separately the cases $\lambda_1 = \lambda_2$ and $\lambda_1 \neq \lambda_2$.

Example 4.1. Continuation of Exercise 4.2. Let

$$A = \begin{bmatrix} -\alpha & \beta \\ \gamma & -\delta \end{bmatrix}, \quad B = \begin{bmatrix} 1 \\ 0 \end{bmatrix}, \quad \beta, \gamma \geq 0. \tag{4.3}$$

§ 4.2. Positive systems

It follows from Theorem 4.3 that system (4.1) is not positively controllable at any $t > 0$. By the same theorem and Exercise 4.2, the system is positively controllable if and only if

$$\lim_{t \uparrow +\infty} \frac{S(t) \begin{bmatrix} 1 \\ 0 \end{bmatrix}}{\left| S(t) \begin{bmatrix} 1 \\ 0 \end{bmatrix} \right|} = \begin{bmatrix} 0 \\ 1 \end{bmatrix}. \tag{4.4}$$

In particular, see Exercise 4.2, the vector $\begin{bmatrix} 0 \\ 1 \end{bmatrix}$ must be an eigenvector of A and so $\beta = 0$. Consequently

$$S(t) = \begin{cases} \begin{bmatrix} e^{-\alpha t} & 0 \\ \gamma t e^{-\alpha t} & e^{-\delta t} \end{bmatrix}, & \text{if } \alpha = \delta, \\ \begin{bmatrix} e^{-\alpha t} & 0 \\ (\gamma e^{-\alpha t} - e^{-\delta t})(\delta - \alpha)^{-1} & e^{-\delta t} \end{bmatrix}, & \text{if } \alpha \neq \delta. \end{cases}$$

We see that (4.4) holds if $\gamma > 0$, $\beta = 0$ and $\delta \leq \alpha$. These conditions are necessary and sufficient for positive controllability of (4.3). Let us remark that the pair (A, B) is controllable under much the weaker condition $\gamma \neq 0$.

The above example shows that positive controllability is a rather rare property. In particular, the system from Example 0.1 is not positively controllable. However in situations of practical interest one does not need to reach arbitrary elements from E_+ but only states in which the system can rest arbitrarily long. This way we are led to the concept of a *positive stationary pair*. We say that $(\bar{x}, \bar{u}) \in E_+ \times U_+$ is a *positive stationary pair* for (4.1) if

$$A\bar{x} + B\bar{u} = 0. \tag{4.5}$$

It follows from the considerations below that a transfer to a state $\bar{x}$ for which (4.5) holds with some $\bar{u} \in U_+$ can be achieved for a rather general class of positive systems.

Theorem 4.4. *Assume that system (4.1) is positive and the matrix A is stable. Then*

(i) *for arbitrary $\bar{u} \in U_+$ such that $B\bar{u} \neq 0$ there exists exactly one vector $\bar{x} \in E_+$ such that*

$$A\bar{x} + B\bar{u} = 0;$$

(ii) *if $(\bar{x}, \bar{u})$ is a positive stationary pair for (4.1) and $\tilde{u}(t) = \bar{u}$ for $t \geq 0$ then, for arbitrary $x \in E_+$*

$$y^{x,\tilde{u}}(t) \longrightarrow \bar{x}, \quad t \uparrow +\infty.$$

Proof. (i) If A is a stable matrix then for some $\omega > 0$ and $M > 0$
$$|e^{At}| \leq M e^{-\omega t}, \quad t \geq 0.$$
Therefore the integral $\int_0^{+\infty} e^{At}\, dt$ is a well defined matrix with nonnegative elements. Since
$$-A^{-1} = \int_0^{+\infty} e^{At}\, dt,$$
the matrix $-A^{-1}$ has nonnegative elements. Consequently the pair
$$(\bar{x}, \bar{u}) = (-A^{-1} B\bar{u}, \bar{u})$$
is the only positive stationary pair corresponding to $\bar{u}$.

(ii) If $\tilde{u}(t) = \bar{u}$ for $t \geq 0$ and
$$\dot{y} = Ay + B\bar{u}$$
$$= Ay - A\bar{x}, \quad y(0) = x,$$
then $\frac{d}{dt}(y(t) - \bar{x}) = A(y(t) - \bar{x})$ and $y(t) - \bar{x} = S(t)(x - \bar{x})$, $t \geq 0$. Since the matrix A is stable we finally have $S(t)(x - \bar{x}) \longrightarrow 0$ as $t \uparrow +\infty$ and $y(t) \longrightarrow \bar{x}$ as $t \uparrow +\infty$. □

Conversely, the existence of positive stationary pairs implies, under rather weak assumptions, the stability of A.

Theorem 4.5. *Assume that for a positive system* (4.1) *there exists a stationary pair* $(\bar{x}, \bar{u})$ *such that vectors* $\bar{x}$ *and* $B\bar{u}$ *have all coordinates positive. Then matrix* A *is stable.*

Proof. It is enough to show that $S(t)\bar{x} \longrightarrow 0$ as $t \uparrow +\infty$. Since
$$S(t+r)\bar{x} = S(t)\bar{x} + \int_r^{t+r} \frac{d}{ds} S(s)\bar{x}\, ds$$
$$= S(t)\bar{x} + \int_r^{t+r} S(s) A\bar{x}\, ds$$
$$= S(t)\bar{x} - \int_r^{t+r} S(s) B\bar{u}\, ds$$
$$\leq S(t)\bar{x}, \quad t, r \geq 0,$$
the coordinates of the function $t \longrightarrow S(t)\bar{x}$ are nondecreasing. Consequently $\lim_{t \uparrow +\infty} S(t)\bar{x} = z$ exists and $S(t)z = z$ for all $t \geq 0$. On the other hand, for arbitrary $t > 0$, the vector $\int_0^t S(r) B\bar{u}\, dr$ has all coordinates positive. Since
$$\bar{x} - z \geq \bar{x} - S(t)\bar{x} = \int_0^t S(r) B\bar{u}\, dr,$$

§4.2. Positive systems

for a number $\mu > 0$,
$$\bar{x} \leq \mu(\bar{x} - z).$$
Finally
$$0 \leq S(t)\bar{x} \leq \mu(S(t)\bar{x} - z) \longrightarrow 0,$$
and $S(t)\bar{x} \longrightarrow 0$ as $t \uparrow +\infty$. This finishes the proof of the theorem. □

The question of attainability of a positive stationary pair is answered by the following basic result:

Theorem 4.6. *Assume that $(\bar{x}, \bar{u})$ is a positive stationary pair for (4.1) such that all coordinates of $\bar{u}$ are positive, and let (A, B) be a controllable pair. For an arbitrary $x \in E_+$ and an arbitrary neighbourhood $V \subset U_+$ of $\bar{u}$ there exists a bounded control $\tilde{u}(\cdot)$ with values in V and a number $t_0 > 0$ such that*
$$y^{x,\tilde{u}}(s) = \bar{x} \quad \text{for } s \geq t_0.$$

Proof. Let $\delta > 0$ be a number such that $\{u;\ |\bar{u} - u| < \delta\} \subset V$. It follows from formula (1.12) that for arbitrary $t_1 > 0$ there exists $\gamma > 0$ and that for arbitrary $b \in \mathbf{R}^n$, $|b| < \gamma$, there exists a control $v^b(\cdot)$, $|v^b(t)| < \delta$, $t \in [0, t_1]$, transferring b to 0 at time t_1. Assume first that $|x - \bar{x}| < \gamma$ and let $x - \bar{x} = b$, $\tilde{u}(t) = v^b(t) + \bar{u}$ and $y(t) = y^{x,\tilde{u}}(t)$, $t \in [0, t_1]$. Then

$$\frac{d}{dt}(y(t) - \bar{x}) = \frac{d}{dt}y(t)$$
$$= Ay(t) + Bv^b(t) + B\bar{u}$$
$$= A(y(t) - \bar{x}) + Bv^b(t), \quad t \in [0, t_1],$$

and $y(0) - \bar{x} = x - \bar{x}$. Hence $y(t_1) - \bar{x} = 0$ or $y(t_1) = \bar{x}$. It is also obvious that $\tilde{u}(t) \in V$ for $t \in [0, t_1]$. Setting $\tilde{u}(t) = \bar{u}$ for $t > t_1$ we get the required control.

If x is an arbitrary element of E_+ then, by Theorem 4.5, there exists $t_2 > 0$ such that the constant control $\tilde{u}(t) = \bar{u}$, $t \in [0, t_2]$ transfers x into the ball $\{z;\ |\bar{x} - z| < \gamma\}$ at the time t_2. By the first part of the proof all points from the ball can be transferred to $\bar{x}$ in the required way. □

Example 4.1. (Continuation.) Let $\alpha, \beta, \gamma, \delta > 0$. The matrix A is stable if and only if $\alpha\delta - \beta\gamma > 0$. A positive stationary pair is proportional to
$$(\bar{x}, \bar{u}) = \left(\frac{1}{\alpha\delta - \beta\gamma}\right)\left(\begin{bmatrix}\delta\\\gamma\end{bmatrix}, 1\right).$$
So if $\alpha\delta > \beta\gamma$ then the state $\begin{bmatrix}\delta\\\gamma\end{bmatrix}$ is attainable from $\begin{bmatrix}0\\0\end{bmatrix}$ using only non-negative controls.

Bibliographical notes

Results on positive systems are borrowed from the paper [53] by T. Schanbacher. In particular, he is the author of Theorem 4.3 and its generalizations to infinite dimensional systems.

PART II

NONLINEAR CONTROL SYSTEMS

Chapter 1
Controllability and observability of nonlinear systems

This chapter begins by recalling basic results of nonlinear differential equations. Controllability and observability of nonlinear control systems are studied next. Two approaches to problems are illustrated: one based on linearization and the other one on concepts of differential geometry.

§1.1. Nonlinear differential equations

The majority of the notions and results of Part I have been generalized to nonlinear systems of the form

$$\dot{y} = f(y, u), \quad y(0) = x \in \mathbf{R}^n, \tag{1.1}$$
$$w = h(y). \tag{1.2}$$

Functions f and h in (1.1) and (1.2) are defined on $\mathbf{R}^n \times \mathbf{R}^m$ and $\mathbf{R}^n$ and take values in $\mathbf{R}^n$ and $\mathbf{R}^k$ respectively.

In the present chapter we discuss generalizations concerned with controllability and observability. The following two chapters are devoted to stability and stabilizability as well as to the problem of realizations.

As in Part I by *control, strategy* or *input* we will call an arbitrary locally integrable function $u(\,\cdot\,)$ from $[0, +\infty)$ into $\mathbf{R}^m$. The corresponding solution $y(\,\cdot\,)$ of the equation (1.1) will be denoted by $y^{x,u}(t)$ or $y^u(t, x)$, $t \geq 0$. The function $h(y(\,\cdot\,))$ is called the *output* or the *response* of the system.

To proceed further we will need some results on general differential equations

$$\dot{z} = f(z(t), t), \quad z(t_0) = x \in \mathbf{R}^n, \tag{1.3}$$

where t_0 is a nonnegative number and f a mapping from $\mathbf{R}^n \times \mathbf{R}$ into $\mathbf{R}^n$. A *solution* $z(t)$, $t \in [0, T]$, of equation (1.3) on the interval $[0, T]$, $t_0 \leq T$, is an arbitrary absolutely continuous function $z(\,\cdot\,)\colon [0, T] \longrightarrow \mathbf{R}^n$ satisfying (1.3) for almost all $t \in [0, T]$. A *local solution* of (1.3) is an absolutely continuous function defined on an interval $[t_0, \tau)$, $\tau > t_0$, satisfying (1.3) for almost all $[t_0, \tau)$. If a local solution defined on $[t_0, \tau)$ cannot be extended to a local solution on a larger interval $[t_0, \tau_1)$, $\tau_1 > \tau$, then it is called a *maximal solution* and the interval $[t_0, \tau)$ is the *maximal interval of existence*. An arbitrary local solution has an extension to a maximal one.

The following classical results on the existence and uniqueness of solutions to (1.3) hold.

Theorem 1.1. *Assume that f is a continuous mapping from $\mathbf{R}^n \times \mathbf{R}$ into $\mathbf{R}^n$. Then for arbitrary $t_0 \geq 0$ and $x \in \mathbf{R}^n$ there exists a local solution to (1.3). If $z(t)$, $t \in [t_0, \tau)$, is a maximal solution and $\tau < +\infty$ then*

$$\lim_{t \uparrow \tau} |z(t)| = +\infty.$$

Theorem 1.2. *Assume that for arbitrary $x \in \mathbf{R}^n$ the function $f(\cdot, x)$: $[0, T] \longrightarrow \mathbf{R}^n$ is Borel measurable and for a nonnegative, integrable on $[0,T]$, function $c(\cdot)$*

$$|f(x,t)| \leq c(t)(|x|+1), \tag{1.4}$$

$$|f(x,t) - f(y,t)| \leq c(t)|x-y|, \quad x,y \in \mathbf{R}^n,\ t \in [0,T]. \tag{1.5}$$

Then equation (1.3) has exactly one solution $z(\cdot, x)$. Moreover, for arbitrary $t \in [0,T]$ the mapping $x \longrightarrow z(t,x)$ is a homeomorphism of $\mathbf{R}^n$ into $\mathbf{R}^n$.

Proofs of the existence and uniqueness of the solutions, similar to those for linear equations (see [46]) will be omitted. The existence of explosion, formulated in the second part of Theorem 1.1, we leave as an exercise.

The above formulated definitions and results can be directly extended to the complex case, when the state space $\mathbf{R}^n$ is replaced by $\mathbf{C}^n$.

The following result is a corollary of Theorem 1.2.

Theorem 1.3. *Assume that the transformation f: $\mathbf{R}^n \times \mathbf{R}^m \longrightarrow \mathbf{R}^n$ is continuous and, for a number $c > 0$,*

$$|f(x,u)| \leq c(|x|+|u|+1), \tag{1.6}$$

$$|f(x,u) - f(y,u)| \leq c|x-y|, \quad x,y \in \mathbf{R}^n,\ u \in \mathbf{R}^m. \tag{1.7}$$

Then for an arbitrary control $u(\cdot)$ there exists exactly one solution of the equation (1.1).

Proof. If $u(\cdot)$: $[0,T] \longrightarrow \mathbf{R}^m$ is an integrable function then $f(x, u(t))$, $x \in \mathbf{R}^n$, $t \in [0,T]$, satisfies the assumptions of Theorem 1.2 and the result follows. □

§ 1.1. Nonlinear differential equations

Corollary 1.1. *If conditions* (1.6), (1.7) *hold then the control* $u(\cdot)$ *and the initial condition x uniquely determine the output* $y^u(\cdot, x)$.

In several proofs of control theory a central role is played by theorems on the regular dependence of solutions on the initial date.

Theorem 1.4. *Assume that the conditions of Theorem 1.2 hold that for arbitrary $t \in [0,T]$ the function $f(\cdot, t)$ has a continuous derivative $f_x(\cdot, t)$ and that the functions $f(\cdot, \cdot)$, $f_x(\cdot, \cdot)$ are bounded on bounded subsets of $[0,T] \times \mathbf{R}^n$. Then the mapping $x \longrightarrow z(\cdot, x)$ acting from $\mathbf{R}^n$ into the space of continuous functions $C(0,T; \mathbf{R}^n)$ is Fréchet differentiable at an arbitrary point x_0 and the directional derivative in the direction $v \in \mathbf{R}^n$ is a solution $\xi(\cdot)$ of the linear equation*

$$\dot{\xi} = f_x(z(t, x_0), t)\xi, \quad \xi(t_0) = v. \tag{1.8}$$

In particular, the function $t \longrightarrow z_x(t, x_0)$ is absolutely continuous and satisfies the equation

$$\frac{d}{dt} z_x(t, x_0) = f_x(z(t, x_0), t) z_x(t, x_0) \text{ for a.a. } t \in [0,T], \tag{1.9}$$

$$z_x(t_0, x_0) = I. \tag{1.10}$$

Theorem 1.4 is often formulated under stronger conditions, therefore we will prove it here. We will need the classical, implicit function theorem, the proof of which, following easily form the contraction principle, will be omitted. See also Lemma IV.4.3.

Lemma 1.1. *Let X and Z be Banach spaces and F a mapping from a neighbourhood $\mathcal{O}$ of a point (x_0, z_0), with the following properties:*

(i) $F(x_0, z_0) = 0$,

(ii) *there exist Gateaux derivatives* $F_x(\cdot, \cdot)$, $F_z(\cdot, \cdot)$ *continuous at $(0,0)$, and the operator $F_z(x_0, z_0)$ has a continuous inverse $F_z^{-1}(x_0, z_0)$.*

Then there exist balls $\mathcal{O}(x_0) \subset X$ and $\mathcal{O}(z_0) \subset Z$ with centres respectively at x_0, z_0 such that for arbitrary $x \in \mathcal{O}(x_0)$ there exists exactly one $z \in \mathcal{O}(z_0)$, denoted by $z(x)$, such that

(iii) $F(x, z) = 0$.

Moreover the function $z(\cdot)$ is differentiable in the neighbourhood $\mathcal{O}(x_0)$ and

$$z_x(x) = -F_z^{-1}(x, z(x)) F_x(x, z(x)).$$

Proof of Theorem 1.4. Let $X = \mathbf{R}^n$, $Z = C(0,T; \mathbf{R}^n)$ and define the mapping $F: X \times Z \longrightarrow Z$ by the formula

$$F(x, z(\cdot))(t) = x + \int_{t_0}^{t} f(z(s), s)\, ds - z(t), \quad t \in [0,T].$$

The mapping $F_x(x,z)$ associates with an arbitrary vector $v \in \mathbf{R}^n$ the function on $[0,T]$ with a constant value v. The Gateaux derivative $F_z(x,z)$ in the direction $\xi \in Z$ is equal to

$$(F_z(x,z)\xi)(t) = \int_{t_0}^t f_x(z(s),s)\xi(s)\,ds - \xi(t), \quad t \in [0,T]. \tag{1.11}$$

To prove (1.11), remark that if $h > 0$ then by Theorem A.6 (the mean value theorem)

$$\sup_{t \in [0,T]} \frac{1}{h}|F(x,z+h\xi)(t) - F(x,z)(t) - \int_{t_0}^t f_x(z(s),s)h\xi(s)\,ds|$$

$$\leq \frac{1}{h}\int_0^T |f(z(s)+h\xi(s),s) - f(z(s),s) - f_x(z(s),s)h\xi(s)|\,ds$$

$$\leq \int_0^T \sup_{\eta \in I(z(s), z(s)+h\xi(s))} |f_x(\eta,s) - f_x(z(s),s)|\,ds.$$

By the assumptions and the Lebesgue dominated convergence theorem we obtain (1.11). From the estimate

$$\|F_z(x,z)\xi - F_z(\bar{x},\bar{z})\xi\| \leq \int_0^T \|f_x(z(s),s) - f_x(\bar{z}(s),s)\|\,ds \leq \|\xi\|$$

valid for $x, \bar{x} \in \mathbf{R}^n$ and $z, \bar{z} \in Z$, $\xi \in Z$ and again by the Lebesgue dominated convergence theorem the continuity of the derivative $F_z(\cdot,\cdot)$ follows.

Assume that

$$x_0 + \int_{t_0}^t f(z_0(s),s)\,ds = z_0(t) = z(t,x_0), \quad t \in [0,T].$$

To show the invertability of $F_z(x_0, z_0)$ assume $\eta(\cdot) \in Z$ and consider the equation

$$x_0 + \int_{t_0}^t f_y(z_0(s),s)\xi(s)\,ds - \xi(t) = \eta(t), \quad t \in [0,T].$$

Denoting $\zeta(t) = \eta(t) + \xi(t)$, $t \in [0,T]$, we obtain the following equivalent differential equation on $\zeta(\cdot)$:

$$\dot{\zeta} = f_z(z_0(t),t)\zeta - f_z(z_0(t),t)\eta(t), \quad t \in [0,T], \tag{1.12}$$
$$\zeta(t_0) = x_0,$$

which, by Theorem I.1.1, has exactly one solution. Therefore, the transformation F satisfies all the assumptions of Lemma 1.1, and, consequently, for all $x \in \mathbf{R}^n$ sufficiently close to x_0, the equation (iii) — equivalent to (1.3) — has the unique solution $z(\cdot, x) \in Z$, with the Fréchet derivative at x_0 given by

$$\xi = z_x(x_0)v = -F_z^{-1}(x_0, z_0)F_x(x_0, z_0)v.$$

Since $\eta(t) = F_x(x_0, z_0)v = v$, $t \in [0, T]$, by (I.1.7) and (1.12):

$$\dot\xi(t) = \dot\zeta(t) = f_z(z_0(t), t)(\xi(t) + v) - f_z(z_0(t), t)v$$
$$= f_z(z_0(t), t)\xi(t), \quad \xi(t_0) = v.$$

$\square$

Let us finally consider equation (1.3) which depends on a parameter $\gamma \in \mathbf{R}^l$ and denote by $z(\cdot, \gamma, x)$, $\gamma \in \mathbf{R}^l$, a solution of the equation

$$\dot z = f(z, \gamma, t), \quad z(t_0) = x. \tag{1.13}$$

We have the following:

Theorem 1.5. *Assume that a function $f(x, \gamma, t)$, $(x, \gamma) \in \mathbf{R}^n \times \mathbf{R}^l$, $t \in [0, T]$, satisfies the assumptions of Theorem 1.4 with $\mathbf{R}^n$ replaced by $\mathbf{R}^n \times \mathbf{R}^l$. Then for arbitrary $t \in [0, T]$, $\gamma_0 \in \mathbf{R}^l$, $x_0 \in \mathbf{R}^n$, the function $\gamma \longrightarrow z(t, \gamma, x_0)$ is Fréchet differentiable at γ_0 and*

$$\frac{d}{dt}z_\gamma(t, \gamma_0, x_0) = f_x(z(t, \gamma_0, z_0), \gamma_0, t)z_\gamma(t, \gamma_0, x_0) \tag{1.14}$$
$$+ f_\gamma(z(t, \gamma_0, x_0), \gamma_0, t), \quad t \in [0, T].$$

Proof. Consider the following system of equations:

$$\dot z = f(z, \gamma, t), \quad \dot\gamma = 0, \quad z(t_0) = x, \quad \gamma(t_0) = \gamma. \tag{1.15}$$

The function $(x, \gamma, t) \longrightarrow (f(x, \gamma, t), 0)$ acting from $\mathbf{R}^n \times \mathbf{R}^l \times [0, T]$ into $\mathbf{R}^n \times \mathbf{R}^l$ satisfies the assumptions of Theorem 1.4 with $\mathbf{R}^n$ replaced by $\mathbf{R}^n \times \mathbf{R}^l$. Consequently the solution $t \longrightarrow (z(t, \gamma, x), \gamma)$ of (1.15) is Fréchet differentiable with respect to (x, γ) and an arbitrary point (x_0, γ_0). It is now enough to apply (1.9). $\square$

§ 1.2. Controllability and linearization

Let $B(a, r)$ denote the open ball of radius r and centre a contained in $\mathbf{R}^n$. We say that the system (1.1) is *locally controllable* at $\bar x$ and at time T if for arbitrary $\varepsilon > 0$ there exists $\delta \in (0, \varepsilon)$ such that for arbitrary

$a, b \in B(\bar{x}, \delta)$ there exists a control $u(\,\cdot\,)$ defined on an interval $[0, t] \subset [0, T]$ for which
$$y^{a,u}(t) = b, \tag{1.16}$$
$$y^{a,u}(s) \in K(\bar{x}, \varepsilon), \quad \text{for all } s \in [0, t]. \tag{1.17}$$

We say also that the point b, for which (1.16) holds, is *attainable* from a at time t.

Exercise 1.1. Assume that the pair (A, B) where $A \in \mathbf{M}(n, n)$, $B \in \mathbf{M}(n, m)$, is controllable, and show that the system
$$\dot{y} = Ay + Bu, \quad y(0) = x, \tag{1.18}$$
is locally controllable at $0 \in \mathbf{R}^n$ at arbitrary time $T > 0$.

Hint. Apply Proposition I.1.1 and Theorem I.1.2.

The result formulated as Exercise 1.1 can be extended to nonlinear systems (1.1) using the concept of linearization. Assume that the mapping f is differentiable at $(\bar{x}, \bar{u}) \in \mathbf{R}^n \times \mathbf{R}^m$; then the system (1.18) with
$$A = f_x(\bar{x}, \bar{u}), \quad B = f_u(\bar{x}, \bar{u}) \tag{1.19}$$
is called the linearization of (1.1) at $(\bar{x}, \bar{u})$.

Theorem 1.6. *Assume that the mapping f is continuously differentiable in a neighbourhood of a point $(\bar{x}, \bar{u})$ for which*
$$f(\bar{x}, \bar{u}) = 0. \tag{1.20}$$

If the linearization (1.19) is controllable then the system (1.1) is locally controllable at the point $\bar{x}$ and at arbitrary time $T > 0$.

Proof. Without any loss of generality we can assume that $\bar{x} = 0$, $\bar{u} = 0$. Let us first consider the case when the initial condition $a = 0$. Controllabity of $(A, B) = (f_x(0, 0), f_u(0, 0))$ implies, see Chapter I.1.1, that there exist smooth controls $u^1(\,\cdot\,), \ldots, u^n(\,\cdot\,)$ with values in $\mathbf{R}^m$ and defined on a given interval $[0, T]$, such that for the corresponding outputs $y^1(\,\cdot\,), \ldots, y^n(\,\cdot\,)$ of the equation (1.1), with the initial condition 0, vectors $y^1(T), \ldots, y^n(T)$ are linearly independent. For an arbitrary $\gamma = (\gamma_1, \ldots, \gamma_n)^* \in \mathbf{R}^n$ we set
$$u(t, \gamma) = u^1(t)\gamma_1 + \ldots + u^n(t)\gamma_n, \quad t \in [0, T].$$

Let $y(t, \gamma, x)$, $t \in [0, T]$, $\gamma \in \mathbf{R}^n$, $x \in \mathbf{R}^n$, be the output of (1.1) corresponding to $u(\,\cdot\,, \gamma)$. Then
$$\frac{d}{dt} y(t, \gamma, x) = f(y(t, \gamma, x), u(t, \gamma)), \quad t \in [0, T], \tag{1.21}$$
$$y(t, \gamma, x) = x.$$

§ 1.2. Controllability and linearization

By Theorem 1.5 applied to (1.21), with $x_0 = 0$, $\gamma_0 = 0$, $t \in [0, T]$, we obtain

$$\frac{d}{dt}y_\gamma(t,0,0) = \frac{\partial f}{\partial x}(y(t,0,0), u(t,0))y_\gamma(t,0,0)$$
$$+ \frac{\partial f}{\partial u}(y(t,0,0), u(t,0))\frac{\partial u}{\partial \gamma}(t,0)$$
$$= Ay_\gamma(t,0,0) + B\left[u^1(t), \ldots, u^n(t)\right].$$

Consequently the columns of the matrix $y_\gamma(t,0,0)$ are identical to the vectors $y^1(t), \ldots, y^n(t)$, $t \in [0, T]$, and the matrix $y_\gamma(T,0,0)$ is nonsingular. By Lemma 1.1, the transformation $\gamma \longrightarrow y(T, \gamma, 0)$ maps an arbitrary neighbourhood of $0 \in \mathbf{R}^n$ onto a neighbourhood of $0 \in \mathbf{R}^n$, so for $\varepsilon > 0$ and $r > 0$ there exists $\delta \in (0, \varepsilon)$ such that for arbitrary $b \in B(0, \delta)$ there exists $\gamma \in \mathbf{R}^n$, $|\gamma| < r$, and

$$y(T, \gamma, 0) = b.$$

This proves local contollability if $a = 0$. To complete the proof let us consider the system

$$\dot{\tilde{y}} = -f(\tilde{y}, \tilde{u}), \quad \tilde{y}(0) = 0, \qquad (1.22)$$

and repeat the above arguments. If

$$u(t) = \tilde{u}(T-t), \quad y(t) = \tilde{y}(T-t), \quad t \in [0,T],$$

then

$$\dot{y}(t) = f(y(t), u(t)), \quad t \in [0,T],$$

and $y(0) = \tilde{y}(T)$. Taking into account that the state $a = y(T)$ could be an arbitrary element from a neighbourhood of 0 we conclude that system (1.1) is locally controllable at 0 and at time $2T$, the required property. □

Condition (1.20) is essential for the validity of Theorem 1.5 even for linear systems.

Theorem 1.7. *System* (1.18) *is locally controllable at* $\bar{x} \in \mathbf{R}^n$ *if and only if*

$$\text{the pair } (A, B) \text{ is controllable, and} \qquad (1.23)$$

$$A\bar{x} + B\bar{u} = 0 \quad \text{for some } \bar{u} \in \mathbf{R}^m. \qquad (1.24)$$

Proof. Taking into account Theorem 1.6 it remains only to prove the necessity of conditions (1.23)-(1.24).

To prove the necessity of (1.23) assume that $A\bar{x} + Bu \neq 0$ for arbitrary $u \in \mathbf{R}^m$. Let v be a vector of the form $A\bar{x} + Bu$ with the minimal norm.

Then $|v| > 0$ and, for arbitrary $u \in \mathbf{R}^m$, $\langle A\bar{x}+Bu-v,v\rangle = 0$. Consequently, for $x \in \mathbf{R}^n$, $u \in \mathbf{R}^m$,

$$\langle Ax+Bu,v\rangle = \langle A\bar{x}+Bu+A(x-\bar{x}),v\rangle$$
$$= |v|^2 + \langle A(x-\bar{x}),v\rangle \geq |v|^2 - |A|\,|x-\bar{x}|\,|v|.$$

If δ is a number such that $0 < \delta < |v|^2$, then

$$\langle Ax+Bu,v\rangle \geq \delta \quad \text{for } x \in B(\bar{x},\bar{r}),\; u \in \mathbf{R}^m,$$

where $\bar{r} = (|v|^2 - \delta)/(|A|\,|v|) > 0$. Therefore for arbitrary control $u(\,\cdot\,)$ and the output $y^{\bar{x},u}$ we have

$$\frac{d}{dt}\langle y^{\bar{x},u}(t),v\rangle \geq \delta, \quad \text{provided } t \in [0,\bar{t}],$$

where $\bar{t} = \inf\{t;\, |y^{\bar{x},u}(t)-\bar{x}| \geq \bar{r}\}$. So an arbitrary state $b = \bar{x} - \beta v/|v|$, $\beta \in (0,\bar{r})$ can be attained from $\bar{x}$ only after the output exits from $B(\bar{x},\bar{r})$. The system (1.1) can not be locally controllable at $\bar{x}$.

Assume now that the system (1.1) is controllable at $\bar{x} \in \mathbf{R}^n$ and (1.24) holds. If $y(\,\cdot\,,\bar{x})$ is the output corresponding to $u(\,\cdot\,)$ then $\tilde{y}(\,\cdot\,,0) = y(\,\cdot\,,\bar{x})-\bar{x}$ is the response of (1.1) to $\tilde{u}(\,\cdot\,) = u(\,\cdot\,) - \bar{u}$. Therefore we obtain that the system (1.1) is locally controllable at 0 and that the pair (A,B) is controllable, see I.1.2. □

Exercise 1.2. Let a function $g(\xi_1,\xi_2,\ldots,\xi_n,\xi_{n+1})$, $\xi_1,\ldots,\xi_{n+1} \in \mathbf{R}$ be of class C^1 and satisfies the condition: $g(0,\ldots,0) = 0$, $\partial g/\partial \xi_{n+1}(0,\ldots,0) \neq 0$. Show that the system

$$\frac{d^n z}{dt^n}(t) = g\left(\frac{d^{n-1}z}{dt^{n-1}}(t),\ldots,z(t),u(t)\right), \quad t \geq 0, \quad (1.25)$$

treated as a system in $\mathbf{R}^n$ is locally controllable at $0 \in \mathbf{R}^n$.

Hint. Show first that the linearization (1.2) is of the form

$$\frac{d^n z}{dt^n} = \frac{\partial g}{\partial \xi_1}(0,\ldots,0)\frac{d^{n-1}z}{dt^{n-1}}(t)+\ldots+\frac{\partial g}{\partial \xi_n}(0,\ldots,0)z(t) \quad (1.26)$$
$$+ \frac{\partial g}{\partial \xi_{n+1}}(0,\ldots,0)u(t), \quad t \geq 0.$$

□

There exist important, locally controllable systems whose linearizations are not controllable. A proper tool to discuss such cases is the Lie bracket introduced in the next section.

§ 1.3. Lie brackets

A vector field of class C^k we define as an arbitrary mapping f, from an open set $D \subset \mathbf{R}^n$ into $\mathbf{R}^n$, whose coordinates are k-times continously differentiable in D. Let f and g be vector fields of class C^1, defined on D, with coordinates $f^1(x),\ldots,f^n(x)$ and $g^1(x),\ldots,g^n(x)$, $x \in D$, respectively. The *Lie bracket* of f, g is a vector field denoted by $[f,g]$ and given by

$$[f,g](x) = f_x(x)g(x) - g_x(x)f(x), \quad x \in D.$$

Thus, the i-th coordinate, $[f,g]^i(x)$, of the Lie bracket is of the form

$$[f,g]^i(x) = \sum_{k=1}^n \left(\frac{\partial f^i}{\partial x_k}(x) g^k(x) - \frac{\partial g^i}{\partial x_k}(x) f^k(x) \right), \tag{1.27}$$

$$x = \begin{bmatrix} x_1 \\ \vdots \\ x_n \end{bmatrix} \in D.$$

A subset $\mathcal{P}$ of $\mathbf{R}^n$ is called a *k-dimensional surface*, $k = 1, 2, \ldots, n$, in short, a *surface*, if there exists an open set $V \subset \mathbf{R}^k$ and a homeomorphism q mapping V onto $\mathcal{P}$ with the properties:

$$q = \begin{bmatrix} q^1 \\ \vdots \\ q^n \end{bmatrix} \text{ is of class } C^1, \tag{1.28}$$

vectors $\dfrac{\partial q}{\partial v_1}(v), \ldots, \dfrac{\partial q}{\partial v_k}(v)$ are linearly independent, $v = \begin{bmatrix} v_1 \\ \vdots \\ v_k \end{bmatrix} \in V.$
$$\tag{1.29}$$

The transformation q is called a *parametrization of the surface* and the linear subspace of $\mathbf{R}^n$ spanned by vectors $\frac{\partial q}{\partial v_1}(v), \ldots, \frac{\partial q}{\partial v_k}(v)$ is called the *tangent space* to the surface $\mathcal{P}$ at the point $q(v)$.

We will need the following proposition.

Proposition 1.1 (i) *If $q(\cdot)$ is a mapping of class C^1 from an open set $V \in \mathbf{R}^k$ into a Banach space E such that the derivative $q_v(\bar{v}): \mathbf{R}^k \longrightarrow E$ at a point $\bar{v}$ is one-to-one then there exists a neighbourhood W of $\bar{v}$ such that the mapping q restricted to W is a homeomorphizm.*

(ii) *If, in addition, $E = \mathbf{R}^k$, then the image $q(W)$ of an arbitrary neighbourhood W of $\bar{v}$ contains a ball with the centre $q(\bar{v})$.*

Proof. (i) Without any loss of generality we can assume that $\bar{v} = 0$, $q(\bar{v}) = 0$. Let $A = q_v(\bar{v})$. Since the linear mapping A is defined on a finite

82   1. Controllability and observability of nonlinear systems

dimensional linear space and is one–to–one, there exists $c > 0$ such that $\|Av\| \geq c|v|$ for $v \in \mathbf{R}^k$. Define $\bar{q}(v) = q(v) - Av$, $v \in \mathbf{R}^k$. Then $\bar{q}_v(\bar{v}) = 0$, and by Theorem A.6 (the mean value theorem), for arbitrary $\varepsilon > 0$ there exists $\delta > 0$ such that

$$\|\bar{q}(u) - \bar{q}(v)\| \leq \varepsilon |u - v|, \quad \text{if } |u|, |v| < \delta.$$

Let $\varepsilon \in (0, c)$ and $W = B(0, \delta)$. For $u, v \in W$

$$c|u - v| - \|q(u) - q(v)\| \leq \|A(u - v)\| - \|q(u) - q(v)\|$$
$$\leq \|\bar{q}(u) - \bar{q}(v)\| \leq \varepsilon |u - v|.$$

So

$$(c - \varepsilon)|u - v| \leq \|q(u) - q(v)\|, \quad u, v \in W,$$

and we see that q restricted to W is one–to–one, and the inverse transformation satisfies the Lipschitz condition with the constant $(c - \varepsilon)^{-1}$. This finishes the proof of (i). The latter part of the proposition follows from Lemma 1.1. □

As far as Lie brackets are concerned, we will need the following basic result.

Theorem 1.8. *Let $q(v)$, $v \in V$, be a parametrization of class C^2 of the surface $\mathcal{P}$ in $\mathbf{R}^n$. Let f and g be vector fields of class C^1 defined in a neighbourhood of $\mathcal{P}$. If for arbitrary $v \in V$ vectors $f(q(v))$, $g(q(v))$ belong to the tangent space to $\mathcal{P}$ at $q(v)$, then for arbitrary $v \in V$ the Lie bracket $[f, g](q(v))$ belongs to the tangent space to $\mathcal{P}$ at $q(v)$.*

Proof. Let $q^1(v), \ldots, q^n(v)$, $f^1(q(v)), \ldots, f^n(q(v))$, $g^1(q(v)), \ldots, g^n(q(v))$ be coordinates respectively of $q(v)$, $f(q(v))$, $g(q(v))$, $v \in V$. By (1.29) and Cramer's formulae, there exist functions $\alpha^1(v), \ldots, \alpha^k(v), \beta^1(v), \ldots, \beta^k(v)$, $v \in V$, of class C^1 such that

$$f(q(v)) = \sum_{j=1}^{k} \frac{\partial q}{\partial v_j}(v) \alpha^j(v), \qquad (1.30)$$

$$g(q(v)) = \sum_{j=1}^{k} \frac{\partial q}{\partial v_j}(v) \beta^j(v), \quad v \in V. \qquad (1.31)$$

Therefore

$$\frac{\partial}{\partial v_l}(f^i(q(v))) = \frac{\partial}{\partial v_l}\left(\sum_{j=1}^{k} \frac{\partial q^i}{\partial v_j}(v) \alpha^j(v)\right) \qquad (1.32)$$

$$= \sum_{j=1}^{k} \frac{\partial^2 q^i}{\partial v_l \partial v_j}(v) \alpha^j(v) + \sum_{j=1}^{k} \frac{\partial q^i}{\partial v_j}(v) \frac{\partial \alpha^j}{\partial v_l}(v),$$

§ 1.3. Lie brackets

$$\frac{\partial}{\partial v_l}(g^i(q(v))) = \sum_{j=1}^{k} \frac{\partial^2 q^i}{\partial v_l \partial v_j}(v)\beta^j(v) + \sum_{j=1}^{k} \frac{\partial q^i}{\partial v_j}(v)\frac{\partial \beta^j}{\partial v_l}(v). \quad (1.33)$$

On the other hand

$$\frac{\partial}{\partial v_l}(f^i(q(v))) = \sum_{r=1}^{n} \frac{\partial f^i}{\partial x_k}(q(v))\frac{\partial q^r}{\partial v_l}(v), \quad (1.34)$$

$$\frac{\partial}{\partial v_l}(g^i(q(v))) = \sum_{r=1}^{n} \frac{\partial g^i}{\partial x_r}(q(v))\frac{\partial q^r}{\partial v_l}(v), \quad v \in V. \quad (1.35)$$

Taking into account definition (1.27)

$$[f,g]^i(q(v)) = \sum_{r=1}^{n} \left(\frac{\partial f^i}{\partial x_r}(q(v))g^r(q(v)) - \frac{\partial g^i}{\partial x_r}(q(v))f^r(q(v)) \right), \quad v \in V.$$

From (1.30), (1.31) and (1.34), (1.35) we have

$$[f,g]^i(q(v)) = \sum_{r=1}^{n} \left[\frac{\partial f^i}{\partial x_r}(q(v)) \left(\sum_{l=1}^{k} \frac{\partial q^r}{\partial v_l}(v)\beta^l(v) \right) \right.$$
$$\left. - \frac{\partial g^i}{\partial x_r}(q(v)) \left(\sum_{l=1}^{k} \frac{\partial q^r}{\partial v_l}(v)\alpha^l(v) \right) \right]$$
$$= \sum_{l=1}^{k} \beta^l(v)\frac{\partial}{\partial v_l}(f^i(q(v))) - \sum_{l=1}^{k} \alpha^l(v)\frac{\partial}{\partial v_l}(g^i(q(v))).$$

Hence by (1.32), (1.33)

$$[f,g]^i(q(v)) = \sum_{l=1}^{k} \beta^l(v) \left[\sum_{j=1}^{k} \left(\frac{\partial^2 q^i}{\partial v_l \partial v_j}(v)\alpha^j(v) + \frac{\partial q^i}{\partial v_j}(v)\frac{\partial \alpha^j}{\partial v_l}(v) \right) \right]$$
$$- \sum_{l=1}^{k} \alpha^l(v) \left[\sum_{j=1}^{k} \left(\frac{\partial^2 q^i}{\partial v_l \partial v_j}(v)\beta^j(v) + \frac{\partial q^i}{\partial v_j}(v)\frac{\partial \beta^j}{\partial v_l}(v) \right) \right].$$

Since functions q^i, $i = 1, \ldots, n$ are of class C^2 we obtain finally

$$[f,g]^i(q(v)) = \sum_{j=1}^{k} \frac{\partial q^i}{\partial v_j}(v) \left(\sum_{l=1}^{k} \frac{\partial \alpha^j}{\partial v_l}(v)\beta^l(v) - \frac{\partial \beta^j}{\partial v_l}(v)\alpha^l(v) \right),$$

$$i = 1, 2, \ldots, n, \quad v \in V.$$

Therefore the vector $[f,g](q(v))$ is in the tangent space to $\mathcal{P}$ at $q(v)$. The proof of the theorem is complete. □

Exercise 1.3. Let $A_1, A_2 \in \mathbf{M}(n,n)$, $a_1, a_2 \in \mathbf{R}^n$ and $f(x) = A_1 x + a_1$, $g(x) = A_2 x + a_2$, $x \in \mathbf{R}^n$. Show that

$$[f,g](x) = (A_1 A_2 - A_2 A_1)x + A_1 a_2 - A_2 a_1, \quad x \in \mathbf{R}^n.$$

§1.4. The openness of attainable sets

Theorem 1.9 of the present section gives algebraic conditions for the set of all attainable points to be open. The property is slightly weaker than that of local controllability. Under additional conditions, formulated in Theorem 1.10, the local controllability follows.

Let U be a fixed subset of $\mathbf{R}^m$ and assume that mappings $f(\cdot, u)$, $u \in U$, from $\mathbf{R}^n$ into $\mathbf{R}^n$, defining system (1.1), are of class C^k. By *elementary control* we understand a right continuous piecewise constant function $u \colon \mathbf{R}_+ \longrightarrow U$, taking on only a finite number of values. Denote by $\mathcal{L}_0$ the set of all vector fields defined on $\mathbf{R}^n$ of the form $f(\cdot, u)$, $u \in U$, and by $\mathcal{L}_j$, $j = 1, 2, \ldots$, the set of all vector fields which can be obtained from vector fields in $\mathcal{L}_0$ by an application of the Lie brackets at most j times:

$$\mathcal{L}_0 = \{f(\cdot, u); \, u \in U\},$$
$$\mathcal{L}_j = \mathcal{L}_{j-1} \cup \{[f,g](\,\cdot\,); \, f(\,\cdot\,) \in \mathcal{L}_{j-1} \text{ and } g(\,\cdot\,) \in \mathcal{L}_0$$
$$\text{or } f(\,\cdot\,) \in \mathcal{L}_0 \text{ and } g(\,\cdot\,) \in \mathcal{L}_{j-1}\}, \quad j = 1, 2, \ldots .$$

Let us define additionally for $x \in \mathbf{R}^n$ and $j = 0, 1, 2, \ldots$

$$\mathcal{L}_j(x) = \{f(x); \, f \in \mathcal{L}_j\},$$

and let $\dim \mathcal{L}_j(x)$ be the maximum number of linearly independent vectors in $\mathcal{L}_j(x)$.

The following result holds.

Theorem 1.9. *Assume that fields $f(\cdot, u)$, $u \in U$, are of class C^k for some $k \geq 2$ and that for some $x \in \mathbf{R}^n$ and $j \leq k$*

$$\dim \mathcal{L}_j(x) = n. \tag{1.36}$$

Then the set of all points attainable from x by elementary controls, which keep the system in a given in advance neighbourhood D of x, has a nonempty interior.

Proof. Since the fields $f(\cdot, u)$, $u \in U$, are smooth, we can assume that condition (1.36) is satisfied for any $x \in D$. For arbitrary parameters $u_1, \ldots, u_l \in U$, for a fixed element $u^+ \in U$ and for positive numbers

§ 1.4. The openness of attaible sets

$v_1, v_2, \ldots, v_l$, $l = 1, 2, \ldots$, denote by $u(t; u_1, \ldots, u_l; v_1, \ldots, v_l)$, $t \geq 0$, a control $u(\cdot)$ defined by:

$$u(t) = \begin{cases} u_1, & \text{if } 0 \leq t < v_1, \\ u_r, & \text{if } v_1 + \ldots + v_{r-1} \leq t < v_1 + \ldots + v_r, \ r = 2, \ldots, l, \\ u^+, & \text{if } t \geq v_1 + \ldots + v_l. \end{cases} \quad (1.37)$$

Let us fix parameters $u_1, \ldots, u_l$ and element $x \in \mathbf{R}^n$ and let

$$q(v_1, \ldots, v_l) = y^{u(\cdot; u_1, \ldots, u_l; v_1, \ldots, v_l)}(v_1 + \ldots + v_l, x), \quad v_1, \ldots, v_l > 0.$$

Note that

$$q(v_1, \ldots, v_l) = z^{u_l}(v_l, z^{u_{l-1}}(v_{l-1}, z^{u_{l-2}}(\ldots, z^{u_1}(v_1, x)\ldots))), \quad (1.38)$$

where $z^u(t, y)$, $t \geq 0$, $y \in \mathbf{R}^n$ denote the solution of the equation

$$\dot{z} = f(z, u), \quad z(0) = y. \quad (1.39)$$

Let $l \leq n$ be the largest natural number for which there exist control parameters $u_1, \ldots, u_l \in U$ and positive numbers $\alpha_1 < \beta_1, \ldots, \alpha_l < \beta_l$ such that the formula (1.38) defines a parametric representation,

$$V = \left\{ \begin{bmatrix} v_1 \\ \vdots \\ v_l \end{bmatrix}; \ \alpha_i < v_i < \beta_i, \ i = 1, 2, \ldots, l \right\},$$

of a surface of class C^k, contained in D.

We show first that $l \geq 1$. Note that there exists $u_1 \in U$ such that $f(x, u_1) \neq 0$. For if $f(x, u) = 0$ for all $u \in U$, then the definition of the Lie bracket implies that $g(x) = 0$ for arbitrary $g \in \mathcal{L}_0$ and more generally $g(x) = 0$ for arbitrary $g \in \mathcal{L}_j$, $j = 1, 2, \ldots$. This contradicts the assumption that $\dim \mathcal{L}_j(x) = n$. If, now, $f(x, u_1) \neq 0$ then, by Proposition 1.1(i), for small $\alpha_1 > 0$ the function $z^{u_1}(v, x)$, $v \in (0, \alpha_1)$, is a parametric representation of a one-dimensional surface contained in D.

Assume that $l < n$ and let $\mathcal{P}$ be a surface given by (1.38). There exist

$$v' = \begin{bmatrix} v'_1 \\ \vdots \\ v'_l \end{bmatrix} \in V, \quad u' \in U,$$

such that the vector $f(q(v'), u')$ is not included in the tangent space to $\mathcal{P}$ at $q(v')$. This follows from Theorem 1.8 and from the assumption that $\dim \mathcal{L}_j(q(v)) = n > l$ for $v \in V$. Since the function $f(\cdot, u')$ is continuous, there exists $\delta > 0$ such that if $\alpha'_j - \delta < v_j < \alpha'_j + \delta$, $j = 1, \ldots, l$ then vector

$f(q(v), u')$ is not in the tangent space to $\mathcal{P}$ at the point $q(v)$. It follows from Proposition 1.1(i) that for some positive numbers $\alpha_{l+1} < \beta_{l+1}$ the transformation $q'(\cdot)$,

$$q'(v_1, \ldots, v_l, v_{l+1}) = z^{u'}(v_{l+1}, q(v_1, \ldots, v_l)),$$

$$\alpha'_j - \delta < v_j < \alpha'_j + \delta, \quad j = 1, \ldots, l, \quad \alpha_{l+1} < v_{l+1} < \beta_{l+1},$$

is a parametric representation of an $(l+1)$-dimensional surface included in D. By the regular dependence of solutions of differential equations on initial conditions (see Theorem 1.4), the transformation $q'(\cdot)$ is of the same class as $q(\cdot)$ and consequently of the same class C^k as fields $f(\cdot, u)$, $u \in U$. This contradicts the definition of the number l, so $l = n$. Hence, by Proposition 1.1(ii), the image $q(v)$ has a nonempty interior. This finishes the proof of the result. $\square$

Example 1.1. Consider the system $\dot{y}_1 = 1$, $\dot{y}_2 = u(y_1)^2$ with the initial condition $x = 0$ and $U = \mathbf{R}$. In this case

$$f(x, u) = \begin{bmatrix} 1 \\ 0 \end{bmatrix} + u \begin{bmatrix} 0 \\ (x_1)^2 \end{bmatrix}, \quad u \in \mathbf{R},$$

and the fields

$$f(x) = \begin{bmatrix} 1 \\ 0 \end{bmatrix}, \quad g(x) = \begin{bmatrix} 1 \\ (x_1)^2 \end{bmatrix}, \quad x = \begin{bmatrix} x_1 \\ x_2 \end{bmatrix} \in \mathbf{R}^2,$$

belong to $\mathcal{L}_0$. By direct calculations,

$$[f, g](x) = -\begin{bmatrix} 0 \\ 2x_1 \end{bmatrix} = f_1(x), \quad [f, f_1](x) = \begin{bmatrix} 0 \\ 2 \end{bmatrix}, \quad x \in \mathbf{R}^2.$$

Hence $\dim \mathcal{L}_2(x) = 2$ for arbitrary $x \in \mathbf{R}^2$. By Theorem 1.7, the set of all attainable points from $\begin{bmatrix} 0 \\ 0 \end{bmatrix}$ has a nonempty interior. However, since $\dot{y}_1 > 0$, points $\begin{bmatrix} x_1 \\ x_2 \end{bmatrix}$, $x_1 < 0$, can not be attained from $\begin{bmatrix} 0 \\ 0 \end{bmatrix}$. Therefore, in general, the conditions of Theorem 1.9 do not imply local controllability at x.

Example 1.2. Let $f(x, u) = Ax + Bu$, $x \in \mathbf{R}^n$, $u \in \mathbf{R}^m$. Then (see Exercise 1.3),

$$[f(\cdot, u), f(\cdot, v)] = AB(v - u).$$

We easily see that

$$\mathcal{L}_0(x) = \{Ax + Bu; \, u \in \mathbf{R}^m\},$$
$$\mathcal{L}_k(x) = \mathcal{L}_0(x) \cup \{A^j B u_j; \, j = 1, \ldots, k, \, u_j \in \mathbf{R}^m\}.$$

§ 1.4. The openness of attaible sets

So $\dim \mathcal{L}_{n-1}(0) = n$ if and only if the pair (A, B) is controllable. If the pair (A, B) is controllable then

$$\dim \mathcal{L}_{n-1}(x) = n \quad \text{for arbitrary } x \in \mathbf{R}^n.$$

This way condition (1.36) generalizes Kalman's condition from Theorem I.1.2.

Under additional assumptions, condition (1.36) does imply local controllability. We say that the system (1.1), with $U \subset \mathbf{R}^m$, is *symmetric* if for arbitrary $u \in U$ there exists $u' \in U$ such that

$$f(x, u) = -f(x, u'), \quad x \in \mathbf{R}^n. \tag{1.40}$$

Theorem 1.10. *If system (1.1) is symmetric and conditions of Theorem 1.9 hold, then system (1.1) is locally controllable at x.*

Proof. Let us fix a neighbourhood D of x. By Theorem 1.7, there exists a ball $B(\tilde{x}, r) \in D$ such that arbitrary $b \in B(\tilde{x}, r)$ can be attained from x at a time $\tau(b)$ using elementary controls $u^b(t)$, $t \in [0, \tau(b)]$, and keeping the system in D. In particular there are numbers $v_i > 0$ and parameters $u_i \in U$, $i = 1, 2, \ldots, l$, such that strategies $u^x(t) = u(t, u_1, \ldots, u_l; v_1, \ldots, v_l)$, $t \geq 0$, transfer x to $\tilde{x}$ at time $\tau(\tilde{x}) = v_1 + v_2 + \ldots + v_l$.

Let $u'_1 \ldots, u'_l \in U$ be parameters such that $f(x, u_i) = -f(x, u'_i)$ and $Z^i \colon \mathbf{R}^n \longrightarrow \mathbf{R}^n$ — transformations given by

$$Z^i(x) = z^{u'_i}(v_i, x), \quad x \in \mathbf{R}^n, \ i = 1, 2, \ldots, l,$$

compare (1.39). It follows from Theorem 1.2 that these transformations are homeomorphic.

For $b \in B(\tilde{x}, r)$ we define new controls

$$\tilde{u}^b(t) = \begin{cases} u^b(t), & \text{if } t \in [0, \tau(b)], \\ u'_i, & \text{if } \tau(b) + v_{i+1} + \ldots + v_l < t \leq \tau(b) + v_i + v_{i+1} + \ldots + v_l, \\ & i = 1, 2, \ldots, l, \ v_{l+1} = 0. \end{cases}$$

Then the set

$$\left\{ y^{\tilde{u}^b}(\tau(b) + v_1 + \ldots + v_l, x); \ b \in K(\tilde{x}, r) \right\}$$

consists of states attainable from x and is equal to the image of the ball $B(\tilde{x}, r)$ by the transformation composed of $Z^l, Z^{l-1}, \ldots, Z^1$. Therefore, this set contains a nonempty neighbourhood of x and local controllability of (1.1) at x follows. □

88   1. Controllability and observability of nonlinear systems

Example 1.3. Consider the following system (compare also § 2.8):

$$\dot{y}_1 = u,$$
$$\dot{y}_2 = v,$$
$$\dot{y}_3 = y_1 v - y_2 u, \quad \begin{bmatrix} u \\ v \end{bmatrix} \in U = \mathbf{R}^2.$$

Let us remark that the linearization of the system at $(x,0)$ is of the form

$$A = \begin{bmatrix} 0 & 0 & 0 \\ 0 & 0 & 0 \\ 0 & 0 & 0 \end{bmatrix}, \quad B = \begin{bmatrix} 1 & 0 \\ 0 & 1 \\ -x_2 & x_1 \end{bmatrix},$$

so the linearization is not controllable and Theorem 1.6 is not applicable. On the other hand, the system is symmetric. Let

$$f(x) = \begin{bmatrix} 1 \\ 0 \\ -x_2 \end{bmatrix}, \quad g(x) = \begin{bmatrix} 0 \\ 1 \\ x_1 \end{bmatrix}.$$

Then $f(x), g(x) \in \mathcal{L}_0(x)$ and

$$[f,g](x) = \begin{bmatrix} 0 \\ 0 \\ -2 \end{bmatrix}, \quad x = \begin{bmatrix} x_1 \\ x_2 \\ x_3 \end{bmatrix} \in \mathbf{R}^3.$$

Therefore $\dim \mathcal{L}_1(x) = 3$, $x \in \mathbf{R}^3$, and the system is locally controllable at arbitrary point of $\mathbf{R}^3$.

§ 1.5. Observability

We will now extend the concept of observability from § I.1.6 to nonlinear systems. Let us assume that the stated equation is independent of the control paremeter

$$\dot{z} = f(z), \quad z(0) = x \in \mathbf{R}^n, \qquad (1.41)$$

and that the observation is of the form

$$w(t) = h(z(t)), \quad t \geq 0. \qquad (1.42)$$

We say that the system (1.41)-(1.42) is *observable at a point* x if there exists a neighbourhood D of x such that for arbitrary $x_1 \in D$, $x_1 \neq x$, there exists $t > 0$ for which

$$h(z(t,x)) \neq h(z(t,x_1)).$$

§ 1.5. Observability

If, in addition, $t \leq T$ then the system (1.41)-(1.42) is said to be *observable at point x and at the time T*.

We start from a sufficient condition for observability at an equilibrium state, based on linearization. To simplify notation we assume that the equilibrium state is $0 \in \mathbf{R}^n$.

Theorem 1.11. *Assume that transformations $f\colon \mathbf{R}^n \longrightarrow \mathbf{R}^n$ and $h\colon \mathbf{R}^n \longrightarrow \mathbf{R}^k$ are of class C^1. If the pair $(f_x(0), h_x(0))$ is observable then the system (1.41)-(1.42) is observable at 0 and at any time $T > 0$.*

Proof. Let us fix $T > 0$ and define a transformation K from $\mathbf{R}^n$ into $C(0, T; \mathbf{R}^n)$ by the formula

$$(Kx)(t) = h(z(t, x)), \quad x \in \mathbf{R}^n,\ t \in [0, T].$$

It follows from Theorem 1.4 that the transformation K is Fréchet differentiable at arbitrary $\bar{x} \in \mathbf{R}^n$. In addition the directional derivative of K at $\bar{x}$ and at the direction $v \in \mathbf{R}^n$ is equal to

$$(K_x(\bar{x}; v))(t) = h_x(z(t, \bar{x})) z_x(t, \bar{x}) v,$$

and

$$\frac{d}{dt} z_x(t, \bar{x}) = f_x(z(t, \bar{x})) z_x(t, \bar{x}), \quad t \in [0, T],$$

$$z_x(0, \bar{x}) = I.$$

In particular, for $\bar{x} = 0$, $v \in \mathbf{R}^n$,

$$K_x(0; v)(t) = h_x(0) e^{f_x(0) t} v, \quad t \in [0, T].$$

Since the pair $(f_x(0), h_x(0))$ is observable therefore the derivative $K_x(0, \cdot)$ is a one-to-one mapping onto a finite dimensional subspace of $C(0, T; \mathbf{R}^n)$. By Proposition 1.1 we have that the transformation K is one-to-one in a neighbourhood of 0 and the system (1.41)-(1.42) is observable at 0 at time T. □

We will now formulate a different sufficient condition for observability, often applicable if the linearization is not observable. Assume that the system (1.41)-(1.42) is of class C^r or equivalently that the mappings f and h are of class at least C^r, $r \geq 1$. Let

$$f(x) = \begin{bmatrix} f^1(x) \\ \vdots \\ f^n(x) \end{bmatrix}, \quad h(x) = \begin{bmatrix} h^1(x) \\ \vdots \\ h^k(x) \end{bmatrix}, \quad x \in \mathbf{R}^n,$$

and let H_0 be the family of functions $h^1, \ldots, h^k$:

$$H_0 = \{h^1, \ldots, h^k\}.$$

Define families H_j, $1 \leq j \leq k$, as follows:

$$H_j = H_{j-1} \cup \left\{ \sum_{i=1}^n f^i \frac{\partial g}{\partial x_i};\, g \in H_{j-1} \right\},$$

and set

$$dH_j(x) = \left\{ \begin{bmatrix} \frac{\partial g}{\partial x_1}(x) \\ \vdots \\ \frac{\partial g}{\partial x_n}(x) \end{bmatrix};\, g \in H_j \right\}, \quad x \in \mathbf{R}^n,\, j = 0,\ldots,r-1.$$

Let $\dim dH_j(x)$ be the maximal number of linearly independent vectors in $dH_j(x)$.

Example 1.4. Assume that $f(x) = Ax$, $h(x) = Cx$, $x \in \mathbf{R}^n$, where $A \in M(n,n)$, $C \in M(k,n)$. Then $f^i(x) = \langle a_i^*, x \rangle$, $h^j(x) = \langle c_j^*, x \rangle$, $i = 1,2,\ldots,n$, $j = 1,2,\ldots,k$, $x \in \mathbf{R}^n$, where $a_1,\ldots,a_n$, $c_1,\ldots,c_k$ are the rows of matrices A and C respectively. Thus for arbitrary $x \in \mathbf{R}^n$

$$dH_0(x) = \{c_1^*,\ldots,c_k^*\},$$
$$dH_1(x) = dH_0(x) \cup \{A^*c_1^*,\ldots,A^*c_k^*\},$$
$$\vdots$$
$$dH_j(x) = dH_{j-1}(x) \cup \{(A^*)^j c_1,\ldots,(A^*)^j c_k,\ j = 1,2,\ldots\}.$$

Theorem 1.12. *Assume that system (1.41)-(1.42) is of class C^r, $r \geq 1$, and*

$$\dim dH_{r-1}(x) = n. \tag{1.43}$$

Then the system is observable at x at arbitrary time $T > 0$.

Proof. Assume that condition (1.43) holds but the system (1.41)-(1.42) is not observable at x and at a time $T > 0$. Then, in an arbitrary neighbourhood D of x there exists a point x_1 such that

$$g(z^x(t)) = g(z^{x_1}(t)), \quad t \in [0,T],\, g \in H_0. \tag{1.44}$$

By an easy induction argument (1.44) holds for arbitrary $g \in H_{r-1}$. Since $\dim dH_{r-1}(x) = n$, there exist in H_{r-1} functions $\tilde{g}^1,\ldots,\tilde{g}^n$ such that the derivative of $G(\,\cdot\,) = \begin{bmatrix} \tilde{g}^1(\,\cdot\,) \\ \vdots \\ \tilde{g}^n(\,\cdot\,) \end{bmatrix}$ at x is nonsingular. In particular, for arbitrary x_1 sufficiently close to x, $G(x) \neq G(x_1)$ and, for some $j < n$, $\tilde{g}^j(z^x(0)) \neq \tilde{g}^j(z^{x_1}(0))$, a contradiction with (1.44). □

Example 1.4. (Continuation.) Condition $\dim dH_{n-1}(x) = n$ is equivalent to the rank condition of Theorem I.1.6.

Bibliographical notes

Theorem 1.6 is due to E. Lee and L. Markus [36] and Theorem 1.9 to H. Sussman and V. Jurdjevic [54]. The proof of Theorem 1.9 follows that of A. Krener. Results of § 1.4 and 1.5 are typical for geometric control theory based on differential geometry. More on this topic can be found in the monograph by A. Isidori [31]. Example 1.3 was introduced by R. Brockett [11].

Chapter 2
Stability and stabilizability

Three types of stability and stabilizability are studied: exponential, asymptotic and Liapunov. Discussions are based on linearization and Liapunov's function approaches. When analysing a relationship between controllability and stabilizability topological methods are used.

§2.1. Differential inequalities

The basic method of studying asymptotic properties of solutions of differential equations in $\mathbb{R}^n$ consists of analysing one dimensional images of the solutions by appropriately chosen transformations from $\mathbb{R}^n$ into $\mathbb{R}$. These images do not satisfy, in general, differential equations but very often are solutions of differential or integral inequalities. This is why we start with such inequalities. We consider first the Gronwall lemma.

Lemma 2.1. *Let k be a nonnegative, bounded, Borel measurable function on an interval $[t_0, t_1]$ and l a nondecreasing one. Let v be a function integrable on $[t_0, t_1]$ such that for almost all $t \in [t_0, t_1]$*

$$v(t) \leq l(t) + \int_{t_0}^{t} k(s)v(s)\,ds. \tag{2.1}$$

Then for those $t \in [t_0, t_1]$ for which (2.1) holds one has

$$v(t) \leq \exp\left(\int_{t_0}^{t} k(s)\,ds\right) l(t). \tag{2.2}$$

In particular, (2.2) holds almost surely on $[t_0, t_1]$.

Proof. Assume first that $l(t) = l(t_0)$ for $t \in [t_0, t_1]$ and define

$$u(t) = l(t_0) + \int_{t_0}^{t} k(s)v(s)\,ds,$$

$$w(t) = \exp\left(-\int_{t_0}^{t} k(s)\,ds\right) u(t), \quad t \in [t_0, t_1].$$

Functions u and w are absolutely continuous and

$$\dot{u}(t) = k(t)v(t) \leq k(t)u(t),$$

$$\dot{w}(t) = \dot{u}(t)\exp\left(-\int_{t_0}^{t} k(s)\,ds\right) - k(t)u(t)\exp\left(-\int_{t_0}^{t} k(s)\,ds\right) \leq 0,$$

§ 2.1. Differential inequalities

almost surely on $[t_0, t_1]$. Therefore,

$$w(t) \leq w(t_0) \leq u(t_0) \leq l(t_0),$$

$$u(t) = w(t) \exp \int_{t_0}^{t} k(s) ds \leq l(t_0) \exp \int_{t_0}^{t} k(s) ds,$$

and, consequently,

$$v(t) \leq u(t) \leq \exp \left(\int_{t_0}^{t} k(s) \, ds \right) l(t_0) \quad \text{almost surely on } [t_0, t_1].$$

If l is an arbitrary nondecreasing function on $[t_0, t_1]$ and $t_2 \in (t_0, t_1)$, then $l(t) \leq l(t_2)$ on $[t_0, t_2]$. Moreover

$$v(t) \leq l(t_2) + \int_{t_0}^{t} k(s) v(s) \, ds \quad \text{almost surely on } [t_0, t_2]. \tag{2.3}$$

Assume additionally that (2.1) holds for $t = t_2$. Then (2.3) holds for $t = t_2$, as well. By the above reasoning,

$$v(t_2) \leq \exp \left(\int_{t_0}^{t_2} k(s) \, ds \right) l(t_2).$$

The proof of the lemma is complete. □

The following result will play an important role in what follows.

Theorem 2.1. *Assume that* $\varphi \colon [t_0, t_1] \times \mathbf{R} \longrightarrow \mathbf{R}$ *is a continuous function such that for some* $M > 0$

$$|\varphi(t, x) - \varphi(t, y)| \leq M |x - y|, \quad t \in [t_0, t_1], \; x, y \in \mathbf{R}. \tag{2.4}$$

If v *is an absolutely continuous function on* $[t_0, t_1]$ *such that*

$$\dot{v}(t) \leq \varphi(t, v(t)) \quad \text{a.s. on } [t_0, t_1], \tag{2.5}$$

or

$$\dot{v}(t) \geq \varphi(t, v(t)) \quad \text{a.s. on } [t_0, t_1], \tag{2.6}$$

then

$$v(t) \leq u(t) \quad \text{for all } t \in [t_0, t_1] \tag{2.7}$$

or, respectively,

$$v(t) \geq u(t) \quad \text{for all } t \in [t_0, t_1], \tag{2.8}$$

where u is a solution of the equation

$$\dot{u}(t) = \varphi(t, u(t)), \quad t \in [t_0, t_1], \tag{2.9}$$

with the initial condition

$$u(t_0) \geq v(t_0)$$

or, respectively,

$$u(t_0) \leq v(t_0).$$

Proof. By Theorem 1.2 there exists exactly one solution of the equation (2.9) on $[t_0, t_1]$. Assume that $\dot{v}(t) \leq \varphi(t, v(t))$ a.s. on $[t_0, t_1]$ and consider a sequence $u_n(\,\cdot\,)$, $n = 1, 2, \ldots$, of functions satisfying

$$\dot{u}_n(t) = \varphi(t, u_n(t)) + \frac{1}{n}, \quad t \in [t_0, t_1],$$

$$u_n(t_0) = u(t_0), \quad n = 1, 2, \ldots.$$

Let $K > 0$ be a number such that $|\varphi(s, 0)| \leq K$, $s \in [t_0, t_1]$. From (2.4),

$$|\varphi(s, x)| \leq |\varphi(s, x) - \varphi(s, 0)| + |\varphi(s, 0)| \leq M|x| + K, \tag{2.10}$$
$$s \in [t_0, t_1], \ x \in \mathbf{R}.$$

Since

$$u_n(t) = u(t_0) + (t - t_0)\frac{1}{n} + \int_{t_0}^t \varphi(s, u_n(s))\, ds, \quad t \in [t_0, t_1], \tag{2.11}$$

therefore

$$|u_n(t)| \leq |u(t_0)| + (K + \frac{1}{n})(t - t_0) + M \int_{t_0}^t |u_n(s)|\, ds, \quad t \in [t_0, t_1].$$

Hence, by Lemma 2.1

$$|u_n(t)| \leq |u(t_0)| + (K + \frac{1}{n})(t_1 - t_0)e^{M(t-t_0)}, \quad t \in [t_0, t_1],$$

and functions $u_n(t)$, $t \in [t_0, t_1]$, $n = 1, 2, \ldots$, are uniformly bounded. They are also equi-continuous as

$$|u_n(t) - u_n(s)| \leq \int_s^t |\varphi(r, u_n(r))|\, dr + \frac{1}{n}(t - s) \leq (L + \frac{1}{n})(t - s),$$
$$t_0 \leq s \leq t \leq t_1,$$

for a constant $L > 0$, independent of n.

It follows from Theorem A.8 (the Ascoli theorem) that there exists a subsequence of $(u_n(\cdot))$ uniformly continuous to a continuous function $\tilde{u}(\cdot)$. Letting n in (2.11) tend to $+\infty$ we see that the function $\tilde{u}(\cdot)$ is an absolutely continuous solution of the equation (2.9) and therefore it is identical with $u(\cdot)$. To prove (2.7) it is sufficient to show that

$$u_n(t) \geq v(t) \quad \text{for } n = 1, 2, \ldots, \ t \in [t_0, t_1].$$

Assume that for a natural number m and for some $t_2 \in (t_0, t_1)$, $u_m(t_2) < v(t_2)$. Then for some $\tilde{t} \in [t_0, t_2)$,

$$u_m(\tilde{t}) = v(\tilde{t}) \quad \text{and} \quad u_m(t) < v(t) \quad \text{for } t \in [\tilde{t}, t_2).$$

Taking into account the definitions of $v(\cdot)$ and $u_m(\cdot)$ and the continuity of $\varphi(\cdot,\cdot)$ we see that for some $\delta > 0$ and arbitrary $t \in (\tilde{t}, \tilde{t} + \delta)$

$$\frac{v(t) - v(\tilde{t})}{t - \tilde{t}} \leq \frac{1}{t - \tilde{t}} \int_{\tilde{t}}^{t} \varphi(s, v(s))\, ds \leq \varphi(\tilde{t}, v(\tilde{t})) + \frac{1}{3m},$$

$$\frac{u_m(t) - v(\tilde{t})}{t - \tilde{t}} \geq \varphi(\tilde{t}, v(\tilde{t})) + \frac{1}{2m}.$$

Therefore $u_m(t) > v(t)$, $t \in (\tilde{t}, \tilde{t}+\delta)$, a contradiction. This proves (2.7). In a similar way (2.6) implies (2.8). □

Corollary 2.1. *If $v(\cdot)$ is an absolutely continuous function on $[t_0, t_1]$ such that for some $\alpha \in \mathbf{R}$*

$$\dot{v}(t) \leq \alpha v(t) \quad \text{a.s. on } [t_0, t_1],$$

then

$$v(t) \leq e^{\alpha(t-t_0)} v(t_0), \quad t \in [t_0, t_1].$$

Similarly if $\dot{v}(t) \geq \alpha v(t)$ a.s. on $[t_0, t_1]$ then

$$v(t) \geq e^{\alpha(t-t_0)} v(t_0), \quad t \in [t_0, t_1].$$

§2.2. The main stability test

In the present section we study the asymptotic behaviour of solutions of the equation

$$\begin{aligned} \dot{z} &= Az + h(t, z), \\ z(0) &= x \in \mathbf{C}^n, \end{aligned} \tag{2.12}$$

where A is a linear transformation from $\mathbf{C}^n$ into $\mathbf{C}^n$, identified with an element of $\mathbf{M}(n,n;C)$, and h is a continuous mapping from $\mathbf{R}_+ \times \mathbf{C}^n$ onto $\mathbf{C}^n$ satisfying

$$\sup_{t\geq 0} \frac{|h(t,x)|}{|x|} \longrightarrow 0, \quad \text{if } |x| \longrightarrow 0. \tag{2.13}$$

As in Part I (see §I.2.7), we set

$$\omega(A) = \max\{\operatorname{Re}\lambda;\, \lambda \in \sigma(A)\},$$

and if x, y are vectors with complex coordinates $\xi_1, \ldots, \xi_n$ and $\eta_1, \ldots, \eta_n$,

$$\langle x, y \rangle = \sum_{i=1}^{n} \xi_i \bar{\eta}_i, \quad |x|^2 = \sum_{i=1}^{n} |\xi_i|^2.$$

Under the imposed conditions, for arbitrary $x \in \mathbf{C}^n$ there exists a maximal solution $z(t)$, $t \in [0,\tau)$, $z(0) = x$ of (2.12).

The following theorem is of basic importance in the theory of nonlinear systems.

Theorem 2.2. (i) *If* (2.13) *holds and* $\omega(A) < 0$, *then for arbitrary* $\omega > \omega(A)$ *there exist* $M > 0, \delta > 0$ *such that an arbitrary maximal solution* $z(\cdot)$ *of* (2.12), *with initial condition* $z(0)$ *satisfying* $|z(0)| < \delta$, *is defined on* $[0, +\infty)$ *and*

$$|z(t)| \leq M e^{\omega t} |z(0)|, \quad t \geq 0.$$

(ii) *If* (2.13) *holds and* $\omega(A) > 0$ *then there exists* $r > 0$ *such that for arbitrary* $\delta > 0$ *one can find a solution* $z(t)$, $t \in [0,\tau)$ *with an initial condition* $z(0)$, $|z(0)| < \delta$, *and a number* $s \in [0,\tau)$, *such that* $|z(s)| > r$. *If, in addition, transformations* A *and* $h(t,\cdot)$, $t \geq 0$, *restricted to* $\mathbf{R}^n$ *take values in* $\mathbf{R}^n$ *then there exists a solution* $z(\cdot)$ *with values in* $\mathbf{R}^n$ *having the described properties.*

Before proving the result we prove a lemma on Jordan blocks. Let us recall (see Theorem I.2.1) that a *complex Jordan block* corresponding to a number $\lambda = \alpha + i\beta$, $\beta \neq 0$, or $\lambda = \alpha$, $\alpha, \beta \in \mathbf{R}$, is a square matrix of arbitrary dimension $m = 1, 2, \ldots$ or of dimension 1, of the form

$$J = \begin{bmatrix} \lambda & \gamma & 0 & \cdots & 0 & 0 & 0 \\ 0 & \lambda & \gamma & \cdots & 0 & 0 & 0 \\ 0 & 0 & \lambda & \cdots & 0 & 0 & 0 \\ \vdots & \vdots & \vdots & \ddots & \vdots & \vdots & \vdots \\ 0 & 0 & 0 & \cdots & \lambda & \gamma & 0 \\ 0 & 0 & 0 & \cdots & 0 & \lambda & \gamma \\ 0 & 0 & 0 & \cdots & 0 & 0 & \lambda \end{bmatrix} \quad \text{or } J = [\lambda],$$

respectively, where γ is a number different from 0.

§ 2.2. The main stability test

Lemma 2.2. *For arbitrary complex Jordan block J and for arbitrary $x \in \mathbf{C}^n$*
$$(\operatorname{Re}\lambda - |\gamma|)|x|^2 \leq \operatorname{Re}\langle Jx, x\rangle \leq (\operatorname{Re}\lambda + |\gamma|)|x|^2.$$

Proof. It is sufficient to consider the case of matrix J with the parameter $\gamma \neq 0$. Then
$$\langle Jx, x\rangle = \lambda|x|^2 + \gamma(\xi_2\bar{\xi}_1 + \ldots + \xi_m\bar{\xi}_{m-1})$$

and therefore
$$\operatorname{Re}\langle Jx, x\rangle = \operatorname{Re}\lambda|x|^2 + \operatorname{Re}(\gamma(\xi_2\bar{\xi}_1 + \ldots + \xi_m\bar{\xi}_{m-1})).$$

By the Schwartz inequality
$$|\operatorname{Re}(\gamma(\xi_2\bar{\xi}_1 + \ldots + \xi_m\bar{\xi}_{m-1}))| \leq |\gamma| \, |x|^2,$$

and the required inequality follows. □

Proof of the theorem. (i) By Theorem I.2.1 there exists a nonsingular matrix P such that the matrix $\tilde{A} = PAP^{-1}$ consists of only complex Jordan blocks $J_1, \ldots, J_r$ corresponding to the eigenvalues of A and to each eigenvalue $\lambda \in \sigma(A)$ corresponds at least one block. Parameters $\gamma \neq 0$ can be chosen in advance.

Function $\tilde{z}(t) = Pz(t)$, $t \in [0, \tau)$, is a solution of the equation
$$\dot{\tilde{z}} = \tilde{A}\tilde{z} + \tilde{h}(t, \tilde{z}), \quad \tilde{z}(0) = Pz(0),$$

in which the transformation $\tilde{h}(t, x) = Ph(t, P^{-1}x)$, $t \geq 0$, $x \in \mathbf{C}^n$, satisfies also the conditions of the theorem. Since the matrix P is nonsingular, therefore, without any loss of generality, we can assume that the matrix A consists of Jordan blocks only with parameters $\gamma \neq 0$ chosen arbitrarily. Since $\omega(A) < 0$ we can assume that $\omega < 0$. It follows from Lemma 2.2 that
$$\operatorname{Re}\langle Ax, x\rangle \leq (\omega(A) + \gamma)|x|^2, \quad x \in \mathbf{C}^n.$$

For arbitrary maximal solution $z(t)$, $t \in I = [0, \tau)$, define
$$v(t) = |z(t)|^2, \quad t \in I.$$

Then
$$\frac{d}{dt}v(t) = 2\operatorname{Re}\langle Az(t), z(t)\rangle + 2\operatorname{Re}\langle h(t, z(t)), z(t)\rangle$$
$$\leq 2(\omega(A) + \gamma)|z(t)|^2 + 2|h(t, z(t))| \, |z(t)|, \quad t \in I.$$

By (2.13) there exists $\delta > 0$ such that

$$|h(t,x)| \leq \varepsilon|x| \quad \text{for } t \geq 0, \ |x| \leq \delta.$$

Let $|z(0)| < \delta$ and $\bar{\tau} = \inf\{t \in [0,\tau); |z(t)| \geq \delta\}$ be the first exit time of the solution from the ball $B(0,\delta)$, $(\inf \emptyset = +\infty)$. Then (see Theorem 1.1) either $\bar{\tau} = +\infty$ or $\bar{\tau} < \tau$ and

$$\frac{d}{dt}v(t) \leq 2(\omega(A) + \gamma + \varepsilon)v(t), \quad t \in \bar{I} = [0,\bar{\tau}).$$

By Corollary 2.1 of Theorem 2.1

$$v(t) \leq e^{2\omega t}v(0), \quad t \in \bar{I},$$

or equivalently

$$|z(t)| \leq e^{\omega t}|z(0)|, \quad t \in \bar{I}. \tag{2.14}$$

If $\bar{\tau} < +\infty$, then

$$\delta \leq \overline{\lim_{t \uparrow \bar{\tau}}}|z(t)| \leq e^{\omega\bar{\tau}}|z(0)| < \delta,$$

a contradiction. Thus $\bar{\tau} = +\infty$ and inequality (2.14) completes the proof of part (i).

(ii) We can assume that the matrix A is of the form

$$A = \begin{bmatrix} B & 0 \\ 0 & C \end{bmatrix}, \tag{2.15}$$

where matrices $B \in \mathbf{M}(k,k,\mathbf{C})$, $C \in \mathbf{M}(l,l,\mathbf{C})$, $k+l = n$, consists of Jordan blocks corresponding to all eigenvalues of A with positive and nonpositive real parts. Let

$$\alpha = \min\{\operatorname{Re}\lambda; \ \lambda \in \sigma(B)\}.$$

Identifying $\mathbf{C}^n$ with the Cartesian product of $\mathbf{C}^k$ and $\mathbf{C}^l$, we denote by $z_1(\cdot)$, $z_2(\cdot)$ and h_1, h_2 projections of the solution $z(\cdot)$ and the function h on $\mathbf{C}^k$ and $\mathbf{C}^l$, respectively. Define

$$v_1(t) = |z_1(t)|^2, \quad v_2(t) = |z_2(t)|^2, \quad t \in [0,\tau) = I.$$

Then for $t \in I$

$$\frac{d}{dt}v_1(t) = 2\operatorname{Re}\langle Bz_1(t), z_1(t)\rangle + 2\operatorname{Re}\langle h_1(t,z), z_1(t)\rangle,$$

$$\frac{d}{dt}v_2(t) = 2\operatorname{Re}\langle Cz_2(t), z_2(t)\rangle + 2\operatorname{Re}\langle h_2(t,z), z_2(t)\rangle.$$

§ 2.2. The main stability test

It follows from Lemma 2.2 and the Schwartz inequality that
$$\frac{d}{dt}v_1(t) \geq 2(\alpha - \gamma)v_1(t) - 2v_1^{1/2}(t)|h_1(t, z(t))|,$$
$$\frac{d}{dt}v_2(t) \leq 2\gamma v_2(t) + 2v_2^{1/2}(t)|h_2(t, z(t))|, \quad t \in I.$$

We can assume that $\alpha > \gamma > 0$. For arbitrary $\varepsilon > 0$ there exists $\delta > 0$, compare (2.13), such that for $x \in \mathbf{C}^n$, $|x| < \delta$,
$$|h_1(t,x)| \leq \varepsilon|x|, \quad |h_2(t,x)| \leq \varepsilon|x|.$$

Assume that $|z(0)| < \delta$ and let
$$\bar\tau = \inf\{t \in [0,\tau); |z(t)| \geq \delta\}.$$

For $t \in \bar{I} = [0, \bar\tau)$
$$|h_1(t, z(t))| \leq \varepsilon(v_1(t) + v_2(t))^{1/2},$$
$$|h_2(t, z(t))| \leq \varepsilon(v_1(t) + v_2(t))^{1/2}, \quad t \in \bar I.$$

This and the inequality $(v_1^{1/2} + v_2^{1/2}) \leq \sqrt{2}(v_1 + v_2)^{1/2}$ imply
$$\frac{d}{dt}(v_1(t) - v_2(t)) \geq 2(\alpha - \gamma)v_1(t) - 2\gamma v_2(t) - \varepsilon(v_1^{1/2}(t)$$
$$+ v_2^{1/2}(t))(v_1(t) + v_2(t))^{1/2}$$
$$\geq 2(\alpha - \gamma)v_1(t) - 2\gamma v_2(t) - 2\varepsilon\sqrt{2}(v_1(t) + v_2(t)), \text{ for } t \in \bar I.$$

Selecting $\varepsilon > 0$ and $\gamma > 0$ properly, we can find $c > 0$ such that for all $t \in \bar I$
$$\frac{d}{dt}(v_1(t) - v_2(t)) \geq c(v_1(t) - v_2(t)).$$

By Theorem 2.1
$$v_1(t) - v_2(t) \geq e^{ct}(v_1(0) - v_2(0)), \quad t \in \bar I.$$

Therefore, if
$$v_1(0) - v_2(0) = |z_1(0)|^2 \geq |z_2(0)|^2 \quad \text{and} \quad |z(0)| < \delta, \qquad (2.16)$$

then $\bar\tau < +\infty$ and $|z(t)| \geq \delta$ for some $t \in (0, \tau)$.

This way part (ii) of the theorem has been proved for the complex case. To show the final part of (ii), we apply the representation theorem from §I.2.1. Since the matrix A is real, one can find invertible matrices $P_0 \in \mathbf{M}(n,n)$, $P_1 \in \mathbf{M}(k,k; \mathbf{C})$, $P_2 \in \mathbf{M}(l,l; \mathbf{C})$ such that $\tilde A = PAP^{-1}$ and $P = \begin{bmatrix} P_1 & 0 \\ 0 & P_2 \end{bmatrix}$, P_0 is of the form (2.15). It is therefore clear, for arbitrary $\delta > 0$, one can find $x \in \mathbf{R}^n$ such that (2.16) holds for $\begin{bmatrix} z_1(0) \\ z_2(0) \end{bmatrix} = Px$. The proof of (ii) is complete. □

Exercise 2.1. Find a matrix $A \in M(n,n)$ with $\omega(A) = 0$, $n \geq 3$ and vectors $x_1, x_2, x_3 \in \mathbf{R}^n$, such that

$$\lim_{t\uparrow+\infty} |e^{At}x_1| = +\infty, \quad 0 < \lim_{t\uparrow+\infty} |e^{At}x_2| < +\infty, \quad \lim_{t\uparrow+\infty} |e^{At}x_3| = 0.$$

Exercise 2.2. A continuous mapping $f\colon \mathbf{R}^n \longrightarrow \mathbf{R}^n$ is called *dissipative* if

$$\langle f(x) - f(y), x - y \rangle \leq 0 \quad \text{for arbitrary } x, y \in \mathbf{R}^n, \tag{2.17}$$

where $\langle \cdot, \cdot \rangle$ is the scalar product on $\mathbf{R}^n$.

Show that if f is a dissipative mapping then an arbitrary maximal solution of the equation

$$\dot{z} = f(z), \quad z(0) = x, \tag{2.18}$$

is defined on $[0, +\infty)$, and for arbitrary $x \in \mathbf{R}^n$ equation (2.12) has exactly one solution $z^x(t), t \geq 0$. Moreover

$$|z^x(t) - z^y(t)| \leq |x - y| \quad \text{for } t \geq 0, \ x, y \in \mathbf{R}^n. \tag{2.19}$$

Conversely, if $f\colon \mathbf{R}^n \longrightarrow \mathbf{R}^n$ is a continuous mapping and equation (2.18) has for arbitrary $x \in \mathbf{R}^n$ exactly one solution $z^x(\cdot)$ for which (2.19) holds, then f is dissipative.

Hint. If $z(\cdot)$ is a solution to (2.19) defined on $[0, \tau)$ then function $v(t) = |z(t)|^2, t \in [0, \tau)$, satisfies inequality $\dot{v} \leq |f(0)|v^{1/2}$ on $[0, \tau)$. For two arbitrary solutions $z_1(t), t \in [0, \tau_1)$, and $z_2(t), t \in [0, \tau_2)$, of (2.18), investigate the monotonic character of the function $|z_1(t) - z_2(t)|^2, t \in [0, \tau_1 \wedge \tau_2)$.

§2.3. Linearization

Let us consider a differential equation:

$$\dot{z} = f(z), \quad z(0) = x \in \mathbf{R}^n. \tag{2.20}$$

Assume that $f(\bar{x}) = 0$ or equivalently that $\bar{x}$ is an equilibrium state for (2.20). We say that the state $\bar{x}$ is *exponentially stable* for (2.20) if there exist $\omega < 0$, $M > 0$, $\delta > 0$ such that arbitrary maximal solution $z(\cdot)$ of (2.20), with the initial condition $z(0)$, $|z(0) - \bar{x}| < \delta$, is defined on $[0, +\infty)$ and satisfies

$$|z(t) - \bar{x}| \leq Me^{\omega t}|z(0) - \bar{x}|, \quad t \geq 0. \tag{2.21}$$

The infinum of all those numbers $\omega < 0$ for which (2.21) holds will be called the *exponent* of $\bar{x}$.

§ 2.3. Linearization

The next theorem gives an effective characterization of exponentially stable equilibria and their exponents.

Theorem 2.3. *Assume that a continuous function f is differentiable at an equilibrium state $\bar{x}$. Then $\bar{x}$ is exponentially stable for (2.20) if and only if the Jacobi matrix*
$$A = f_x(\bar{x})$$
is stable. Moreover the exponent of $\bar{x}$ is $\omega(A)$.

Proof. We can assume, without any loss of generality, that $\bar{x} = 0$. Define
$$h(x) = f(x) - Ax, \quad x \in \mathbf{R}^n.$$
It follows from the assumptions that
$$\frac{|h(x)|}{|x|} \longrightarrow 0, \quad \text{when } |x| \to 0.$$
If matrix A is stable then the equation
$$\dot{z} = f(z) = Az + h(z), \quad z(0) = x,$$
satisfies the conditions of Theorem 2.2 (i). Therefore the state 0 is exponentially stable and its exponent is at least equal to $\omega(A)$. The following lemma completes the proof of the theorem.

Lemma 2.3. *Assume that condition (2.21) holds for some $\omega < 0$, $\delta > 0$, $M > 0$ and $\bar{x} = 0$. Then for arbitrary $N > M$ and $\gamma > \omega$*
$$|e^{At}| \leq Ne^{\gamma t}, \quad t \geq 0. \tag{2.22}$$

Proof. We show first that (2.22) holds for some and then for arbitrary $N > M$. We can assume that $\gamma < 0$. Let η be an arbitrary number from the interval (ω, γ) and let
$$y(t) = e^{-\eta t} z(t), \quad t \geq 0.$$
Then
$$\dot{y}(t) = (-\eta + A)y(t) + e^{-\eta t} h(e^{\eta t} y(t)), \quad t \geq 0.$$
Since
$$\sup_{t \geq 0} \frac{|e^{-\eta t} h(e^{\eta t} x)|}{|x|} = \sup_{t \geq 0} \frac{|h(e^{\eta t} x)|}{|e^{\eta t} x|} \longrightarrow 0, \quad \text{if } |x| \to 0,$$
therefore, by Theorem 2.2 (ii), $\omega(-\eta I + A) \leq 0$ and consequently $\omega(A) \leq \eta$. There exists $N_1 > 0$ such that, for arbitrary $t \geq 0$,
$$|e^{At}| \leq N_1 e^{\gamma t}.$$

Moreover, for the solution $z(\,\cdot\,)$ of (2.20),
$$z(t) = e^{At}z(0) + \int_0^t e^{A(t-s)}h(z(s))\,ds, \quad t \geq 0,$$
and, assuming that $x = z(0)$,
$$|e^{At}x| \leq |z(t)| + \int_0^t |e^{A(t-s)}|\,|h(z(s))|\,ds, \quad t \geq 0.$$
For arbitrary $\varepsilon > 0$ there exists $\delta_1 > 0$ such that if $|x| = |z(0)| < \delta_1$ then
$$|h(z(t))| \leq \varepsilon |z(t)|, \quad t \geq 0.$$
Hence, for $|x| < \delta_1$
$$\begin{aligned}|e^{At}x| &\leq Me^{\omega t}|x| + MN_1\varepsilon \left(\int_0^t e^{\gamma(t-s)}e^{\omega s}\,ds\right)|x| \\ &\leq \left(Me^{\omega t} + MN_1\varepsilon \frac{1}{\gamma-\omega}e^{\gamma t}\right)|x| \\ &\leq \left(M + \frac{MN_1\varepsilon}{\gamma-\omega}\right)e^{\gamma t}|x|, \quad t \geq 0.\end{aligned}$$
Number $\varepsilon > 0$ was arbitrary and the lemma follows. □

Theorem 2.3 reduces the problem of stability of an equilibrium state $\bar{x} \in \mathbf{R}^n$ to the question of stability of the matrix A obtained by the linearization of the mapping f at $\bar{x}$. The content of Theorem 2.3 is called the *linearization method* or the *first method of Liapunov* of stability theory. The practical algorithm due to Routh allowing one to determine whether a given matrix A is stable was given in §I.2.3.

Exercise 2.3. Watt's regulator (see Example 0.3) is modelled by the equations
$$\begin{aligned}\dot{x} &= a\cos y - b, \quad x(0) = x_0, \\ \ddot{y} &= cx^2 \sin y \cos y - d\sin y - e\dot{y}, \quad y(0) = y_0,\ \dot{y}(0) = z_0,\end{aligned}$$
where a,b,c,d,e are some positive constants, $a > b$. Show that there exists exactly one equilibrium state for the system, with coordinates $\bar{x} > 0$, $\bar{y} \in (0,\tfrac{1}{2}\pi)$, $\bar{z} = 0$. Prove that if $e\sqrt{da} > 2\sqrt{cb^3}$ then the equilibrium state is exponentially stable and if $e\sqrt{da} < 2\sqrt{cb^3}$ then it is not stable in the Liapunov sense; see §2.4.

Hint. Apply Theorems 2.3, 2.2 and I.2.4.

§ 2.4. The Liapunov function method

We say that a state $\bar{x}$ is *stable in the Liapunov sense* for (2.20), or that $\bar{x}$ is *Liapunov stable*, if for arbitrary $r > 0$ there is $\delta > 0$ such that if $|z(0) - \bar{x}| < \delta$ then $|z(t) - \bar{x}| < r$ for $t \geq 0$. It is clear that exponentially stable equilibria are stable in the Liapunov sense.

Assume that a mapping $f \colon \mathbf{R}^n \longrightarrow \mathbf{R}^n$ is continuous and $G \subset \mathbf{R}^n$ is a neighbourhood of a state $\bar{x}$ for which $f(\bar{x}) = 0$. A real function V differentiable on G is said to be a *Liapunov function* at the state $\bar{x}$ for equation (2.20) if

$$V(\bar{x}) = 0, \quad V(x) > 0 \quad \text{for } x \in G,\ x \neq \bar{x}, \tag{2.23}$$

$$V_f(x) = \sum_{j=1}^{n} \frac{\partial V}{\partial x_j}(x) f^j(x) \leq 0, \quad x \in G. \tag{2.24}$$

The function V_f defined by (2.24) is called the *Liapunov derivative* of V with respect to f.

Theorem 2.4. (i) *If there exists a Liapunov function at $\bar{x}$ for (2.20), then the state $\bar{x}$ is Liapunov stable for (2.20).*

(ii) *If $\bar{x}$ is exponentially stable at $\bar{x}$ and f is differentiable at $\bar{x}$, then there exists a Liapunov function at $\bar{x}$ for (2.20), being a quadratic form.*

For the proof of part (i) it is convenient to introduce the concept of an invariant set. We say that a set $K \subset \mathbf{R}^n$ is *invariant* for (2.20), if for arbitrary solution $z(t)$, $t \in [0, \tau)$, then $z(0) \in K$, $z(t) \in K$ for all $t \in [0, \tau)$.

We will need the following lemma:

Lemma 2.4. *If for an open set $G_0 \subset G$ and for an $\alpha > 0$ the set*

$$K_0 = \{x \in G_0;\ V(x) \leq \alpha\}$$

is closed, then it is also invariant for (2.20).

Proof. Assume that the lemma is not true and that there exists a point $x \in K_0$ and a solution $z(t)$, $t \in [0, \tau)$, of the equation (2.20), with values in G, $z(0) = x$ such that for some $s \in [0, \tau)$, $z(s) \in K_0^c$.

Let

$$t_0 = \inf\{s \geq 0;\ s < \tau,\ z(s) \in K_0^c\} < +\infty.$$

Since the set K_0 is closed $z(t_0) \in K_0$ and $t_0 < \tau$. For some $s \in (t_0, \tau)$, $V(z(s)) > V(z(t_0))$. Since

$$\frac{dV(z(t))}{dt} = V_f(z(t)) \leq 0 \quad \text{for } t < \tau,$$

the function $V(z(\cdot))$ is nonincreasing on $[0, \tau)$, a contradiction. $\square$

Proof of the theorem. (i) Let $r > 0$ be a number such that the closure of $G_0 = B(\bar{x}, r) = \{x; |x - \bar{x}| < r\}$ is contained in G. Let β be the minimal value of the function V on the boundary S of the ball $B(\bar{x}, r)$. Let us fix $\alpha \in (0, \beta)$ and define

$$K_0 = \{x \in B(\bar{x}, r); V(x) \leq \alpha\}.$$

Let $\hat{x}$ be the limit of a sequence (x_m) of elements from K_0. Then $V(\hat{x}) \leq \alpha < \beta$ and consequently $\hat{x} \in B(\bar{x}, r)$ and $\hat{x} \in K_0$. The set K_0 is closed and, by Lemma 2.4, is also invariant. Since the function V is continuous, $V(\bar{x}) = 0$ and there exists $\delta \in (0, r)$ such that $V(x) \leq \alpha$ for $x \in B(\bar{x}, \delta)$. This implies (i).

(ii) Matrix $A = f_x(\bar{x})$ is stable by Theorem 2.3. It follows from Theorem I.2.7 that there exists a positive definite matrix Q satisfying the Liapunov equation

$$A^*Q + QA = -I.$$

To simplify notation we set $\bar{x} = 0$ and define

$$V(x) = \langle Qx, x \rangle, \quad h(x) = f(x) - Ax, \quad x \in \mathbf{R}^n.$$

It is clear that the function V satisfies (2.23). To show that (2.24) holds as well remark that

$$\begin{aligned} V_f(x) &= 2\langle Qx, f(x) \rangle = 2\langle Qx, Ax + h(x) \rangle \\ &= \langle (A^*Q + QA)x, x \rangle + 2\langle Qx, h(x) \rangle \\ &\leq -|x|^2 + 2|Q||x||h(x)|, \quad x \in \mathbf{R}^n. \end{aligned}$$

Since f is differentiable at 0, there exists a number $\delta > 0$ such that $|h(x)| \leq \frac{1}{4|Q|}|x|$, provided $|x| < \delta$. Consequently

$$V_f(x) < -\frac{1}{2}|x|^2, \quad \text{for } |x| < \delta.$$

Hence the function V is the required Liapunov function in the ball $G = B(0, \delta)$.

□

Exercise 2.4. Let $f: \mathbf{R}^n \longrightarrow \mathbf{R}^n$, $f(0) = 0$, be a continuous mapping differentiable at 0 such that all eigenvalues of $A = f_x(0)$ have positive real parts. Show that there exists a number $r > 0$ such that, for arbitrary maximal solution $z(t)$, $t \in [0, \tau)$, of the equation (2.20), one can find $t_0 \in [0, \tau)$ such that, for all $t \in [t_0, \tau)$, $|z(t)| \geq r$.

Hint. Matrix $-A$ is stable. Follow the proof of part (ii) of Theorem 2.4.

Exercise 2.5. Let a matrix $A \in M(n, n)$ be stable and a continuous mapping $F: \mathbf{R}^n \longrightarrow \mathbf{R}^n$ be bounded. Then, for arbitrary solution $z(t)$,

$t \in [0, \tau)$, of the equation $\dot z = Az + F(z)$, there exists $M > 0$ such that $|z(t)| \leq M$ for $t \in [0, \tau)$.

Hint. Examine the behaviour of $V(x) = \langle Qx, x \rangle$, $z \in \mathbf{R}^n$, where $A^*Q + QA = -I$, along the trajectories of the equation.

A closed invariant set K for the equation (2.20) is called *nonattainable* if one cannot find a solution $z(t)$, $t \in [0, \tau)$, of (2.20) such that $z(0) \notin K$ and $z(t) \in K$ for some $t \in [0, \tau)$.

Exercise 2.6. (i) Assume that the mapping $f \colon \mathbf{R}^n \longrightarrow \mathbf{R}^n$ satisfies the Lipschitz condition in a neighbourhood of a point $\bar x \in \mathbf{R}^n$, $f(\bar x) = 0$. Show that the set $K = \{\bar x\}$ is nonattainable for (2.20).

(ii) Construct a mapping $f \colon \mathbf{R}^n \longrightarrow \mathbf{R}^n$ satisfying the Lipschitz condition and an invariant compact set K for (2.20) which is not nonattainable.

(iii) Assume that $f \colon \mathbf{R}^n \longrightarrow \mathbf{R}^n$ satisfies the Lipschitz condition and K is a closed invariant set for (2.20). Let moreover for arbitrary $y \in K$ there be $x \in K$ and a solution $z(t)$, $t \geq 0$, of (2.20) such that $z(0) = x$ and $z(s) = y$ for some $s > 0$. Prove that K is nonattainable for (2.20).

(iv) Let $z(t)$, $t \geq 0$, be a solution of the following *prey-predator* equation in $\mathbf{R}^2$:

$$\dot z_1 = \alpha z_1 - \beta z_1 z_2, \quad \dot z_2 = -\gamma z_2 + \delta z_1 z_2,$$

in which α, β, γ, δ are positive numbers. Show that if $z_1(0) > 0$, $z_2(0) > 0$ then $z_1(t) > 0$, $z_2(t) > 0$ for all $t > 0$.

§2.5. La Salle's theorem

Besides exponential stability and stability in the sense of Liapunov it is also interesting to study asymptotic stability. An equilibrium state $\bar x$ for (2.20) is *asymptotically stable* if it is stable in the sense of Liapunov and

there exists $\delta > 0$ such that, for any maximal solution $z(t)$, $t \in [0, \tau)$, of (2.20), with $|z(0) - \bar x| < \delta$, one has $\tau = \infty$ and $\lim\limits_{t \uparrow +\infty} z(t) = \bar x$. (2.25)

For the following linear system on $\mathbf{R}^2$

$$\dot z_1 = z_2, \quad \dot z_2 = -z_1 \qquad (2.26)$$

the origin is stable in the sense of Liapunov but it is not asymptotically stable. The function $V(x) = |x|^2$, $x \in \mathbf{R}^2$, is a Liapunov function for (2.26) and $V_f(x) = 0$, $x \in \mathbf{R}^2$. Hence the existence of a Liapunov function does not necessarily imply asymptotic stability. Additional conditions, which we discuss now, are needed.

We will limit our considerations to equations (2.26) with the function $f\colon \mathbf{R}^n \longrightarrow \mathbf{R}^n$ satisfying the *local Lipschitz condition*. Let us recall that a function $f\colon G \longrightarrow \mathbf{R}^n$ satisfies the *Lipschitz condition* on a set $G \subset \mathbf{R}^n$ if there exists $c > 0$ such that

$$|f(x) - f(y)| \leq c|x-y| \quad \text{far all } x, y \in G.$$

If the Lipschitz condition is satisfied on an arbitrary bounded subset of $\mathbf{R}^n$ then we say that it is satisfied locally.

Exercise 2.7. If a function $f\colon \mathbf{R}^n \longrightarrow \mathbf{R}^n$ satisfies the local Lipschitz condition then for arbitrary $r > 0$ the function

$$\hat{f}(x) = \begin{cases} f(x), & \text{for } |x| \leq r, \\ f\left(\frac{x}{|x|}r\right), & \text{for } |x| \geq r, \end{cases}$$

satisfies the Lipschitz condition on $\mathbf{R}^n$.

It follows from Theorem 1.1 and Theorem 1.2 that if a mapping $f\colon \mathbf{R}^n \longrightarrow \mathbf{R}^n$ satisfies the local Lipschitz condition then for arbitrary $x \in \mathbf{R}^n$ there exists exactly one maximal solution $z(t,x) = z^x(t)$, $t \in [0, \tau(x))$, of the equation (2.20). The maximal solution (see Exercise 2.7 and Theorem 2.2) depends continuously on the initial state: If $\lim_{k \uparrow +\infty} x_k = x$ and $t \in [0, \tau(x))$ then for all large k, $\tau(x_k) > t$ and $\lim_{k \uparrow +\infty} z(t, x_k) = z^x(t, x)$.

If $\tau(x) = +\infty$, then the *orbit* $\mathcal{O}(x)$ and the *limit set* $K(x)$ of x are given by the formulae,

$$\mathcal{O}(x) = \{y \in \mathbf{R}^n;\ y = z(t,x),\ t \geq 0\},$$
$$K(x) = \{y \in \mathbf{R}^n;\ \lim_k z(t_k, x) = y \text{ for a sequence } t_k \uparrow +\infty\}.$$

It is clear that the limit set $K(x)$ is contained in the closure of $\mathcal{O}(x)$.

Lemma 2.5. *If $\mathcal{O}(x)$ is a bounded set then*
  (i) *the set $K(x)$ is invariant for* (2.20),
  (ii) *the set $K(x)$ is compact,*
  (iii) $\lim_{t \uparrow +\infty} \varrho(z(t,x), K(x)) = 0$, *where $\varrho(\cdot, \cdot)$ denotes the distance between a point and a set.*

Proof. (i) Assume that $y \in K(x)$ and $t > 0$. There exists a sequence $t_m \uparrow +\infty$ such that $y_m = z(t_m x) \longrightarrow y$. The uniqueness and the continuous dependence of solutions on initial data imply

$$z(t + t_m, x) = z(y_m, t) \longrightarrow z(y, t), \quad \text{when } m \longrightarrow +\infty.$$

Hence $z(y,t) \in K(x)$.

(ii) Assume that (y_m) is a sequence of elements from $K(x)$ converging to $y \in \mathbf{R}^n$. There exists a sequence $t_m \uparrow +\infty$ such that $|z(t_m, x) - y_m| \longrightarrow 0$. Consequently $z(t_m, x) \longrightarrow y$ and $y \in K(x)$.

§ 2.5. La Salle's theorem

(iii) Assume that $\varrho(z(t,x), K(x)) \not\to 0$ when $t \uparrow +\infty$. Then there exists $\varepsilon > 0$ and a sequence (t_m), $t_m \uparrow +\infty$, such that

$$|z(t_m, x) - y)| \geq 0 \quad \text{for arbitrary } y \in K(x), \ m = 1, 2, \ldots . \tag{2.27}$$

Since the set $\mathcal{O}(x)$ is bounded, one can find a subsequence $(\tilde{t}_k)$ of (t_m) and an element $\tilde{y} \in G$ such that $z(\tilde{t}_k, x) \longrightarrow \tilde{y}$. Hence $\tilde{y} \in K(x)$ and, by (2.27), $|y - \tilde{y}| \geq \varepsilon$ for arbitrary $y \in K(x)$. In particular for $y = \tilde{y}$, $0 = |\tilde{y} - \tilde{y}| \geq \varepsilon$, a contradiction. This way the proof of (iii) is complete. □

We are now in a position to prove the La Salle theorem.

Theorem 2.5. *Assume that a mapping* $f: \mathbf{R}^n \longrightarrow \mathbf{R}^n$ *is locally Lipschitz,* $f(\bar{x}) = 0$ *and there exists a Liapunov function V for (2.20) in a neighbourhood G of $\bar{x}$. If, for an $x \in G$, $\mathcal{O}(x) \subset G$, then*

$$K(x) \subset \{y \in G; \ V_f(y) = 0\}.$$

Proof. Let $v(t) = V(z(t, x))$, $t \geq 0$. Then $v(\cdot)$ is a nonnegative, decreasing function and

$$\lim_{t \uparrow +\infty} v(t) = \hat{v} \geq 0.$$

If $y \in K(x)$ then for a sequence $t_m \uparrow +\infty$, $z(t_m, x) \longrightarrow y$. Thus $V(y) = \hat{v}$. It follows from Lemma 2.5(i) that for an arbitrary $y \in K(x)$ and $t \geq 0$, $V(z(t, y)) = V(y)$. Hence

$$V_f(y) = \lim_{t \downarrow 0} \frac{1}{t}(V(y) - V(z(t, y))) = 0.$$

□

We will now formulate a version of Theorem 2.5 useful in specific applications.

Let

$$L = \{y \in G; \ V_f(y) = 0\} \tag{2.28}$$

and

$$K \text{ be the maximal, invariant for (2.20), subset of } L \tag{2.29}$$

Since the sum of invariant subsets for (2.20) is also invariant for (2.20) as well as the set $\{\bar{x}\}$ is invariant, therefore K is a well defined nonempty subset of G.

Theorem 2.6. (i) *Under conditions of Theorem 2.5,*

$$\varrho(z(t, x), K) \longrightarrow 0, \quad \text{for } t \uparrow +\infty.$$

(ii) *If, in addition, $V_f(y) < 0$ for $y \in G \setminus \{\bar{x}\}$, then $\bar{x}$ is an asymptotically stable equilibrium for* (2.20).

Proof. (i) Since $K(x) \subset K$, it is enough to apply Lemma 2.5(iii).
(ii) In this case $L = \{\bar{x}\} = K$. □

Exercise 2.8. Show that maximal solutions of the Liénard equation $\ddot{x} + (2a+(\dot{x})^2)\dot{x}+\nu^2 x = 0$, $\nu \neq 0$, exist on $[0,+\infty)$. Find the unique equilibrium state for this system and prove that
  (i) if $a \geq 0$, then it is asymptotically stable, and
  (ii) if $a < 0$, then it is not stable.

Hint. Let $V(x,y) = \frac{1}{2}(\nu^2 x^2 + y^2)$, $x,y \in \mathbf{R}$. Show that $\frac{d}{dt}V(x,\dot{x}) \leq |a|^2$. To prove (i) apply Theorem 2.5. Base the proof of (ii) on Theorem 2.3.

§2.6. Topological stability criteria

We proceed now to stability criteria based on the concept of the *degree* of a vector field. They will be applied to stabilizability of control systems in §2.8.

Let S be the unit sphere, i.e., the boundary of the unit ball $B(0,1) = \{y \in \mathbf{R}^n; |y| < 1\}$. With an arbitrary continuous mapping $F\colon S \longrightarrow S$ one associates, in topology, an integer called the degree of F, denoted by $\deg F$, see [25]. Intuitively, if $n = 1$, the degree of F is the number of times the image $F(x)$ rotates around S when x performs one oriented rotation. We will not give here a precise definition of the degree of a map but gather its basic properties in the following proposition. For the proofs we refer to [29].

Let us recall that if F_0 and F_1 are two continuous mappings from S into S and there exists a continuous function $H\colon [0,1] \times S \longrightarrow S$ such that

$$H(0,x) = F_0(x), \quad H(1,x) = F_1(x), \quad x \in S,$$

then F_0 and F_1 are called *homotopic* and H is a *homotopy* which deforms F_0 to F_1. If F_0 and F_1 are homotopic we symbolically write $F_0 \simeq F_1$.

Proposition 2.1. (i) *For arbitrary continuous mappings F_0, F_1 from S onto S:*

If $F_0 \simeq F_1$, then $\deg F = \deg F_1$.

(ii) *If a mapping F is constant: $F(x) = \tilde{x}$ for some $\tilde{x} \in S$ and all $x \in S$, then $\deg F = 0$.*

(iii) *If F is an antipodal mapping: $F(x) = -x$, $x \in S$, then $\deg F = (-1)^n$.*

§ 2.6. Topological stability criteria

Assume that f is a continuous mapping from a ball $B(\bar{x}, \delta) \subset \mathbf{R}^n$ onto $\mathbf{R}^n$, $f(\bar{x}) = 0$ and $f(x) \neq 0$ for $x \neq \bar{x}$. For arbitrary $r \in (0, \delta)$ define a mapping F_r from S onto S by the formula

$$F_r(x) = \frac{f(rx + \bar{x})}{|f(rx + \bar{x})|}, \quad x \in S.$$

If $0 < r_0 \leq r_1 < \delta$ then the transformation

$$H(s, x) = \frac{f((r_0 + s(r_1 - r_0))x + \bar{x})}{|f((r_0 + s(r_1 - r_0))x + \tilde{x}|}, \quad s \in [0, 1], \ x \in S,$$

defines a homotopy deforming F_{r_0} to F_{r_1}. By Proposition 2.1(i), all the mappings F_r, $r \in (0, \delta)$, have the same degree, which is called the *index of* f *at* $\bar{x}$ and denoted by $\mathrm{Ind}_{\bar{x}} f$. Thus

$$\mathrm{Ind}_{\bar{x}} f = \deg F_r, \quad 0 < r < \delta.$$

The following theorem concerns asymptotically stable equilibria and *attracting* points. A point $\bar{x}$ is attracting for (2.20) if an arbitrary maximal solution $z(\,\cdot\,)$ of (2.20) is defined on $[0, +\infty)$ and $\lim_{t \uparrow +\infty} z(t) = \bar{x}$.

Exercise 2.9. Construct an equation with an attracting point which is not asymptotically stable.

Hint. See [15, page 59, picture 1.7.9].

Theorem 2.7. *If a point $\bar{x}$ is either asymptotically stable or attracting for equation (2.20) with the right hand side satisfying the local Lipschitz condition then*

$$\mathrm{Ind}_{\bar{x}} f = (-1)^n.$$

Proof. Without any loss of generality we can assume that $\bar{x} = 0$. It follows from the assumptions that $f(0) = 0$ and $f(x) \neq 0$ for all $x \neq 0$ in a neighbourhood of 0.

We show first that there exist numbers $R > r > 0$ and $t_0 > 0$ such that for all solutions $z(\cdot, x)$, $|x| = R$, of (2.20)

$$|z(t_0, x)| < r.$$

Moreover, in the case of an asymptotically stable point, the numbers $R > 0$ and $r > 0$ can be chosen arbitrarily small.

It follows from the assumptions of the theorem that there exist numbers $R_1 > r_1 > 0$ such that for all $x \in \mathbf{R}^n$, $|x| = R_1$:

$$T_{r_1}(x) = \inf\{t \geq 0;\ |z(t, x)| \leq r_1\} < +\infty.$$

Fix $r_2 \in (r_1, R_1)$. If $|x| = R_1$, then $T_{r_2}(x) < T_{r_1}(x)$. Since the solutions of (2.20) depend continuously on initial data, for arbitrary x, $|x| = R_1$ one can find $\delta > 0$ such that $T_{r_2}(y) \leq T_{r_1}(x) < +\infty$ provided $|y| = R_1$ and $|y - x| < \delta$. Consequently the function $T_{r_2}(\cdot)$ is locally bounded on a compact set $\{y; |y| = R_1\}$ and therefore it is globally bounded:

$$\hat{T} = \sup\{T_{r_2}(y); |y| = R_1\} < +\infty. \tag{2.30}$$

If $\mathcal{O}$ is an asymptotically stable equilibrium, numbers $R = R_1 > r_2 > r_1$ and $r \in (r_2, R)$ can be chosen arbitrarily small and such that

$$|z(t, x)| < r \quad \text{for } t \geq 0 \text{ and } |x| = r_2.$$

It is therefore enough to take as t_0 an arbitrary number greater than $\hat{T}$.

If $\mathcal{O}$ is an attracting point then for arbitrary $R_1 > 0$ the set-theoretic sum K of all orbits $\mathcal{O}(x)$, $|x| = R_1$, is bounded and invariant. To see this choose a number $T > \tilde{T}$. Then

$$K \subset \{y; |y| \leq r_2\} \cup \{y; y = z(s, x), \ s \leq T, \ |x| = R_1\}.$$

Since the sets K_1 and K_2 are compact, the boundedness of K follows. It is also obvious that the set K is invariant. Let R be an arbitrary number such that

$$\overline{K} \subset \{y; |x| < R\}$$

and let

$$r = \sup\{|y|; y \in \overline{K}\}.$$

Then

$$\hat{T} = \sup\{T_{R_1}(x); |x| = R\} < +\infty.$$

It is therefore clear that it is enough to choose as t_0 an arbitrary number greater than $\hat{T}$.

The proof of the theorem follows immediately from the following lemma.

Lemma 2.6. *Assume that for some $R > r > 0$, $t_0 > 0$,*

$$\lim_{t\uparrow+\infty} |z(t, x)| < R, \quad \text{where } |x| \leq R, \tag{2.31}$$

$$|z(t_0, x)| < r, \quad \text{where } |x| = R. \tag{2.32}$$

Then

$$\text{Ind}_0 f = (-1)^n.$$

Proof of the lemma. Define for $t \in (0, t_0]$

$$F_t(y) = \frac{z(t, Ry) - Ry}{|z(t, Ry) - Ry|}, \quad y \in S. \tag{2.33}$$

§ 2.6. Topological stability criteria

It follows from (2.31) that solutions $z(\cdot, x)$, $|x| = R$, are not periodic functions. Consequently the formula (2.33) defines continuous transformations from S onto S which are clearly homotopic. Taking into account the following homotopy

$$H(\alpha, y) = \frac{\alpha z(t_0, Ry) - Ry}{|\alpha z(t_0, Ry) - Ry|}, \quad \alpha \in [0,1], \ y \in S,$$

we see that the transformation $H(1, \cdot) = F_{t_0}$ is homotopic to the antipodal map $H(0, y) = -y$, $y \in S$. So $\deg F_{t_0} = (-1)^n$. On the other hand, $\frac{1}{t}(z(t, Ry) - Ry) \longrightarrow f(Ry)$ uniformly on S, when $t \downarrow 0$ so for sufficiently small $t > 0$, $F_t \simeq f(R\cdot)/|f(R\cdot)|$. Consequently by Proposition 2.1 the result follows. $\square$

As a corollary from Theorem 2.7 we obtain that the vector field f defining a stable system is "rich in directions".

Theorem 2.8. *Assume that the conditions of Theorem 2.7 are satisfied. Then for sufficiently small $r > 0$, the transformation $F_r \colon S \longrightarrow S$ given by*

$$F_r(x) = \frac{f(rx + \bar{x})}{|f(rx + \bar{x})|}, \quad x \in S, \ r > 0,$$

transforms S onto S. In addition, the mapping f transforms an arbitrary neighbourhood of $\bar{x}$ onto a neighbourhood of 0.

Proof. It follows from Theorem 2.7 that $\deg F_r \neq 0$ for sufficiently small $r > 0$. Assume that the transformation F_r is not onto S and that a point $\tilde{x} \in S$ is not in the image of F_r. One can easily construct a homotopy deforming $S \setminus \{\tilde{x}\}$ to the antipodal point $\tilde{x}$. Consequently the transformation F_r is homotopic to a constant transformation and, by Proposition 2.1.(ii), $\deg F_r = 0$, a contradiction. Thus F_r is a transformation onto S. To prove the final statement of the theorem assume that $\bar{x} = 0$. Let $r > 0$ be a number such that $f(x) \neq 0$ for all $x \in B(0, r) \setminus \{0\}$. If $0 < \delta < \min(|f(x)|, |x| = r)$ then for arbitrary $\lambda \in [0,1]$ and $y \in \mathbf{R}^n$, $|y| < \delta$,

$$|\lambda(f(x) - y) + (1 - \lambda)f(x)| = |f(x) - \lambda y| \geq |f(x)| - |y| \geq 0.$$

Let us fix $y \in \mathbf{R}^n$, $0 < |y| < \delta$ and define

$$H(\lambda, x) = \frac{(1 - \lambda)f(rx) + \lambda(f(rx) - y)}{|(1 - \lambda)f(rx) + \lambda(f(rx) - y)|}, \quad \lambda \in [0,1], \ x \in S.$$

Then H is a homotopy deforming $G(x) = \frac{f(rx)}{|f(rx)|}$ to $G_r(x) = \frac{f(rx) - y}{|f(rx) - y|}$, $x \in S$. Since $\deg G \neq 0$, $\deg G_r \neq 0$ as well. Assume that for arbitrary $s \in (0, r]$ and arbitrary x, $|x| \leq s$, $f(x) \neq y$. Then the transformation G_r is homotopic to all, defined analogically, transformations G_s, $s \in (0, r]$. Since $f(0) = 0$ and f is continuous, for sufficiently small $s > 0$ the image G_s is included in a given in advance neighbourhood of $-y/|y|$. Hence for such $s > 0$ the transformations G_s cannot be onto S and therefore $\deg G_s = 0$, contrary to $\deg G_s = \deg G_r \neq 0$. This finishes the proof. □

§2.7. Exponential stabilizability and the robustness problem

In this section we begin our discussion of the stabilizability of nonlinear systems

$$\dot{y} = f(y, u), \quad y(0) = x, \tag{2.34}$$

Let $U \subset \mathbf{R}^m$ be the set of control parameters and assume that the continuous function $f: \mathbf{R}^n \times U \longrightarrow \mathbf{R}^n$ is such that $f(\bar{x}, \bar{u}) = 0$ for some $\bar{x} \in \mathbf{R}^n$ and $\bar{u} \in U$. Then a *feedback* is defined as an arbitrary continuous function $v: \mathbf{R}^n \longrightarrow U$ such that $v(\bar{x}) = \bar{u}$. Feedbacks determine *closed loop systems*

$$\dot{z} = g(z), \quad z(0) = x \in \mathbf{R}^n, \tag{2.35}$$

where

$$g(x) = f(x, v(x)), \quad x \in \mathbf{R}^n. \tag{2.36}$$

A feedback v is *stabilizing* if $\bar{x}$ is a stable equilibrium for (2.35). A feedback v *stabilizes* (2.34) *exponentially* or *asymptotically* or *in the sense of Liapunov* if the state $\bar{x}$ is respectively exponentially stable or asymptotically stable or stable in the sense of Liapunov for the closed loop system. If for system (2.34) there exists a feedback with one of the specified properties, then system (2.34) is called *stabilizable exponentially, asymptotically* or *in the sense of Liapunov*.

The stabilization problem consists of formulating checkable conditions on the right hand side of (2.34) implying stabilizability and of constructing stabilizing feedback.

To simplify notation we assume that $\bar{x} = 0$, $\bar{u} = 0$.

We first examine exponential stabilizability and restrict considerations to the case when $0 \in \mathbf{R}^m$ is an interior point of U and the function f as well as admissible *feedbacks* are differentiable respectively at $(0,0) \in \mathbf{R}^n \times \mathbf{R}^m$ and $0 \in \mathbf{R}^n$. According to our definitions, system (2.34) is *exponentially stabilizable* if there exists a feedback v and numbers $\omega < 0$, $\delta > 0$ and M such that for arbitrary solution $z(t)$, $t \in [0, \tau)$, of (2.34), with $|z(0)| < \delta$,

$$|z(t)| \leq M e^{\omega t} |z(0)|, \quad t \in [0, \tau). \tag{2.37}$$

The infimum of all $\omega < 0$ from (2.37) is called the *exponent of stabilizability* of (2.34). Let us recall that the linearization of the system (2.34) is of the form

$$\dot{y} = Ay + Bu, \tag{2.38}$$

§ 2.7. Exponential stabilizability and the robustness problem

where
$$A = f_x(0,0), \quad B = f_u(0,0). \tag{2.39}$$

The following theorem gives a complete solution to the question of exponential stabilizability.

Theorem 2.9. (i) *System (2.34) is exponentially stabilizable if and only if the linear system (2.38)-(2.39) is stabilizable.*
(ii) *An exponentially stabilizing feedback can always be found to be linear.*
(iii) *System (2.34) and its linearization (2.38)-(2.39) have the same stabilizability exponents.*

Proof. Assume that v is a feedback such that (2.37) holds. It follows from the differentiability of f and v that

$$g(x) = (f_x(0,0) + f_u(0,0)v_x(0))x + h(x), \quad x \in \mathbf{R}^n,$$

where
$$\frac{|h(x)|}{|x|} \longrightarrow 0, \quad \text{if } |x| \longrightarrow 0.$$

By Lemma 2.3 the matrix $\tilde{A} = f_x(0,0) + f_u(0,0)v_x(0)$ is stable, and, for arbitrary $N > M$ and $\omega < \gamma < 0$,

$$|e^{\tilde{A}t}| \leq Ne^{\gamma t}, \quad t \geq 0. \tag{2.40}$$

Therefore the linearization (2.38)-(2.39) is exponentially stabilizable by the linear feedback $x \longrightarrow v_x(0)x$ and the stabilizability exponent of the linearization is not greater than the stabilizability exponent of (2.34).

Assume that a feedback v is of the form $v(x) = Kx$, $x \in K$, and that (2.40) holds for the matrix $\tilde{A} = f_x(0,0) + f_u(0,0)K$. Then the derivative of $\tilde{g}(\,\cdot\,) = f(\,\cdot\,, v(\,\cdot\,))$ at 0 is identical with $\tilde{A}$. By Theorem 2.3 the state 0 has identical exponent stability with respect to the following linear and nonlinear systems:

$$\dot{z} = \tilde{A}z \quad \text{and} \quad \dot{z} = \tilde{g}(z).$$

Consequently the stabilizability exponent for (2.34) is not greater than the stabilizability exponent of its linearization, and a linear feedback which stabilizes the linearization of (2.34) also stabilizes the system (2.34). This way the proof of the theorem is complete. □

It follows from the proof of Theorem 2.9 that

Corollary 2.2. *If the pair $(f_x(0,0), f_u(0,0))$ is stabilizable and for a matrix K the matrix $f_x(0,0) + f_u(0,0)K$ is stable then the linear feedback $v(x) = Kx$, $x \in \mathbf{R}^n$, stabilizes (2.34) exponentially.*

Corollary 2.3. *If the pair $(f_x(0,0)), f_u(0,0))$ is not stabilizable then for arbitrary (differentiable at 0) feedback there exist trajectories of the closed loop system which start arbitrarily close to 0 but do not tend to 0 exponentially fast.*

We will now discuss the related problem of the *robustness* of exponentially stabilizable systems.

Assume that (2.37) holds for the closed–loop system (2.35)-(2.36) and let r be a transformation from $\mathbf{R}^n$ into $\mathbf{R}^n$ of class C^1 and such that $r(0) = 0$. We ask for what "perturbations" r the system

$$\dot{z} = g(z) + r(z), \quad z(0) = x, \tag{2.41}$$

remains exponentially stable. The following lemma holds.

Lemma 2.7. *Assume that*

$$\lim_{x \to 0} \frac{|r(x)|}{|x|} < \frac{|\omega|}{M}. \tag{2.42}$$

Then system (2.41) is exponentially stable.

Proof. Let $\widetilde{A} = g_x(0)$, $C = r_x(0)$. The linearization of (2.41) is of the form

$$\dot{z} = (\widetilde{A} + C)z. \tag{2.43}$$

Note that $|C| \leq \lim_{x \to 0} \frac{|r(x)|}{|x|}$ and therefore

$$|C| < \frac{|\beta|}{N}, \tag{2.44}$$

where $\beta > \omega$ and $N > M$ are numbers sufficiently close to ω and M respectively. By Lemma 2.3

$$|e^{\widetilde{A}t}| \leq Ne^{\beta t} \quad \text{for } t \geq 0. \tag{2.45}$$

It follows from (2.44) and (2.45) that for a solution $z(\cdot)$ of (2.43)

$$z(t) = e^{\widetilde{A}t}z(0) + \int_0^t e^{\widetilde{A}(t-s)}Cz(s)\,ds, \quad t \geq 0,\ x \in \mathbf{R}^n,$$

and

$$|z(t)| \leq Ne^{\beta t}|z(0)| + N|C|\int_0^t e^{\beta(t-s)}|z(s)|\,ds,$$

$$e^{-\beta t}|z(t)| \leq N|z(0)| + N|C|\int_0^t e^{-\beta s}|z(s)|\,ds, \quad t \geq 0.$$

§ 2.7. Exponential stabilizability and the robustness problem

Taking into account Lemma 2.1 we see that
$$e^{-\beta t}|z(t)| \le Ne^{N|C|t}|z(0)|,$$
$$|z(t)| \le Ne^{(\beta+N|C|)t}|z(0)|, \quad t \ge 0.$$
Since $\beta + N|C| < 0$, system (2.43) is exponentiably stable.

The supremum of $|\omega|/M$, where ω and M are arbitrary numbers from the definition (2.37) of exponential stabilizability, will be called the *robustness index* of system (2.34).

The following theorem gives an upper bound on the robustness index. In its formulation,
$$\delta = \sup\{\varrho(x, \operatorname{Im} A); \ x \in \operatorname{Im} B, \ |x| = 1\},$$
where $\operatorname{Im} A$ and $\operatorname{Im} B$ are the images of the linear transformations $A = f_x(0,0)$ and $B = f_u(0,0)$.

Theorem 2.10. *If, for a feedback v, relation (2.37) holds, then*
$$\frac{|\omega|}{M} \le \frac{|A|}{\delta}.$$

Proof. Taking into account Lemma 2.3, we can assume that system (2.34) and the feedback v are linear, $v(x) = Kx, \ x \in \mathbf{R}^n$. The theorem is trivially true if $\delta = 0$. Assume that $\delta \ne 0$ and let $\bar{x}$ be a vector such that $A\bar{x} + Bu \ne 0$ for all $u \in \mathbf{R}^m$. Define the control $\bar{u}(\cdot)$ by the formula
$$\bar{u}(t) = K(y(t) - \bar{x}), \quad t \ge 0,$$
where
$$\dot{y} = Ay + B\bar{u}, \quad y(0) = \bar{x}.$$
Let $\bar{y}(t) = y(t) - \bar{x}$ and $\widetilde{A} = A + BK$. Then
$$\dot{\bar{y}} = (A + BK)\bar{y} + A\bar{x} \tag{2.46}$$
$$= \widetilde{A}\bar{y} + A\bar{x}, \quad \bar{y}(0) = 0.$$
It follows from (2.37) and (2.46) that
$$|y(t) - \bar{x}| = |\bar{y}(t)| = \left|\int_0^t e^{\widetilde{A}(t-s)}|A\bar{x}|\,ds\right|$$
$$\le \frac{M}{|\omega|}|A\bar{x}|, \quad t \ge 0. \tag{2.47}$$
On the other hand, repeating the arguments from the first part of the proof of Theorem 1.7, we obtain that for an arbitrary control $u(\cdot)$ and the corresponding output $y^u(\cdot)$ satisfying
$$\dot{y} = Ay + Bu, \quad y(0) = \bar{x},$$
one has
$$\sup_{t \ge 0}|y^u(t) - \bar{x}| \ge \left(|a| - \frac{\gamma}{|a|}\right)|A|^{-1}, \tag{2.48}$$

where a is the vector of the form $A\bar{x} + Bu$ with the minimal norm and $\gamma \in (0, |a|^2)$. Comparing formulae (2.47) and (2.48) we have

$$\left(|a| - \frac{\gamma}{|a|}\right) \le \frac{M}{|\omega|}|A\bar{x}||A|.$$

Taking into account constraints on γ and $\bar{x}$ and the definition of δ we easily obtain the required estimate. □

It follows from the theorem that the robustness index is in general finite. Hence, if all solutions of the closed–loop system tend to zero fast then some of the solutions will deviate far from the equilibrium on initial time intervals.

Corollary 2.4. *If* $\operatorname{Im} A$ *is not contained in* $\operatorname{Im} B$ *then*

$$\frac{|\omega|}{M} \le \frac{|A|}{\delta} < +\infty.$$

Corollary 2.5. *The robustness index of* (2.34) *is bounded from above by* $|A|/\delta$.

§2.8. Necessary conditions for stabilizability

We will now deduce necessary conditions for stabilizability. The following theorem is a direct consequence of the result from §2.6 and is concerned with systems (2.34), (2.35) with f and v satisfying local Lipschitz conditions.

Theorem 2.11. *If system* (2.34) *is asymptotically stabilizable then the mapping* $f(\cdot, \cdot)$ *transforms arbitrary neighbourhoods of* $(0,0) \in \mathbf{R}^n \times \mathbf{R}^m$ *onto neighbourhoods* $0 \in \mathbf{R}^n$.

Proof. If v is a feedback stabilizing (2.34) asymptotically then the closed-loop system (2.35)-(2.36) satisfies all the assumptions of Theorem 2.8. Hence the mapping $x \longrightarrow f(x, v(x))$ transforms an arbitrary neighbourhood of $0 \in \mathbf{R}^n$ onto a neighbourhood of $0 \in \mathbf{R}^n$. □

Taking into account Theorem 2.11 one can show that for nonlinear systems local controllability does not imply, in general, asymptotic stabilizability. Such implication is of course true for linear systems, see Theorem I.2.9.

Example 2.1. Consider again the system

$$\dot{y}_1 = u,$$
$$\dot{y}_2 = v,$$
$$\dot{y}_3 = y_1 v - y_2 u, \quad \begin{bmatrix} u \\ v \end{bmatrix} \in U = \mathbf{R}^2.$$

We know, see § 1.4, Example 1.3, that the system is locally controllable at an arbitrary point, in particular, at $0 \in \mathbf{R}^3$. Let us remark, however, that the image of $f(\cdot, \cdot)$ does not contain vectors of the form $\begin{bmatrix} 0 \\ 0 \\ \gamma \end{bmatrix}$, $\gamma \neq 0$, so the necessary condition for stabilizability is not satisfied. Hence the system is not asymptotically stabilizable. This is a suprising result as it concerns a system which differs only slightly from a linear one.

The following theorem is also an immediate consequence of Theorem 2.8.

Theorem 2.12. *Assume that there exists a feedback which either asymptotically stabilizes (2.34) or for which $0 \in \mathbf{R}^n$ is an attracting point for the closed-loop system (2.35)-(2.36). Then, for sufficiently small $r > 0$, transformations $F_r \colon S \times U \longrightarrow S$ given by*

$$F_r(x, u) = \frac{f(rx, u)}{|f(rx, u)|}, \quad x \in S, \ u \in U, \tag{2.49}$$

are onto S.

Example 2.2. It is not difficult show that the following system on $\mathbf{R}^2$

$$\dot{y}_1 = (4 - y_2^2)u, \quad \dot{y}_2 = y_2 - e^{-y_1}v, \tag{2.50}$$

with $U = \left\{ \begin{bmatrix} u \\ v \end{bmatrix} \in \mathbf{R}^2;\ u \geq 0,\ -1 \leq v \leq 1 \right\}$ is controllable to $0 \in \mathbf{R}^2$ from an arbitrary state in $\mathbf{R}^2$. Let us also remark that the first coordinate of the right side of (2.50) is nonnegative for state vectors close to $0 \in \mathbf{R}^2$. Therefore, for sufficiently small $r > 0$, transformation (2.49) cannot be onto S. Thus, although system (2.50) is exactly controllable to $0 \in \mathbf{R}^2$, one cannot find a feedback v such that the state 0 is attracting for the closed-loop system determined by v.

§ 2.9. Stabilization of the Euler equations

We will now apply the obtained results to an analysis of an important control system described by the Euler equations from Example 0.2.

Let us recall that the stated equation was of the form

$$J\dot{\omega} = S(\omega)J\omega + Bu, \quad \omega(0) \in \mathbf{R}^3, \tag{2.51}$$

where

$$J = \begin{bmatrix} I_1 & 0 & 0 \\ 0 & I_2 & 0 \\ 0 & 0 & I_3 \end{bmatrix}, \quad S(\omega) = \begin{bmatrix} 0 & \omega_3 & -\omega_2 \\ -\omega_3 & 0 & \omega_1 \\ \omega_2 & -\omega_1 & 0 \end{bmatrix}, \quad \omega \in \mathbf{R}^3,$$

$$B = [b_1, \quad b_2, \quad b_3], \quad b_i = \begin{bmatrix} b_{1i} \\ b_{2i} \\ b_{3i} \end{bmatrix} \in \mathbf{R}^3, \quad i = 1, 2, 3.$$

The control set is $U = \mathbf{R}^3$. We assume that $I_i \neq 0$, $I_i \neq I_j$, $i, j = 1, 2, 3$, $i \neq j$. Vectors b_i, $i = 1, 2, 3$ will be called the *control axes*.

We will look for conditions under which system (2.51) is stabilizable and for formulae defining stabilizing feedbacks $v \colon \mathbf{R}^3 \to \mathbf{R}^3$, $v(0) = 0$.

The closed–loop system corresponding to v is given by the equation

$$\dot{\omega} = F(\omega), \tag{2.52}$$

where

$$F(\omega) = J^{-1} S(\omega) J \omega + J^{-1} B v(\omega).$$

The question of exponential stabilizability has, in the present case, a simple solution. The linearization of (2.51) is given by

$$\dot{\omega} = J^{-1} B u. \tag{2.53}$$

Since the pair $(0, J^{-1}B)$ is stabilizable if and only if the matrix $J^{-1}B$ is invertible therefore, by Theorem 2.9, system (2.51) is exponentially stabilizable if and only if $\det B \neq 0$.

Let

$$W(\omega) = \frac{1}{2} \omega^* J \omega = \frac{1}{2} \left(I_1 \omega_1^2 + I_2 \omega_2^2 + I_3 \omega_3^2 \right), \quad \omega \in \mathbf{R}^3.$$

Then W is the kinetic energy of the system and the Liapunov derivative W_F of W, see (2.24), is given by

$$\begin{aligned} W_F(\omega) &= \omega^* J J^{-1} S(\omega) J \omega + \omega^* J J^{-1} B v(\omega) \\ &= \det[J\omega, \omega, \omega] + \omega^* B v(\omega) \\ &= \omega^* B v(\omega), \quad \omega \in \mathbf{R}^3. \end{aligned}$$

If $v \equiv 0$ then $W_F \leq 0$ and W is a Liapunov function for (2.52). By Theorem 2.4, the constant feedback $v \equiv 0$ stabilizes system (2.51) in the sense of Liapunov.

Let us remark that if a feedback v is given by

$$v(\omega) = -B^* \omega, \quad \omega \in \mathbf{R}^3. \tag{2.54}$$

then

$$W_F(\omega) = -\omega^* B B^* \omega = -|B^* \omega|^2 \leq 0, \quad \omega \in \mathbf{R}^3 \tag{2.55}$$

and W is a Liapunov function for (2.52) also in this case.

§ 2.9. Stabilization of the Euler equations

The following theorem shows that for a large class of systems (2.51) feedback (2.54) is asymptotically stabilizing.

Theorem 2.13 *Feedback $v(\omega) = -B^*\omega$, $\omega \in \mathbf{R}^3$, stabilizes system (2.51) asymptotically if and only if every row of the matrix B has a non-zero element.*

Proof. By (2.55) the closed-loop system

$$\dot\omega = J^{-1}S(\omega)J\omega - J^{-1}BB^*\omega \qquad (2.56)$$

is stable in the sense of Liapunov. It follows from the La Salle theorem that an arbitrary trajectory (2.56) converges, as $t \uparrow +\infty$, to the maximal invariant set K contained in

$$L = \{\omega \in \mathbf{R}^3;\ W_F(\omega) = 0\} = \ker B^*.$$

In particular the set M of all stationary points for (2.56) is in L:

$$M = \{\omega \in \mathbf{R}^3;\ F(\omega) = 0\} = \{\omega \in L;\ S(\omega)J\omega = 0\} = N \cap L,$$

where $N = \{\omega \in \mathbf{R}^3;\ S(\omega)J\omega = 0\}$.

Let us denote by q_i the i-th coordinate axis and by P_j the hyperplane orthogonal to the j-th control axis b_j, $i,j = 1,2,3$. Then $L = \ker B^* = \bigcap_{j=1}^{3} P_j$ and

$$M = \bigcap_{j=1}^{3} P_j \cap \left(\bigcup_{i=1}^{3} q_i\right) = \bigcap_{j=1}^{3} \bigcup_{i=1}^{3} P_j \cap q_i.$$

Moreover

$$P_j \cap q_i = \begin{cases} \{0\}, & \text{if } b_{i,j} \neq 0, \\ q_i, & \text{if } b_{i,j} = 0, \end{cases}$$

Therefore, if $b_{i,j} = 0$ for $j = 1,2,3$, then $M = q_i$. Consequently $M = \{0\}$ if and only if for arbitrary $i = 1,2,3$ there exists $j = 1,2,3$ such that $b_{i,j} \neq 0$.

If system (2.56) is asymptotically stable then $M = \{0\}$ and therefore every row of B has a non-zero element. This proves the theorem in one direction.

To prove the converse implication assume that $M = \{0\}$. We will consider three cases.

(i) rank $B = 3$. Then $L = \{0\}$ and since $K \subset L$ we have $K = \{0\}$.

(ii) rank $B = 2$. In this case L is a straight line. If $\omega(\cdot)$ is an arbitrary trajectory of (2.56), completely contained in K, then, taking into account that $K \subset L$,

$$\frac{d}{dt}W(\omega(t)) = -|B^*\omega(t)|^2 = 0, \quad t \geq 0,$$

and, consequently, for a constant c

$$W(\omega(t)) = c = W(\omega(0)), \quad t \geq 0.$$

Since the trajectory $\omega(\,\cdot\,)$ is contained in the line L and in the ellipsoid $\{\omega \in \mathbf{R}^3;\ W(\omega) = c\}$, it has to be a stationary solution: $\omega(t) = \omega(0)$, $t \geq 0$. Finally we get that $c = 0$ and

$$M \subset K \subset L \cap \{0\} = \{0\}.$$

We see that $K = \{0\}$.

(iii) rank $B = 1$. Without any loss of generality we can assume that $b_1 \neq 0$, $b_2 = b_3 = 0$. Note that

$$\begin{aligned}
K \subset P &= \{\omega \in L;\ F(\omega) \in L\} \\
&= \{\omega \in L;\ b_1^*(J^{-1}S(\omega)J\omega - J^{-1}BB^*\omega) = 0\} \\
&= \{\omega \in \mathbf{R}^3;\ b_1^*J^{-1}S(\omega)J\omega = 0 \text{ and } b_1^*\omega = 0\}.
\end{aligned}$$

Therefore the set P is an intersection of a cone, defined by the equation

$$b_{11}I_{23}\omega_2\omega_3 + b_{21}I_{31}\omega_3\omega_1 + b_{31}I_{12}\omega_1\omega_2 = 0,$$

where $I_{23} \leq (I_2 - I_3)/I_1$, $I_{31} = (I_3 - I_1)/I_2$, $I_{12} = (I_1 - I_2)/I_3$ and of the plane

$$b_{11}\omega_1 + b_{21}\omega_2 + b_{31}\omega_3 = 0, \quad \omega \in \mathbf{R}^3.$$

Consequently P can be either a point or a sum of two, not necessarily different, straight lines. As in case (ii), we show that

$$K \subset P \cap \{\omega \in \mathbf{R}^3;\ W(\omega) = c\} \quad \text{for } c = 0,$$

and finally $K = \{0\}$. □

Bibliographical notes

The main stability test is taken from E. Coddington and N. Levinson [15]. The content of Theorem 2.4 is sometimes called the second Liapunov method of studying stability. The first one is the linearization. Theorem 2.7 is due to M. Krasnoselski and P. Zabreiko [35]; see also [70]. A similar result was obtained independently by R. Brockett, and Theorem 2.11 is due to him. He applied it to show that, in general, for nonlinear systems, local controllability does not imply asymptotic stabilizability. Theorem 2.9 is taken from [70], as is Theorem 2.10. Theorem 2.13 is due to M. Szafrański [55] and Exercise 2.9 to H. Sussman.

Chapter 3
Realization theory

It is shown in this chapter how, by knowing the input-output map of a control system, one can determine the impulse-response function of its linearization. A control system with a given input-output map is also constructed.

§ 3.1. Input-output maps

Let us consider a control system

$$\dot{y} = f(y, u), \quad y(0) = x \in \mathbf{R}^n, \tag{3.1}$$
$$w = h(y), \tag{3.2}$$

with an open set of control parameters $U \subset \mathbf{R}^m$. Assume that the function f satisfies the conditions of Theorem 1.3 and that h is a continuous function with values in $\mathbf{R}^k$. Let us fix T and denote by $B(0, T; U)$ and $C(0, T; \mathbf{R}^k)$ the spaces of bounded, Borel and continuous functions respectively defined on $[0, T]$ and with values in U and in $\mathbf{R}^k$. For an arbitrary function $u(\,\cdot\,) \in B(0, T; U)$ there exists exactly one solution $y^u(t, x)$, $t \in [0, T]$, of the equation (3.1), and thus for arbitrary $\bar{x} \in \mathbf{R}^n$ the formula

$$\mathcal{R}u(t) = w(t) = h(y^u(t, \bar{x})), \quad t \in [0, T], \tag{3.3}$$

defines an operator from $B(0, T; U)$ into $C(0, T; \mathbf{R}^k)$. As in the linear case (compare § I.3.1, formula (3.2)), the transformation $\mathcal{R}$ will be called the *input-output map of the system* (3.1)-(3.2). Realization theory is concerned with the construction of a system (3.1)-(3.2) assuming the knowledge of its input-output map. The linear theory was discussed in Chapter I.3. The nonlinear case is much more complicated, and we limit our presentation to two typical results.

The same input-output map can be defined by different linear systems (3.1)-(3.2). On the other hand, input-output maps determine uniquely several important characteristics of control systems. One of them is the impulse-response function of the linearization we now show.

Assume that $U = \mathbf{R}^m$ and that a state $\bar{x}$ and a parameter $\bar{u} \in \mathbf{R}^m$ form a stationary pair, $f(\bar{x}, \bar{u}) = 0$. Let us recall (see (1.19) and Theorem 1.11) that if the mappings f and h are differentiable at $(\bar{x}, \bar{u})$ and $\bar{x}$ respectively, then the linearization of (3.1)-(3.2) at $(\bar{x}, \bar{u})$ is of the form

$$\dot{y} = Ay + Bu, \quad y(0) = x, \qquad (3.4)$$
$$w = Cy, \qquad (3.5)$$

where $A = f_x(\bar{x}, \bar{u})$, $B = f_u(\bar{x}, \bar{u})$, $C = h_x(\bar{x})$.

The function

$$\Psi(t) = Ce^{tA}B = h_x(\bar{x})e^{tf_x(\bar{x},\bar{u})}f_u(\bar{x}, \bar{u}), \quad t \geq 0, \qquad (3.6)$$

is the impulse-response function of the linearization (3.4)-(3.5).

Theorem 3.1. *If mappings $f: \mathbf{R}^n \times \mathbf{R}^m \longrightarrow \mathbf{R}^n$, $h: \mathbf{R}^n \longrightarrow \mathbf{R}^k$ are of class C^1 and f satisfies the linear growth condition (1.6), then the input-output map $\mathcal{R}$ of (3.1)-(3.2) is well defined and uniquely determines the impulse-response function (3.6) of the linearization (3.4)-(3.5).*

Proof. Let $u: [0, T] \longrightarrow \mathbf{R}^m$ be a continuous function such that $u(0) = 0$. For an arbitrary number γ and $t \in [0, T]$

$$\mathcal{R}(\bar{u} + \gamma u)(t) = h(y^{\gamma u}(t, \bar{x})) = h(z(t, \gamma)),$$

where $z(t) = z(t, \gamma)$, $t \in [0, T]$, is a solution to the equation

$$\dot{z}(t) = f(z(t), \bar{u} + \gamma u(t)), \quad t \in [0, T], \quad z(0) = \bar{x}.$$

By Theorem 1.5, solution $z(\cdot, \cdot)$ is differentiable with respect to γ. Moreover derivatives $z_\gamma(t) = z_\gamma(t, 0)$, $t \in [0, T]$, satisfy the linear equation

$$\frac{d}{dt}z_\gamma(t) = f_x(\bar{x}, \bar{u})z_\gamma(t) + f_u(\bar{x}, \bar{u})u(t), \quad z_\gamma(0) = 0.$$

Consequently

$$\lim_{\gamma \to 0} \frac{1}{\gamma}(\mathcal{R}(\bar{u} + \gamma u)(t) - \mathcal{R}(\bar{u})(t)) = g_x(\bar{x})z_\gamma(t)$$
$$= \int_0^t h_x(\bar{x})e^{(t-s)f_x(\bar{x},\bar{u})}f_u(\bar{x}, \bar{u})\,u(s)\,ds, \quad t \in [0, T],$$

and we see that $\mathcal{R}$ determines $\Psi(\cdot)$ uniquely. $\square$

§3.2. Partial realizations

Given an input-output map $\mathcal{R}$, we construct here a system of the type (3.1)-(3.2), which generates, at least on the same time intervals, the mapping $\mathcal{R}$. To simplify the exposition, we set $k = 1$.

§ 3.2. Partial realizations

Let
$$u(t) = u(t; v_1, \ldots, v_l; u_1, \ldots, u_l), \quad (3.7)$$
$$u'(t) = u(t; v'_1, \ldots, v'_l; u'_1, \ldots, u'_l), \quad t \geq 0, \quad (3.8)$$

be inputs defined by (1.37); see §1.4. We define a new control function $u' \circ u$ setting

$$u' \circ u(t) = \begin{cases} u(t) & \text{for } t \in [0, v_1 + \ldots + v_l), \\ u'(t - (v_1 + \ldots + v_l)) & \text{for } t \geq v_1 + \ldots + v_l. \end{cases}$$

Hence
$$u' \circ u(t) = u(t; v_1, \ldots, v_l, v'_1, \ldots, v'_l; u_1, \ldots, u_l, u'_1, \ldots, u'_l), \quad t \geq 0.$$

Let us fix parameters $(u_1, \ldots, u_l)$ and inputs

$$u^j(\cdot) = u(\cdot; v^j_1, \ldots, v^j_{l_j}; u^j_1, \ldots, u^j_{l_j}), \quad j = 1, \ldots, r. \quad (3.9)$$

Define, on the set $\Delta = \{(v_1, \ldots, v_l)^*; v_1 + \ldots + v_l < T, v_1 > 0, \ldots, v_l > 0\}$,
a mapping $G = \begin{bmatrix} g_1 \\ \vdots \\ g_r \end{bmatrix}$ with values in $\mathbf{R}^r$ in the following way:

$$g_j(v_1, \ldots, v_l) = \mathcal{R}(u^j \circ u)(v_1 + \ldots + v_l + v^j_1 + \ldots + v^j_{l_j})$$
$$= h(y^{u^j \circ u}(v_1 + \ldots + v_l + v^j_1 + \ldots + v^j_{l_j}, \bar{x})),$$

$$v = \begin{bmatrix} v_1 \\ \vdots \\ v_l \end{bmatrix} \in \Delta, \quad j = 1, \ldots, r.$$

The mapping G is a composition $G = G^2 G^1$ of transformations $G^1: \Delta \longrightarrow \mathbf{R}^n$ and $G^2: \mathbf{R}^n \longrightarrow \mathbf{R}^r$, given by

$$G^1(v) = y^{u(\cdot, v_1, \ldots, v_l; u_1, \ldots, u_l)}(v_1 + \ldots + v_l, \bar{x}), \quad v = \begin{bmatrix} v_1 \\ \vdots \\ v_l \end{bmatrix} \in \Delta, \quad (3.10)$$

$$G^2(x) = \left(h(y^{u^j}(v^j_1 + \ldots + v^j_{l_j}, x)) \right)_{j=1,2,\ldots,r}, \quad x \in \mathbf{R}^n. \quad (3.11)$$

The following lemma holds.

Lemma 3.1. *If h and f are of class C^1, then*

$$\operatorname{rank} G_v(v) \leq n, \quad v = \begin{bmatrix} v_1 \\ \vdots \\ v_l \end{bmatrix} \in \Delta.$$

Proof. The derivative $G_v(v)$ of G at $v \in \Delta$ is the composition of the derivatives $G^1_v(v)$ and $G^2_x(G^1(v))$ which have ranks not greater than the dimension of $\mathbf{R}^n$. □

An input-output map $\mathcal{R}$ is said to be *regular* if there exist parameters $u_1, \ldots, u_n \in U$, controls $u^1, \ldots, u^n$ of the form (3.9) and $\tilde{v} \in \mathbf{R}^n$ such that

$$\operatorname{rank} G_v(\tilde{v}) = n.$$

In addition, coordinates $\tilde{v}_1, \ldots, \tilde{v}_n$ of $\tilde{v}$ should satisfy the constraints

$$\tilde{v}_1 + \ldots + \tilde{v}_n + v_1^j + \ldots + v_{l_j}^j < T, \quad j = 1, \ldots, n.$$

Controls $\tilde{u}(\,\cdot\,) = u(\,\cdot\,; \tilde{v}_1, \ldots, \tilde{v}_n; u_1, \ldots, u_n)$, $u^1(\,\cdot\,), \ldots, u^n(\,\cdot\,)$ and the sequence $(\tilde{v}_1, \ldots, \tilde{v}_n)$ are called *reference inputs* and a *reference sequence*.

Let D be a bounded open subset of $\mathbf{R}^n$, $\tilde{v} \in \mathbf{R}^n$, $\tilde{f}$ a continuous mapping from $D \times U$ in $\mathbf{R}^n$ and $\tilde{h}$ a real function on D. Assume that there exists a constant $c > 0$ such that $|\tilde{f}(v,u)| \le c(|v| + |u| + 1)$, $|\tilde{f}(v,u) - \tilde{f}(w,u)| \le c|v-w|$, $v, w \in D$, $u \in U$. Let $\widetilde{T} > 0$ be a positive number. Then for an arbitrary bounded control $u(t)$, $t \ge 0$, there exists a unique maximal in D solution $\tilde{y}(t) = \tilde{y}^u(t, v)$, $t \in [\widetilde{T}, \tau(u))$, of

$$\begin{aligned}\dot{\tilde{y}} &= \tilde{f}(\tilde{y}(t), u(t)), \quad t \in [\widetilde{T}, \tau(u)), \\ \tilde{y}(\widetilde{T}) &= v;\end{aligned} \qquad (3.12)$$

see Theorem 1.3 and Theorem 1.1.

If there exist a bounded control $\tilde{u}(t)$, $t \ge 0$, a vector $\tilde{v} \in D$ and a number $\widetilde{T} < T$ such that, for all controls $u(\,\cdot\,) \in B(0, T; U)$ which coincide with $\tilde{u}(\,\cdot\,)$ on $[0, \widetilde{T}]$, one has

$$\mathcal{R}(u)(t) = \tilde{h}(\tilde{y}^u(t, \tilde{v})), \quad t \in [\widetilde{T}, T \wedge \tau(u)), \qquad (3.13)$$

then one says that the system

$$\dot{\tilde{y}} = \tilde{f}(\tilde{y}, u)$$

$$\tilde{w}(t) = \tilde{h}(\tilde{y}(t)), \quad t \in [0, \tau(u)), \qquad (3.14)$$

is a *partial realization* of the input-output map $\mathcal{R}$.

If a partial realization of an input-output map $\mathcal{R}$ can be constructed, then one says that the map $\mathcal{R}$ is *partially realizable*.

The following theorem holds.

Theorem 3.2. *Assume that all transformations $f(\,\cdot\,, u)$, $u \in U$, are of class C^2 and that an input-output map $\mathcal{R}$ is regular. Then the map $\mathcal{R}$ is partially realizable.*

§ 3.2. Partial realizations

Proof. Assume that $\mathcal{R}$ is regular. Let $\tilde{u}(\,\cdot\,)$, $u^1, \ldots, u^n$ and $(\tilde{v}_1, \ldots, \tilde{v}_n)$ be the corresponding reference inputs and the reference sequence. Define

$$\tilde{v} = \begin{bmatrix} \tilde{v}_1 \\ \vdots \\ \tilde{v}_n \end{bmatrix}, \quad \tilde{T} = \tilde{v}_1 + \ldots + \tilde{v}_n$$

and fix a neighbourhood D of $\tilde{v}$ in which the mapping G is a homeomorphism. For $x = G^1(v)$, $v \in D$, $u \in U$ and sufficiently small $\eta > 0$ the equation

$$y^u(t,x) = G^1(\tilde{y}^u(t,v)), \quad t \in [0, \eta),$$

determines a function $\tilde{y}^u(\cdot, v)$ which satisfies the differential equation

$$\frac{d}{dt}\tilde{y}^u(t,v) = [G_v^1(\tilde{y}^u(t,v))]^{-1} f(\tilde{y}^u(t,v), u), \quad t \in [0, \eta). \qquad (3.15)$$

We claim that the formulae

$$\tilde{f}(v,u) = [G_v^1(v)]^{-1} f(G^1(v), u), \qquad (3.16)$$

$$\tilde{h}(v) = h(G^1(v)), \quad v \in D, \ u \in U, \qquad (3.17)$$

define a partial realization of $\mathcal{R}$.

To show this note first that

$$h(G^1(v)) = h(y^{u(\cdot;v_1,\ldots,v_n;u_1,\ldots,u_n)}(v_1 \ldots + v_n, \bar{x}))$$

$$= \mathcal{R}(u(\cdot; v_1, \ldots, v_n; u_1, \ldots, u_n))(v_1 + \ldots + v_n), \quad v = \begin{bmatrix} v_1 \\ \vdots \\ v_n \end{bmatrix}.$$

Hence the function $\tilde{h}$ can be defined in terms of the mapping $\mathcal{R}$ only.

Let $u^j(v; s; u)(\,\cdot\,)$ be a control identical to $u(\cdot; u_1, \ldots, u_n; v_1, \ldots, v_n)$ on the interval $[0, v_1 + \ldots + v_n)$, and to $u, u_1^j, \ldots, u_{l_j}^j$, $j = 1, 2, \ldots, n$, on consecutive intervals of lengths $s, v_1^j, \ldots, v_{l_j}^j$ respectively. Let us remark that

$$\mathcal{R}(u^j(v; s; u))(v_1 + \ldots + v_n + s + v_1^j + \ldots + v_{l_j}^j)$$

$$= h(y^{u^j}(v_1^j + \ldots + v_{l_j}^j, y^u(s, y^{u(\cdot; v_1, \ldots, v_n; u_1, \ldots, u_n)}(v_1 + \ldots + v_n, \bar{x})))).$$

Therefore, for sufficiently small $s > 0$ and all $u \in U$, the transformations $G^2(y^u(s, G^1(v)))$, $v \in D$, are also given in terms of $\mathcal{R}$. Moreover, denoting by G^{-1}, $(G^1)^{-1}$ the inverses of G and G^1 respectively, we obtain that

$$G^{-1}(G^2(y^u(s, G^1(v)))) = (G^1)^{-1}(y^u(s, G^1(v))). \qquad (3.18)$$

The derivative at $s = 0$ of the function defined by (3.18) is identical with $\tilde{f}$. Hence the function $\tilde{f}$ is expressible in terms of the transformation $\mathcal{R}$. It follows from (3.15) and (3.17) that the functions $\tilde{f}$ and $\tilde{h}$ define a partial realization of $\mathcal{R}$. □

We finally show that the family of all regular input-output maps is rather large.

Proposition 3.1. *Assume that $h(\cdot)$ and $f(\cdot, u)$, $u \in U$, are of class C^1 and C^k, $k \geq 2$, respectively, and that for some $j \leq k$, $\dim \mathcal{L}_j(\bar{x}) = n$. If $f(\bar{x}, \bar{u}) = 0$ for some $\bar{u} \in U$ and the pair $(f_x(\bar{x}, \bar{u}), h_x(\bar{x}))$ is observable, then the input-output map $\mathcal{R}$ corresponding to the system (3.1)-(3.2), with $x = \bar{x}$, is regular.*

Proof. We see, using similar arguments to those in the proof of Theorem 1.9, that for arbitrary neighbourhood V of the point $\bar{x}$, there exist parameters $u_1, \ldots, u_n \in U$ and a point $v = \begin{bmatrix} v_1 \\ \vdots \\ v_n \end{bmatrix}$, such that the derivative $G_v^1(v)$ of the transformation G^1 defined by (3.10) is nonsingular and $G^1(v) \in V$. Define $u^j(t) = u(t; \bar{u}; v^j)$, $t \geq 0$ and $j = 1, \ldots, n$. It is enough to show that the derivative of $G^2(\cdot)$ at $\bar{x}$ is nonsingular as well. However this derivative is of the form

$$G_x^2(\bar{x}) = \begin{bmatrix} h_x(\bar{x}) e^{v_1 f_x(\bar{x}, \bar{u})} \\ \vdots \\ h_x(\bar{x}) e^{v_n f_x(\bar{x}, \bar{u})} \end{bmatrix}.$$

Consequently, to complete the proof of the theorem, it remains to establish the following lemma.

Lemma 3.2. *If a pair (A, C), $A \in \mathbf{M}(n, n)$, $C \in \mathbf{M}(n, k)$, is observable, then for arbitrary $\delta > 0$ there exist numbers v^j, $0 < v^j < \delta$, $j = 1, \ldots, n$ such that*

$$\operatorname{rank} \begin{bmatrix} C e^{v^1 A} \\ \vdots \\ C e^{v^n A} \end{bmatrix} = n.$$

Proof. Taking into account that the pair (A^*, C^*) is controllable, it is enough to show that for an arbitrary controllable pair (A, B), $A \in \mathbf{M}(n, n)$, $B \in \mathbf{M}(n, m)$, there exist numbers v^j, $0 < v^j < \delta$, $j = 1, \ldots, n$ such that $\operatorname{rank} \begin{bmatrix} e^{v^1 A} B, \ldots, e^{v^n A} B \end{bmatrix} = n$. Denote by $b_1, \ldots, b_m$ the columns of the matrix B. It easily follows from the controllability of (A, B) that the smallest linear subspace containing $\{e^{sA} b_r; s \in (0, \delta), r = 1, \ldots, m\}$ is identical with $\mathbf{R}^n$. Hence this set contains n linearly independent vectors. □

Bibliographical notes

Theorem 3.2 is a special case of a theorem due to B. Jakubczyka [32]. The construction introduced in its proof applies also to nonregular systems, however, that proof is much more complicated.

PART III

OPTIMAL CONTROL

Chapter 1
Dynamic programming

This chapter starts from a derivation of the dynamic programming equations called Bellman's equations. They are used to solve the linear regulator problem on a finite time interval. A fundamental role is played here by the Riccati-type matrix differential equations. The stabilization problem is reduced to an analysis of an algebraic Riccati equation.

§ 1.1. Introductory comments

Part III is concerned with controls which are optimal, according to a given criterion. Our considerations will be devoted mainly to control systems

$$\dot{y} = f(y, u), \quad y(0) = x, \tag{1.1}$$

and to *criteria*, called also *cost functionals*,

$$J_T(x, u(\,\cdot\,)) = \int_0^T g(y(t), u(t))\, dt + G(y(T)), \tag{1.2}$$

when $T < +\infty$. If the control interval is $[0, +\infty]$, then the *cost functional*

$$J(x, u(\,\cdot\,)) = \int_0^{+\infty} g(y(t), u(t))\, dt. \tag{1.3}$$

Our aim will be to find a control $\hat{u}(\,\cdot\,)$ such that for all admissible controls $u(\,\cdot\,)$

$$J_T(x, \hat{u}(\,\cdot\,)) \leq J_T(x, u(\,\cdot\,)) \tag{1.4}$$

or

$$J(x, \hat{u}(\,\cdot\,)) \leq J(x, u(\,\cdot\,)). \tag{1.5}$$

In Chapter 2 and § 3.3 we will also consider the so-called impulse control problems.

There are basically two methods for finding controls minimizing cost functionals (1.2) or (1.3). One of them *embeds* a given minimization problem into a parametrized family of similar problems. The embedding should be such that the minimal value, as a function of the parameter, satisfies an analytic relation. If the selected parameter is the initial state and the length of the control interval, then the minimal value of the cost functional is called the value function and the analytical relation, Bellman's equation. Knowing the solutions to the Bellman equation one can find the optimal strategy in the form of a closed loop control.

The other method leads to necessary conditions on the optimal, open-loop, strategy formulated in the form of the so-called maximum principle discovered by L. Pontriagin and his collaborators. They can be obtained (in the simplest case) by considering a parametrized family of controls and the corresponding values of the cost functional (1.2) and by an application of classical calculus.

§1.2. Bellman's equation and the value function

Assume that the state space E of a control system is an open subset of $\mathbf{R}^n$ and let the set U of control parameters be included in $\mathbf{R}^m$. We assume that the functions f, g and G are continuous on $E \times U$ and E respectively and that g is nonnegative.

Theorem 1.1. *Assume that a real function $W(\cdot,\cdot)$, defined and continuous on $[0,T] \times E$, is of class C^1 on $(0,T) \times E$ and satisfies the equation*

$$\frac{\partial W}{\partial t}(t,x) = \inf_{u \in U}(g(x,u) + \langle W_x(t,x), f(x,u)\rangle), \quad (t,x) \in (0,T) \times E, \quad (1.6)$$

with the boundary condition

$$W(0,x) = G(x), \quad x \in E. \tag{1.7}$$

(i) *If $u(\cdot)$ is a control and $y(\cdot)$ the corresponding absolutely continuous, E-valued, solution of (1.1) then*

$$J_T(x, u(\cdot)) \geq W(T,x). \tag{1.8}$$

(ii) *Assume that for a certain function $\hat{v}: [0,T] \times E \longrightarrow U$:*

$$g(x, \hat{v}(t,x)) + \langle W_x(t,x), f(x, \hat{v}(t,x))\rangle \tag{1.9}$$
$$\leq g(x,u) + \langle W_x(t,x), f(x,u)\rangle, \quad t \in (0,T), \; x \in E, \; u \in U,$$

§1.2. Bellman's equation and the value function

and that $\hat{y}$ is an absolutely continuous, E-valued solution of the equation

$$\frac{d}{dt}\hat{y}(t) = f(\hat{y}(t), \hat{v}(T-t, \hat{y}(t))), \quad t \in [0,T], \qquad (1.10)$$
$$\hat{y}(0) = x.$$

Then, for the control $\hat{u}(t) = \hat{v}(T-t, \hat{y}(t))$, $t \in [0,T]$,

$$J_T(x, \hat{u}(\,\cdot\,)) = W(x, T).$$

Proof. (i) Let $w(t) = W(T-t, y(t))$, $t \in [0,T]$. Then $w(\,\cdot\,)$ is an absolutely continuous function on an arbitrary interval $[\alpha, \beta] \subset (0, T)$ and

$$\begin{aligned}\frac{dw}{dt}(t) &= -\frac{\partial W}{\partial t}(T-t, y(t)) + \langle W_x(T-t, y(t)), \frac{dy}{dt}(t)\rangle \qquad (1.11)\\ &= -\frac{\partial W}{\partial t}(T-t, y(t)) + \langle W_x(T-t, y(t)), f(y(t), u(t))\rangle\end{aligned}$$

for almost all $t \in [0,T]$. Hence, from (1.6) and (1.7)

$$\begin{aligned}W(T-\beta, y(\beta)) - W(T-\alpha, y(\alpha)) &= w(\beta) - w(\alpha) = \int_\alpha^\beta \frac{dw}{dt}(t)\,dt \\ &= \int_\alpha^\beta \left[-\frac{\partial W}{\partial t}(T-t, y(t)) + \langle W_x(T-t, y(t)), f(y(t), u(t))\rangle\right] dt \\ &\geq -\int_\alpha^\beta g(y(t), u(t))\,dt.\end{aligned}$$

Letting α and β tend to 0 and T respectively we obtain

$$G(y(T)) - W(T, x) \geq -\int_0^T g(y(t), u(t))\,dt.$$

This proves (i).

(ii) In a similar way, taking into account (1.9), for the control $\hat{u}$ and the output $\hat{y}$,

$$\begin{aligned}G(\hat{y}(T)) - W(T, x) &= \int_0^T \left[-\frac{\partial W}{\partial t}(T-t, \hat{y}(t)) + \langle W_x(T-t, \hat{y}(t))\rangle\right] dt \\ &= \int_0^T g(\hat{y}(t), \hat{u}(t))\,dt.\end{aligned}$$

Therefore

$$G(\hat{y}(T)) + \int_0^T g(\hat{y}(s), \hat{u}(s))\,ds = W(T, x),$$

the required identity. $\square$

Remark. Equation (1.6) is called *Bellman's equation*. It follows from Theorem 1.1 that, under appropriate conditions, $W(T, x)$ is the minimal value of the functional $J_T(x, \cdot)$. Hence W is the *value function* for the problem of minimizing (1.2).

Let $U(t, x)$ be the set of all control parameters $u \in U$ for which the infimum on the right hand side of (1.6) is attained. The function $\hat{v}(\cdot, \cdot)$ from part (ii) of the theorem is a *selector* of the multivalued function $U(\cdot, \cdot)$ in the sense that

$$\hat{v}(t, x) \in U(t, x), \quad (t, x) \in [0, T] \times E.$$

Therefore, for the conditions of the theorem to be fulfilled, such a selector not only should exist, but the closed loop equation (1.10) should have a well defined, absolutely continuous, solution.

Remark. A similar result holds for a more general cost functional

$$J_T(x, u(\cdot)) = \int_0^T e^{-\alpha t} g(y(t), u(t))\, dt + \bar{e}^{\alpha T} G(y(T)). \tag{1.12}$$

In this direction we propose to solve the following exercise.

Exercise 1.1. Taking into account a solution $W(\cdot, \cdot)$ of the equation

$$\frac{\partial W}{\partial t}(t, x) = \inf_{u \in U}(g(x, u) - \alpha W(t, x) + \langle W_x(t, x), f(x, u) \rangle),$$
$$W(0, x) = G(x), \quad x \in E,\ t \in (0, T),$$

and a selector $\hat{v}$ of the multivalued function

$$U(t, x) = \left\{ u \in U;\, g(x, u) + \langle W_x(t, x), f(x, u) \rangle \right.$$
$$\left. = \inf_{u \in U}(g(x, u) + \langle W_x(t, x), f(x, u) \rangle) \right\},$$

generalize Theorem 1.1 to the functional (1.12).

We will now describe an intuitive derivation of equation (1.6). Similar reasoning often helps to guess the proper form of the Bellman equation in situations different from the one covered by Theorem 1.1.

Let $W(t, x)$ be the minimal value of the functional $J_t(x, \cdot)$. For arbitrary $h > 0$ and arbitrary parameter $v \in U$ denote by $u^v(\cdot)$ a control which is constant and equal v on $[0, h]$ and is identical with the optimal strategy for the minimization problem on $[h, t + h]$. Let $z^{x,v}(t)$, $t \geq 0$, be the solution of the equation $\dot{z} = f(z, v)$, $z(0) = x$. Then

$$J_{t+h}(x, u^v(\cdot)) = \int_0^h g(z^{x,v}(s), v)\, ds + W(t, z^{x,v}(h))$$

§ 1.2. Bellman's equation and the value function

and, approximately,

$$W(t+h,x) \approx \inf_{v \in U} J_{t+h}(x, u^v(\,\cdot\,)) \approx \inf_{v \in U} \int_0^h g(z^{x,v}(s), v)\, ds + W(t, z^{x,v}(h)).$$

Subtracting $W(t,x)$ we obtain that

$$\frac{1}{h}(W(t+h,x) - W(t,x))$$
$$\approx \inf_{u \in U} \left[\frac{1}{h} \int_0^h g(z^{x,v}(s), v)\, ds + \frac{1}{h}(W(t, z^{x,v}(h)) - W(t,x)) \right].$$

Assuming that the function W is differentiable and taking the limits as $h \downarrow 0$ we arrive at (1.6). □

For the control problem on the infinite time interval we have the following easy consequence of Theorem 1.1.

Theorem 1.2. *Let g be a nonnegative, continuous function and assume that there exists a nonnegative function W, defined on E and of class C^1, which satisfies the equation*

$$\inf_{u \in U}(g(x,u) + \langle W_x(x), f(x,u)\rangle) = 0, \quad x \in E. \tag{1.13}$$

If for a strategy $u(\,\cdot\,)$ and the corresponding output y, $\lim_{t \uparrow +\infty} W(y(t)) = 0$, then

$$J(x, u(\,\cdot\,)) \geq W(x). \tag{1.14}$$

If $\hat{v} \colon E \longrightarrow U$ is a mapping such that

$$g(x, \hat{v}(x)) + \langle W_x(x), f(x, \hat{v}(x))\rangle = 0 \quad \text{for } x \in E,$$

and $\hat{y}$ is an absolutely continuous, E-valued, solution of the equation

$$\frac{d}{dt}\hat{y}(t) = f(\hat{y}(t), \hat{v}(\hat{y}(t))), \quad t \geq 0, \tag{1.15}$$

for which $\lim_{t \uparrow +\infty} W(\hat{y}(t)) = 0$, then

$$J(x, \hat{u}(\,\cdot\,)) = W(x),$$

Proof. We apply Theorem 1.1 with

$$G(x) = W(x), \quad x \in E.$$

Since the function $\widetilde{W}(t,x) = W(x)$, $t \in [0,T]$, $x \in E$, satisfies all the conditions of Theorem 1.1 we see that, for an arbitrary admissible control $u(\,\cdot\,)$ and the corresponding output,

$$\int_0^T g(y(t), u(t))\, dt + W(y(T)) \geq W(x), \quad T > 0.$$

Letting T tend to $+\infty$ and taking into account that $\lim_{T\uparrow+\infty} W(y(T)) = 0$ we obtain the required inequality. On the other hand, for the strategy $\hat{u}(\,\cdot\,)$,

$$\int_0^T g(\hat{y}(t), \hat{u}(t))\, dt + W(\hat{y}(T)) = W(x), \quad T \geq 0.$$

Hence, if $\lim_{T\uparrow+\infty} W(\hat{y}(T)) = 0$, then

$$J(x, \hat{u}) = W(x).$$

$\square$

Let us remark that if a function W is a nonnegative solution of equation (1.13) then function $W + c$, where c is a nonnegative constant, is also a nonnegative solution to (1.13). Therefore, without additional conditions of the type $\lim_{t\uparrow+\infty} W(y(t)) = 0$, the estimate (1.14) cannot hold.

Theorem 1.2 can be generalized (compare Exercise 1.1) to the cost functional

$$J(x, u(\,\cdot\,)) = \int_0^{+\infty} e^{-\alpha t} g(y(t), u(t))\, dt.$$

The corresponding Bellman equation (1.13) is then of the form

$$\inf_{u\in U}((g(x,u) + \langle W_x(x), f(x,u)\rangle)) = \alpha W(x), \quad x \in E.$$

Exercise 1.2. Show that the solution of the Bellman equation corresponding to the optimal consumption model of Example 0.6, with $\alpha \in (0,1)$, is of the form

$$W(t,x) = p(t)x^\alpha, \quad t \geq 0,\ x \geq 0,$$

where the function $p(\,\cdot\,)$ is the unique solution of the following differential equation:

$$\dot{p} = \begin{cases} 1, & \text{for } p \leq 1, \\ \alpha p + (1-\alpha)\left(\dfrac{1}{p}\right)^{\alpha/(1-\alpha)}, & \text{for } p \geq 1, \end{cases}$$

$$p(0) = a.$$

Find the optimal strategy.

Hint. First prove the following lemma.

Lemma 1.1. *Let $\psi_p(u) = \alpha u p + (1-u)^\alpha$, $p \geq 0$, $u \in [0,1]$. The maximal value $m(p)$ of the function $\psi_p(\cdot)$ is attained at*

$$u(p) = \begin{cases} 0, & \text{if } p > 1, \\ \left(\frac{1}{p}\right)^{1/(1-\alpha)}, & \text{if } p \in [0,1]. \end{cases}$$

Moreover

$$m(p) = \begin{cases} 1, & \text{if } p \geq 1, \\ \alpha p + (1-\alpha)\left(\frac{1}{p}\right)^{\alpha/(1-\alpha)}, & \text{if } p \in [0,1]. \end{cases}$$

§1.3. The linear regulator problem and the Riccati equation

We now consider a special case of Problems (1.1) and (1.4) when the system equation is linear

$$\dot{y} = Ay + Bu, \quad y(0) = x \in \mathbf{R}^n, \tag{1.16}$$

$A \in \mathbf{M}(n,n)$, $B \in \mathbf{M}(n,m)$, the state space $E = \mathbf{R}^n$ and the set of control parameters $U = \mathbf{R}^m$. We assume that the cost functional is of the form

$$J_T = \int_0^T (\langle Qy(s), y(s)\rangle + \langle Ru(s), u(s)\rangle)\, ds + \langle P_0 y(T), y(T)\rangle, \tag{1.17}$$

where $Q \in \mathbf{M}(n,n)$, $R \in \mathbf{M}(m,m)$, $P_0 \in \mathbf{M}(n,n)$ are symmetric, nonnegative matrices and the matrix R is positive definite; see § I.1.1. The problem of minimizing (1.17) for a linear system (1.16) is called the *linear regulator problem* or the *linear-quadratic problem*.

The form of an optimal solution to (1.16) and (1.17) is strongly connected with the following *matrix Riccati equation*:

$$\dot{P} = Q + PA + A^*P - PBR^{-1}B^*P, \quad P(0) = P_0, \tag{1.18}$$

in which $P(s)$, $s \in [0,T]$, is the unknown function with values in $\mathbf{M}(n,n)$. The following theorem takes place.

Theorem 1.3. *Equation (1.18) has a unique global solution $P(s)$, $s \geq 0$. For arbitrary $s \geq 0$ the matrix $P(s)$ is symmetric and nonnegative definite.*

The minimal value of the functional (1.17) is equal to $\langle P(T)x, x\rangle$ and the optimal control is of the form

$$\hat{u}(t) = -R^{-1}B^*P(T-t)\hat{y}(t), \quad t \in [0,T], \tag{1.19}$$

where

$$\dot{\hat{y}}(t) = (A - BR^{-1}B^*P(T-t))\hat{y}(t), \quad t \in [0,T], \quad \hat{y}(0) = x. \tag{1.20}$$

Proof. The proof will be given in several steps.

Step 1. For an arbitrary symmetric matrix P_0 equation (1.18) has exactly one local solution and the values of the solution are symmetric matrices.

Equation (1.18) is equivalent to a system of n^2 differential equations for elements $p_{ij}(\cdot)$, $i,j = 1, 2, \ldots, n$ of the matrix $P(\cdot)$. The right hand sides of these equations are polynomials of order 2, and therefore the system has a unique local solution being a smooth function of its argument. Let us remark that the same equation is also satisfied by $P^*(\cdot)$. This is because matrices Q, R and P_0 are symmetric. Since the solution is unique, $P(\cdot) = P^*(\cdot)$, and the values of $P(\cdot)$ are symmetric matrices.

Step 2. Let $P(s)$, $s \in [0, T_0)$, be a symmetric solution of (1.18) and let $T < T_0$. The function $W(s,x) = \langle P(s)x, x\rangle$, $s \in [0,T]$, $x \in \mathbf{R}^n$, is a solution of the Bellman equation (1.6)-(1.7) associated with the linear regular problem (1.16)-(1.17).

The condition (1.7) follows directly from the definitions. Moreover, for arbitrary $x \in \mathbf{R}^n$ and $t \in [0,T]$

$$\inf_{u \in \mathbf{R}^n} (\langle Qx, x\rangle + \langle Ru, u\rangle + 2\langle P(t)x, Ax + Bu\rangle) \tag{1.21}$$

$$= \langle Qx,x\rangle + \langle (A^*P(t)+P(t)A)x,x\rangle + \inf_{u \in \mathbf{R}^m}(\langle Ru,u\rangle + \langle u, 2B^*P(t)x\rangle).$$

We need now the following lemma, the proof of which is left as an exercise.

Lemma 1.2. *If a matrix $R \in \mathbf{M}(m,m)$ is positive definite and $a \in \mathbf{R}^m$, then for arbitrary $u \in \mathbf{R}^m$*

$$\langle Ru, u\rangle + \langle a, u\rangle \geq -\frac{1}{4}\langle R^{-1}a, a\rangle.$$

Moreover, the equality holds if and only if

$$u = -\frac{1}{2}R^{-1}a.$$

It follows from the lemma that the expression (1.21) is equal to

$$\langle Q + A^*P(t) + P(t)A^* - P(t)BR^{-1}B^*P(A)x, x\rangle$$

§1.3. The linear regulator problem and the Riccati equation

and that the infimum in formula (1.21) is attained at exactly one point given by
$$-R^{-1}B^*P(t)x, \quad t \in [0,T].$$
Since $P(t)$, $t \in [0,T_0)$, satisfies the equation (1.18), the function W is a solution to the problem (1.6)-(1.7).

Step 3. The control $\hat{u}$ given by (1.19) on $[0,T]$, $T < T_0$, is optimal with respect to the functional $J_T(x,\cdot)$.

This fact is a direct consequence of Theorem 1.1.

Step 4. For arbitrary $t \in [0,T]$, $T < T_0$, the matrix $P(t)$ is nonnegative definite and

$$\langle P(t)x, x\rangle \leq \int_0^t \langle Q\tilde{y}^x(s), \tilde{y}^x(s)\rangle\, ds + \langle P_0\tilde{y}^x(t), \tilde{y}^x(t)\rangle, \qquad (1.22)$$

where $\tilde{y}^x(\cdot)$ is the solution to the equation

$$\dot{\tilde{y}} = A\tilde{y}, \quad \tilde{y}(0) = x.$$

Applying Theorem 1.1 to the function $J_t(x,\cdot)$ we see that its minimal value is equal to $\langle P(t)x, x\rangle$. For arbitrary control $u(\cdot)$, $J_t(x,u) \geq 0$, the matrix $P(t)$ is nonnegative definite. In addition, estimate (1.22) holds because its right hand side is the value of the functional $J_t(x,\cdot)$ for the control $u(s) = 0$, $s \in [0,t]$.

Step 5. For arbitrary $t \in [0,T_0)$ and $x \in \mathbf{R}^n$

$$0 \leq \langle P(t)x, x\rangle \leq \langle \left(\int_0^t S^*(r)QS(r)\, dr + S^*(t)P_0 S(t)\right)x, x\rangle,$$

where $S(r) = e^{Ar}$, $r \geq 0$.

This result is an immediate consequence of the estimate (1.22).

Exercise 1.3. Show that if, for some symmetric matrices $P = (p_{ij}) \in \mathbf{M}(n,n)$ and $S = (s_{ij}) \in \mathbf{M}(n,n)$,

$$0 \leq \langle Px, x\rangle \leq \langle Sx, x\rangle, \quad x \in \mathbf{R}^n,$$

then

$$-\frac{1}{2}(s_{ii} + s_{jj}) \leq p_{ij} \leq s_{ij} + \frac{1}{2}(s_{ii} + s_{jj}), \quad i,j = 1,\ldots,n.$$

It follows from Step 5 and Exercise 1.3 that solutions of (1.18) are bounded in $\mathbf{M}(n,n)$ and therefore an arbitrary maximal solution $P(\cdot)$ in $\mathbf{M}(n,n)$ exists for all $t \geq 0$; see Theorem II.1.1.

The proof of the theorem is complete. □

Exercise 1.4. Solve the linear regulator problem with a more general cost functional

$$\int_0^T \left(\langle Q(y(t) - a), y(t) - a\rangle + \langle Ru(t), u(t)\rangle\right) dt + \langle P_0 y(T), y(T)\rangle,$$

where $a \in \mathbf{R}^n$ is a given vector.
Answer. Let $P(t)$, $q(t)$, $r(t)$, $t \geq 0$, be solutions of the following matrix, vector and scalar equations respectively,

$$\dot{P} = Q + A^*P + PA - PBR^{-1}B^*P, \quad P(0) = P_0,$$
$$\dot{q} = A^*q - PBR^{-1}q - 2Qa, \quad q(0) = 0,$$
$$\dot{r} = -\frac{1}{4}\langle R^{-1}q, q\rangle + \langle Qa, a\rangle, \quad r(0) = 0.$$

The minimal value of the functional is equal

$$r(T) + \langle q(T), x\rangle + \langle P(T)x, x\rangle,$$

and the optimal, feedback strategy is of the form

$$u(t) = -\frac{1}{2}R^{-1}q(T-t) - R^{-1}B^*P(T-t)y(t), \quad t \in [0,T].$$

§1.4. The linear regulator and stabilization

The obtained solution of the linear regulator problem suggests an important way to stabilize linear systems. It is related to the *algebraic Riccati equation*

$$Q + PA + A^*P - PBR^{-1}B^*P = 0, \quad P \geq 0, \tag{1.23}$$

in which the unknown is a nonnegative definite matrix P. If $\widetilde{P}$ is a solution to (1.23) and $\widetilde{P} \leq P$ for all the other solutions P, then $\widetilde{P}$ is called a *minimal solution* of (1.23). For arbitrary control $u(\cdot)$ defined on $[0, +\infty)$ we introduce the notation

$$J(x, u) = \int_0^{+\infty} \left(\langle Qy(s), y(s)\rangle + \langle Ru(s), u(s)\rangle\right) ds. \tag{1.24}$$

Theorem 1.4. *If there exists a nonnegative solution P of equation (1.23) then there also exists a unique minimal solution $\widetilde{P}$ of (1.23), and the control $\tilde{u}$ given in the feedback form*

$$\tilde{u}(t) = -R^{-1}B^*\widetilde{P}y(t), \quad t \geq 0,$$

§1.4. The linear regulator and stabilization

minimizes functional (1.24). *Moreover the minimal value of the cost functional is equal to*
$$\langle \widetilde{P}x, x \rangle.$$

Proof. Let us first remark that if $P_1(t)$, $P_2(t)$, $t \geq 0$, are solutions of (1.18) and $P_1(0) \leq P_2(0)$ then $P_1(t) \leq P_2(t)$ for all $t \geq 0$. This is because the minimal value of the functional

$$J_t^1(x,u) = \int_0^t (\langle Qy(s), y(s) \rangle + \langle Ru(s), u(s) \rangle)\, ds + \langle P_1(0)y(t), y(t) \rangle$$

is not greater than the minimal value of the functional

$$J_t^2(x,u) = \int_0^t (\langle Qy(s), y(s) \rangle + \langle Ru(s), u(s) \rangle)\, ds + \langle P_2(0)y(t), y(t) \rangle,$$

and by Theorem 1.3 the minimal values are $\langle P_1(t)x, x \rangle$ and $\langle P_2(t)x, x \rangle$ respectively.

If, in particular, $P_1(0) = 0$ and $P_2(0) = P$ then $P_2(t) = P$ and therefore $P_1(t) \leq P$ for all $t \geq 0$. It also follows from Theorem 1.3 that the function $P_1(\cdot)$ is nondecreasing with respect to the natural order existing in the space of symmetric matrices; see §I.1.1. This easily implies that for arbitrary $i,j = 1, 2, \ldots, n$ there exist finite limits $\tilde{p}_{ij} = \lim\limits_{t \uparrow +\infty} \tilde{p}_{ij}(t)$, where $(\tilde{p}_{ij}(t)) = P_1(t)$, $t \geq 0$. Taking into account equation (1.18) we see that there exist finite limits

$$\lim_{t \uparrow +\infty} \frac{d}{dt} \tilde{p}_{ij}(t) = \gamma_{ij}, \quad i, j = 1, \ldots, n.$$

These limits have to be equal to zero, for if $\gamma_{i,j} > 0$ or $\gamma_{i,j} < 0$ then $\lim\limits_{t \uparrow +\infty} \tilde{p}_{ij}(t) = +\infty$. But $\lim\limits_{t \uparrow +\infty} \tilde{p}_{ij}(t) = -\infty$, a contradiction. Hence the matrix $\widetilde{P} = (\tilde{p}_{ij})$ satisfies equation (1.23). It is clear that $\widetilde{P} \leq P$.

Now let $\tilde{y}(\cdot)$ be the output corresponding to the input $\tilde{u}(\cdot)$. By Theorem 1.3, for arbitrary $T \geq 0$ and $x \in \mathbf{R}^n$,

$$\langle \widetilde{P}x, x \rangle = \int_0^T (\langle Q\tilde{y}(t), \tilde{y}(t) \rangle + \langle R\tilde{u}(t), \tilde{u}(t) \rangle)\, dt + \langle \widetilde{P}\tilde{y}(T), \tilde{y}(T) \rangle, \quad (1.25)$$

and

$$\int_0^T (\langle Q\tilde{y}(t), \tilde{y}(t) \rangle + \langle R\tilde{u}(t), \tilde{u}(t) \rangle)\, dt \leq \langle \widetilde{P}x, x \rangle.$$

Letting T tend to $+\infty$ we obtain

$$J(x, \tilde{u}) \leq \langle \widetilde{P}x, x \rangle.$$

On the other hand, for arbitrary $T \geq 0$ and $x \in \mathbf{R}^m$,

$$\langle P_1(T)x, x\rangle \leq \int_0^T (\langle Q\tilde{y}(t), \tilde{y}(t)\rangle + \langle R\tilde{u}(t), \tilde{u}(t)\rangle)\, dt \leq J(x, \tilde{u}),$$

consequently, $\langle \tilde{P}x, x\rangle \leq J(x, \tilde{u})$ and finally

$$J(x, \tilde{u}) = \langle \tilde{P}x, x\rangle.$$

The proof is complete. $\square$

Exercise 1.5. For the control system

$$\ddot{y} = u,$$

find the strategy which minimizes the functional

$$\int_0^{+\infty} (y^2 + u^2)\, dt$$

and the minimal value of this functional.

Answer. The solution of equation (1.23) in which $A = \begin{bmatrix} 0 & 1 \\ 0 & 0 \end{bmatrix}$, $B = \begin{bmatrix} 0 \\ 1 \end{bmatrix}$, $Q = \begin{bmatrix} 1 & 0 \\ 0 & 0 \end{bmatrix}$, $R = [1]$, is matrix $P = \begin{bmatrix} \sqrt{2} & 1 \\ 1 & \sqrt{2} \end{bmatrix}$. The optimal strategy is of the form $u = -y - \sqrt{2}(\dot{y})$ and the minimal value of the functional is $\sqrt{2}(y(0))^2 + 2y(0)\dot{y}(0) + \sqrt{2}(\dot{y}(0))^2$.

For stabilizability the following result is essential.

Theorem 1.5. (i) *If the pair (A, B) is stabilizable then equation (1.23) has at least one solution.*

(ii) *If $Q = C^*C$ and the pair (A, C) is detectable then equation (1.23) has at most one solution, and if P is the solution then the matrix $A - BR^{-1}B^*P$ is stable.*

Proof. (i) Let K be a matrix such that the matrix $A + BK$ is stable. Consider a feedback control $u(t) = Ky(t)$, $t \geq 0$. It follows from the stability of $A + BK$ that $y(t) \longrightarrow 0$, and therefore $u(t) \longrightarrow 0$ exponentially as $t \uparrow +\infty$. Thus for arbitrary $x \in \mathbf{R}^n$,

$$J(x, u(\,\cdot\,)) = \int_0^{+\infty} (\langle Qy(t), y(t)\rangle + \langle Ru(t), u(t)\rangle)\, dt < +\infty.$$

Since

$$\langle P_1(T)x, x\rangle \leq J(x, u(\,\cdot\,)) < +\infty, \quad T \geq 0,$$

§ 1.4. The linear regulator and stabilization

for the solution $P_1(t)$, $t \geq 0$, of (1.18) with the initial condition $P_1(0) = 0$, there exists $\lim_{T \uparrow +\infty} P_1(T) = P$ which satisfies (1.23). (Compare the proof of the previous theorem.)

(ii) We prove first the following lemma.

Lemma 1.3. (i) *Assume that for some matrices $M \geq 0$ and K of appropriate dimensions,*

$$M(A - BK) + (A - BK)^* M + C^* C + K^* RK = 0. \tag{1.26}$$

If the pair (A, C) is detectable, then the matrix $A - BK$ is stable.

(ii) *If, in addition, P is a solution to (1.23), then $P \leq M$.*

Proof. (i) Let $S_1(t) = e^{(A-BK)t}$, $S_2(t) = e^{(A-LC)t}$, where L is a matrix such that $A - LC$ is stable and let $y(t) = S_1(t)x$, $t \geq 0$. Since

$$A - BK = (A - LC) + (LC - BK),$$

therefore

$$y(t) = S_2(t)x + \int_0^t S_2(t-s)(LC - BK)y(s)\, ds. \tag{1.27}$$

We show now that

$$\int_0^{+\infty} |Cy(s)|^2\, ds < +\infty \quad \text{and} \quad \int_0^{+\infty} |Ky(s)|^2\, ds < +\infty. \tag{1.28}$$

Let us remark that, for $t \geq 0$,

$$\dot{y}(t) = (A - BK)y(t) \quad \text{and} \quad \frac{d}{dt}\langle My(t), y(t)\rangle = 2\langle M\dot{y}(t), y(t)\rangle.$$

It therefore follows from (1.26) that

$$\frac{d}{dt}\langle My(t), y(t)\rangle + \langle Cy(t), Cy(t)\rangle + \langle RKy(t), Ky(t)\rangle = 0.$$

Hence, for $t \geq 0$,

$$\langle My(t), y(t)\rangle + \int_0^t |Cy(s)|^2\, ds + \int_0^t \langle RKy(s), Ky(s)\rangle\, ds = \langle Mx, x\rangle. \tag{1.29}$$

Since the matrix R is positive definite, (1.29) follows from (1.28). By (1.29),

$$|y(t)| \leq |S_2(t)x| + N \int_0^t |S_2(t-s)|(|Cy(s)| + |Ky(s)|)\, ds,$$

where $N = \max(|L|, |B|)$, $t \geq 0$. By Theorem A.8 (Young's inequality) and by (1.28),

$$\int_0^{+\infty} |y(s)|^2 \, ds \leq N \int_0^{+\infty} |S_2(s)| \, ds \left(\int_0^{+\infty} (|Cy(s)| + |Ky(s)|)^2 \, ds \right)^{1/2}$$
$$+ \left(\int_0^{+\infty} |S_2(s)|^2 \, ds \right)^{1/2} |x| < +\infty.$$

It follows from Theorem I.2.3(iv) that $y(t) \to 0$ as $t \to \infty$. This proves the required result. Let us also remark that

$$M = \int_0^{+\infty} S_1^*(s)(C^*C + K^*RK)S_1(s) \, ds. \qquad (1.30)$$

(ii) Define $K_0 = R^{-1}B^*P$ then $RK_0 = -B^*P$, $PB = -K_0^*R$. Consequently,

$$P(A - BK) + (A - BK)^*P + K^*RK = -C^*C + (K - K_0)^*R(K - K_0)$$

and

$$M(A - BK) + (A - BK)^*M + K^*RK = -C^*C.$$

Hence if $V = M - P$ then

$$V(A - BK) + (A - BK)^*V + (K - K_0)^*R(K - K_0) = 0.$$

Since the matrix $A - BK$ is stable,

$$V = \int_0^{+\infty} S_1^*(s)(K - K_0)^*R(K - K_0)S_1(s) \, ds \geq 0,$$

by Theorem I.2.7, and therefore $M \geq P$. The proof of the lemma is complete. □

To prove part (ii) of Theorem 1.5 assume that matrices $P \geq 0$, $P_1 \geq 0$ are solutions of (1.23). Define $K = R^{-1}B^*P$. Then

$$P(A - BK) + (A - BK)^*P + C^*C + K^*RK \qquad (1.31)$$
$$= PA + A^*P + C^*C - PBR^{-1}B^*P = 0.$$

Therefore, by Lemma 1.3(ii), $P_1 \leq P$. In the same way $P_1 \geq P$. Hence $P_1 = P$. Identity (1.31) and Lemma 1.3(i) imply the stability of $A - BK$. □

Remark. Lemma 1.3(i) implies Theorem I.2.7(1). It is enough to define $B = 0$, $K = 0$ and remark that observable pairs are detectable; see § I.2.6.

As a corollary from Theorem 1.5 we obtain

Theorem 1.6. *If the pair (A, B) is controllable, $Q = C^*C$ and the pair (A, C) is observable, then equation (1.23) has exactly one solution, and if P is this unique solution, then the matrix $A - BR^{-1}B^*P$ is stable.*

Theorem 1.6 indicates an effective way of stabilizing linear system (1.16). Controllability and observability tests in the form of the corresponding rank conditions are effective, and equation (1.23) can be solved numerically using methods similar to those for solving polynomial equations. The uniqueness of the solution of (1.23) is essential for numerical algorithms.

The following examples show that equation (1.23) does not always have a solution and that in some cases it may have many solutions.

Example 1.1. If, in (1.23), $B = 0$, then we arrive at the Liapunov equation

$$PA + A^*P = Q, \quad P \geq 0. \tag{1.32}$$

If Q is positive definite, then equation (1.32) has at most one solution, and if, in addition, matrix A is not stable, then it does not have any solutions; see §I.2.4.

Exercise 1.6. If Q is a singular matrix then equation (1.23) may have many solutions. For if P is a solution to (1.23) and

$$\widetilde{A} = \begin{bmatrix} 0 & 0 \\ 0 & A \end{bmatrix}, \quad \widetilde{Q} = \begin{bmatrix} 0 & 0 \\ 0 & Q \end{bmatrix}, \quad \widetilde{A} \in \mathbf{M}(k, k), \ k > n,$$

then, for an arbitrary nonnegative matrix $R \in \mathbf{M}(k - n, k - n)$, matrix

$$\widetilde{P} = \begin{bmatrix} R & 0 \\ 0 & P \end{bmatrix}$$

satisfies the equation

$$\widetilde{P}\widetilde{A} + \widetilde{A}^*\widetilde{P} = \widetilde{Q}.$$

Bibliographical notes

Dynamic programming ideas are presented in the monograph by R. Bellmann [5].

The results of the linear regulator problem are classic. Theorem 1.5 is due to W.M. Wonham [61]. In the proof of Lemma 1.3(i) we follow [65].

Chapter 2
Dynamic programming for impulse control

The dynamic programming approach is applied to impulse control problems. The existence of optimal impulse strategy is deduced from general results on fixed points for monotonic and concave transformations.

§2.1. Impulse control problems

Let us consider a differential equation

$$\dot{z} = f(z), \quad z(0) = x \in \mathbf{R}^n, \tag{2.1}$$

with the right hand side f satisfying the Lipschitz condition, and let us denote the solution to (2.1) by $z^x(t)$, $t \geq 0$. Assume that for arbitrary $x \in \mathbf{R}^n$ a nonempty subset $\Gamma(x) \subset \mathbf{R}^n$ is given such that

$$\Gamma(w) \subset \Gamma(x) \quad \text{for arbitrary } w \in \Gamma(x). \tag{2.2}$$

Let, in addition, $c(\cdot,\cdot)$ be a positive bounded function defined on the graph $\{(x,v); x \in \mathbf{R}^n, v \in \Gamma(x)\}$ of the multifunction $\Gamma(\cdot)$, satisfying the following conditions:

$$c(x,v) + c(v,w) \geq c(x,w), \quad x \in \mathbf{R}^n, \ v \in \Gamma(x), \ w \in \Gamma(v), \tag{2.3}$$

there exists $\gamma > 0$, such that $c(x,w) \geq \gamma$ for $x \in \mathbf{R}^n$, $w \in \Gamma(x)$. (2.4)

Elements of the set $\Gamma(x)$ can be interpreted as those states of $\mathbf{R}^n$ to which one can shift immediately the state x and $c(x,v)$, the cost of the shift from x to v.

An *impulse strategy* π will be defined as a set of three sequences: a nondecreasing sequence (t_m) of numbers from $[0,\infty)$ and two sequences (x_m) and (w_m) of elements of $\mathbf{R}^n \cup \{\Delta\}$ (where Δ is an additional point outside of $\mathbf{R}^n$), satisfying the following conditions:

$$\text{if } t_m < +\infty, \quad \text{then } t_{m+1} > t_m; \tag{2.5}$$

$$\text{if } t_m = +\infty, \quad \text{then } x_m = z_m = \Delta; \tag{2.6}$$

$$\text{if } t_1 = 0, \quad \text{then } x_1 = x \text{ and } w_1 \in \Gamma(x); \tag{2.7}$$

$$\text{if } 0 < t_1 < +\infty, \quad \text{then } x_1 = z^x(t_1), \ w_1 \in \Gamma(x_1) \text{ and} \tag{2.8}$$

if $m > 1$, and $t_m < +\infty$, then

§ 2.1. Impulse control problem 143

$$x_m = z^{w_{m-1}}(t_m - t_{m-1}), \quad w_m \in \Gamma(x_m).$$

The number t_m should be interpreted as the moment of the m-th impulse, identical with an immediate shift from x_m to w_m. An arbitrary impulse trajectory π determines, together with the initial state x, the output strategy $y^{\pi,x}(t)$, $t \geq 0$, by the formulae

$$y^{\pi,x}(t) = z^x(t) \quad \text{for } t \in [0, t_1),$$
$$y^{\pi,x}(t) = z^{w_{m-1}}(t - t_{m-1}) \quad \text{for } t \in [t_{m-1}, t_m), \quad m = 2, 3, \ldots,$$
$$y^{\pi,x}(+\infty) = \Delta.$$

With the strategy π we associate the following *cost functional*

$$J_T(\pi, x) = \int_0^T e^{-\alpha t} g(y^{\pi,x}(t)) dt + \sum_{t_m \leq T} e^{-\alpha t_m} c(x_m, w_m), \quad \alpha > 0,$$

where g is a nonnegative function, α is a positive constant and $T \leq +\infty$. We assume also that $c(\Delta, \Delta) = 0$.

It is our aim to find a strategy minimizing $J_T(\cdot, x)$. We consider first the case of the infinite control interval, $T = +\infty$. The case of finite $T < +\infty$ will be analysed in §3.3 by the maximum principle approach. Instead of J_T we write simply J.

An important class of impulse strategies form the *stationary strategies*. Optimal strategies often are stationary. To define a stationary strategy we start from a closed set $K \subset \mathbf{R}^n$ and a mapping $S: K \longrightarrow \mathbf{R}^n$ such that $S(x) \in \Gamma(x) \setminus K$, $x \in K$, and define sequences (t_m), (x_m) and (w_m) as follows. Let the starting point be x. If $x \in K$ then $t_1 = 0$, $x_1 = x$ and $w_1 = S(x_1)$. If $x \notin K$ then $t_1 = \inf\{t \geq 0; z^x(t) \in K\}$ and if $t_1 < +\infty$ then $x_1 = z^x(t_1)$ and $w_1 = S(x_1)$; if $t_1 = +\infty$ then we define $x_1 = \Delta$, $w_1 = \Delta$. Suppose that sequences $t_1, \ldots, t_m$, $x_1, \ldots, x_m$ and $w_1, \ldots, w_m$ have been defined. We can assume that $t_m < +\infty$. Let us define $\tilde{t}_m = \inf\{t \geq 0; z^{w_m}(t) \in K\}$, $t_{m+1} = t_m + \tilde{t}_m$, and if $\tilde{t}_m < +\infty$ then $x_{m+1} = z^{w_m}(\tilde{t}_m)$, $w_{m+1} = S(x_{m+1})$; if $\tilde{t}_m = +\infty$ then we set $t_{m+1} = +\infty$, $x_{m+1} = \Delta = w_{m+1}$.

To formulate the basic existence result it is convenient to introduce two more conditions.

For an arbitrary nonnegative and bounded continuous function v,

$$Mv(x) = \inf\{v(z) + c(x, z); z \in \Gamma(x)\}, \; x \in \mathbf{R}^n, \text{ is continuous.} \quad (2.9)$$

For arbitrary $x \in \mathbf{R}^n$, the set $\Gamma(x)$ is compact and the function $c(x, \cdot)$ is continuous on $\Gamma(x)$. (2.10)

Theorem 2.1 *Assume that g is a nonnegative, continuous and bounded function and that f satisfies the Lipschitz condition and conditions (2.9) and (2.10). Then*

(i) *The equation*

$$v(x) = \inf_{t \geq 0} \int_0^t e^{-\alpha s} g(z^x(s))\, ds + e^{-\alpha t} M v(z^x(t)), \quad x \in \mathbf{R}^n, \qquad (2.11)$$

has exactly one nonnegative, continuous and bounded solution $v = \hat{v}$.

(ii) *For arbitrary x in the set*

$$\hat{K} = \{x;\, \hat{v}(x) = M\hat{v}(x)\},$$

there exists an element $\hat{S}(x) \in \Gamma(\hat{x}) \setminus \hat{K}$ such that $M\hat{v}(x) = \hat{v}(x) = c(x, \hat{S}(x)) + \hat{v}(\hat{S}(x))$, and the stationary strategy $\hat{\pi}$ determined by $\hat{K}$ and $\hat{S}$ is an optimal one.

Remark. Equation (2.11) is the Bellman equation corresponding to the problem of the impulse control on the infinite time interval. In a heuristic way one derives it as follows.

Assume that $v(x)$ is the minimal value of the function $J(x, \cdot)$, and consider a strategy which consists first of the shift at moment $t \geq 0$ from the current state $z^x(t)$ to a point $w \in \Gamma(z^x(t))$ and then of an application of the optimal strategy corresponding to the new initial state w. The total cost associated with this strategy is equal to

$$\int_0^t e^{-\alpha s} g(z^x(s))\, ds + e^{-\alpha t} [c(z^x(t), w) + v(w)].$$

Minimizing the above expression with respect to t and w we obtain $v(x)$, hence equation (2.11) follows. These arguments are incomplete because in particular they assume the existence of optimal strategies for an arbitrary initial condition $x \in \mathbf{R}^n$.

To prove the theorem we will need several results of independent interest.

§2.2. An optimal stopping problem

Let us assume that a continuous and bounded function φ on $\mathbf{R}^n$ is given, and consider, for arbitrary $x \in \mathbf{R}^n$, the question of finding a number $t \geq 0$, at which the infimum

$$\inf \left\{ e^{-\alpha t} \varphi(z^x(t));\, t \geq 0 \right\} = \Phi(x), \qquad (2.12)$$

is attained in addition to the minimal value $\Phi(x)$.

The following result holds.

Lemma 2.1. *Assume that f is Lipschitz continuous, φ continuous and bounded and α a positive number. Then the function Φ defined by (2.12) is continuous. Let $K = \{x \in \mathbf{R}^n;\ \Phi(x) = \varphi(x)\}$ and*

$$t(x) = \begin{cases} \inf\{t \geq 0;\ z^x(t) \in K\}, \\ +\infty,\ if\ z^x(t) \notin K\ for\ all\ t \geq 0. \end{cases}$$

If $t(x) < +\infty$ then $\Phi(x) = e^{-\alpha t(x)}\varphi(z^x(t(x)))$; if $t(x) = +\infty$ then $\varphi(z^x(t)) > 0$ for $t > 0$, and $\Phi(x) = 0$.

Proof. Note that for $x, w \in \mathbf{R}^n$

$$|\Phi(x) - \Phi(w)| \leq \sup_{t \geq 0}\left(e^{-\alpha t}\left|\varphi(z^x(t)) - \varphi(z^w(t))\right|\right).$$

If $x_m \longrightarrow w$, then $z^{x_m} \longrightarrow z^w$ uniformly on an arbitrary finite interval $[0,T]$, $T < +\infty$. The continuity of φ and positivity of α imply that $\Phi(x_m) \longrightarrow \Phi(w)$. Therefore Φ is a continuous function and the set K is closed. Note also that if $t \leq t(x) < +\infty$ then

$$\Phi(x) = e^{-\alpha t}\Phi(z^x(t)). \tag{2.13}$$

This follows from the identities

$$\Phi(z^x(t)) = \inf_{s \geq 0} e^{-\alpha s}\Phi(z^x(s+t))$$
$$= e^{\alpha t}\inf_{s \geq 0} e^{-\alpha(s+t)}\Phi(z^x(s+t)) = e^{\alpha t}\Phi(x).$$

The formula (2.13) implies the remaining parts of the lemma. $\square$

Lemma 2.1 states that the solution to the problem (2.12) is given by the first hitting time of the *coincidence set*

$$K = \{x;\ \Phi(x) = \varphi(x)\}.$$

§ 2.3. Iterations of convex mappings

The existence of a solution to equation (2.11) will be a corollary of the general results on convex transformations.

Let $\mathcal{K}$ be a *convex cone* included in a linear space $\mathcal{L}$. Hence if $x, y \in \mathcal{K}$ and $\alpha, \beta \geq 0$ then $\alpha x + \beta y \in \mathcal{K}$. In $\mathcal{L}$ we introduce a relation $\geq$:

$$x \geq y\ \text{if and only if}\ x - y \in \mathcal{K}.$$

Let $\mathcal{A}$ be a transformation from $\mathcal{K}$ into $\mathcal{K}$. We say that $\mathcal{A}$ is *monotonic* if for arbitrary $x \geq y$, $\mathcal{A}(x) \geq \mathcal{A}(y)$. In a similar way the transformation $\mathcal{A}$ is called *concave* if for arbitrary $x, y \in \mathcal{K}$ and $\alpha, \beta \geq 0$, $\alpha + \beta = 1$,
$$\mathcal{A}(\alpha x + \beta y) \geq \alpha \mathcal{A}(x) + \beta \mathcal{A}(y).$$

Lemma 2.2. *Assume that a mapping* $\mathcal{A}\colon \mathcal{K} \longrightarrow \mathcal{K}$ *is monotonic and concave. If for an element* $h \in \mathcal{K}$, $h \geq \mathcal{A}(h)$, *and for some* $\delta > 0$, $\delta h \leq \mathcal{A}(0)$, *then there exists* $\gamma \in (0,1)$ *such that*
$$0 \leq \mathcal{A}^{m-1}(h) - \mathcal{A}^m(h) \leq \gamma^m \mathcal{A}^{m-1}(h) \quad \text{for } m = 1, 2, \ldots \quad (2.14)$$

Proof. We can assume that $\delta \in (0,1)$. We show that (2.14) holds with $\gamma = 1 - \delta$. For $m = 1$ inequalities (2.14) take the forms
$$0 \leq h - \mathcal{A}(h) \leq \gamma h,$$
and therefore (2.14) follows from the assumptions. If (2.14) holds for some m then also
$$(1 - \gamma^m)\mathcal{A}^{m-1}(h) + \gamma^m 0 \leq \mathcal{A}^m(h).$$
It follows, from the monotonicity and concavity of $\mathcal{A}$, that
$$\mathcal{A}^{m+1}(h) \geq (1 - \gamma^m)\mathcal{A}^m(h) + \gamma^m \mathcal{A}(0)$$
or, equivalently, that
$$\mathcal{A}^m(h) - \mathcal{A}^{m+1}(h) \leq \gamma^m \mathcal{A}^m(h) - \gamma^m \mathcal{A}(0).$$
It suffices therefore to show that
$$\gamma^m \mathcal{A}^m(h) - \gamma^m \mathcal{A}(0) \leq \gamma^{m+1} \mathcal{A}^m(h),$$
or that
$$(1 - \gamma)\mathcal{A}^m(h) \leq \mathcal{A}(0). \quad (2.15)$$
Since $1 - \gamma = \delta$ and $\mathcal{A}^m(h) \leq h$, the inequality (2.15) holds. This finishes the proof of the lemma. $\square$

§2.4. The proof of Theorem 2.1

(i) We have to prove the existence of a solution to (2.11) in the class of all nonnegative continuous and bounded functions v. For this purpose, define $\mathcal{L}$ as the space of all continuous and bounded functions on $\mathbf{R}^n$ and

§2.4. The proof of Theorem 2.1

$\mathcal{K} \subset \mathcal{L}$ as the cone of all nonnegative functions from $\mathcal{L}$. For arbitrary $v \in \mathcal{K}$ we set

$$\mathcal{A}v(x) = \inf_{t \geq 0} \left(\int_0^t e^{-\alpha s} g\left(z^x(s)\right) ds + e^{-\alpha t} Mv\left(z^x(t)\right) \right).$$

Then equation (2.11) is equivalent to

$$v(x) = \mathcal{A}v(x), \quad x \in \mathbf{R}^n.$$

Since $g \geq 0$, the solution v has to be nonnegative, and a function h defined by the formula

$$h(x) = \int_0^{+\infty} e^{-\alpha s} g\left(z^x(s)\right) ds, \quad x \in \mathbf{R}^n,$$

is in $\mathcal{K}$.

We check easily that, for $x \in \mathbf{R}^n$,

$$\mathcal{A}v(x) = h(x) + \inf_{t \geq 0} \left(e^{-\alpha t} (Mv - h)(z^x(t)) \right). \tag{2.16}$$

Therefore, by (2.9) and Lemma 2.1, the function $\mathcal{A}(v)$ is bounded and continuous for arbitrary $v \in \mathcal{K}$. Hence $\mathcal{A}$ transforms $\mathcal{K}$ into $\mathcal{K}$. Monotonicity of M implies that $\mathcal{A}$ is monotonic.

To show that the transformation $\mathcal{A}$ is concave, remark that for non-negative numbers $\beta, \gamma, \beta + \gamma = 1$ and for functions $v_1, v_2 \in \mathcal{K}$,

$$M(\beta v_1 + \gamma v_2)(x) \geq \beta M v_1(x) + \gamma M v_2(x), \quad x \in \mathbf{R}^n.$$

Therefore, for arbitrary $t \geq 0$

$$\int_0^t e^{-\alpha s} g\left(z^x(s)\right) ds + e^{-\alpha t} M(\beta v_1 + \gamma v_2)(z^x(t))$$

$$\geq \beta \left[\int_0^t e^{-\alpha s} g\left(z^x(s)\right) ds + e^{-\alpha t} M v_1\left(z^x(t)\right) \right]$$

$$+ \gamma \left[\int_0^t e^{-\alpha s} g\left(z^x(s)\right) ds + e^{-\alpha t} M v_2\left(z^x(t)\right) \right]$$

$$\geq \beta \mathcal{A} v_1(x) + \gamma \mathcal{A} v_2(x), \quad x \in \mathbf{R}^n.$$

By the definition of $\mathcal{A}$,

$$\mathcal{A}(\beta v_1 + \gamma v_2)(x) \geq \beta \mathcal{A} v_1(x) + \gamma \mathcal{A} v_2(x), \quad x \in \mathbf{R}^n,$$

the required concavity.

It follows, from the definition of h, that $\mathcal{A}(h) \leq h$. On the other hand, for $v = 0$ and arbitrary $t \geq 0$,

$$\int_0^t e^{-\alpha s} g\left(z^x(s)\right) ds + e^{-\alpha T} M(0)(x)$$

$$\geq \int_0^t e^{-\alpha s} g\left(z^x(s)\right) ds + e^{-\alpha t} \gamma \geq h(x) + e^{-\alpha t}(\gamma - h)\left(z^x(t)\right),$$

where γ is a positive lower bound for $c(\cdot, \cdot)$. Let $\delta \in (0,1)$ be such that $\delta h \leq \gamma$. Since $h(x) \geq e^{-\alpha t} h\left(z^x(t)\right)$, $t \geq 0$, so finally

$$\int_0^t e^{-\alpha s} g\left(z^x(s)\right) ds + e^{-\alpha t} M(0)(x) \geq h(x) + (\delta - 1) e^{-\alpha t} h\left(z^x(t)\right) \geq \delta h(x),$$

and $\mathcal{A}(0) \geq \delta h$.

All the conditions of Lemma 2.2 are satisfied and therefore the sequence $(\mathcal{A}^m(h))$ is uniformly convergent to a continuous, bounded and nonnegative function $\hat{v}$. It is not difficult to see that if $v_2 \geq v_1 \geq 0$ then

$$0 \leq \mathcal{A}(v_2) - \mathcal{A}(v_1) \leq \|v_2 - v_1\|.$$

In a similar way, for $m = 1, 2, \ldots$,

$$0 \leq \mathcal{A}^{m+1}(h) - \mathcal{A}(\hat{v}) \leq \|\mathcal{A}^m h - \hat{v}\|,$$

and therefore $\hat{v} = \mathcal{A}(\hat{v})$. This way we have shown that there exists a solution of (2.11) in the set $\mathcal{K}$. This finishes the proof of (i).

(ii) Since $\hat{v} \in \mathcal{K}$, the set

$$\hat{K} = \{x;\ \hat{v}(x) = M\hat{v}(x)\}$$

is closed by condition (2.9).

We show first that there exists a function $\hat{S}(\cdot)$ with the properties required by the theorem. By (2.10), function $c(x, \cdot) + \hat{v}(\cdot)$ attains its minimum at the point $w_1 \in \Gamma(x)$. Hence

$$c(x, w_1) + \hat{v}(w_1) = \hat{v}(x).$$

If $w_1 \in \hat{K}$, then there exists an element $w_2 \in \Gamma(w_1)$ such that

$$c(w_1, w_2) + \hat{v}(w_2) = \hat{v}(w_1).$$

More generally, assume that $w_2, \ldots, w_m$ are elements such that

$$c(w_i, w_{i+1}) + \hat{v}(w_{i+1}) = \hat{v}(w_i)$$

§ 2.4. The proof of Theorem 2.1

and $w_{i+1} \in \Gamma(w_i) \cap \hat{K}$, $i = 1, 2, \ldots, m-1$. Adding the above identities we obtain

$$c(x, w_1) + c(w_1, w_2) + \ldots + c(w_{m-1}, w_m) + \hat{v}(w_m) = \hat{v}(w_1).$$

Since $\hat{v}$ is nonnegative and $c(\cdot, \cdot) \geq \gamma > 0$,

$$\gamma m \leq \hat{v}(w_1).$$

Therefore for some m, $w_m \in \Gamma(w_{m-1})$ and $w_m \notin \hat{K}$. In addition,

$$c(w_{m-1}, w_m) + \hat{v}(w_m) = \hat{v}(w_{m-1}).$$

Since $w_m \in \Gamma(x)$, by (2.3),

$$c(x, w_1) + \ldots + c(w_{m-1}, w_m) \geq c(x, w_m),$$
$$c(x, w_m) + \hat{v}(w_m) \leq \hat{v}(x)$$

and

$$c(x, w_m) + \hat{v}(w_m) = \hat{v}(x).$$

Hence a function $\hat{S}$ with the required properties exists.

We finally show that if $\hat{v} \in \mathcal{K}$ is a solution to (2.11) then, for arbitrary $x \in \mathbf{R}^n$ and the stationary strategy $\hat{\pi}$ determined by $\hat{K}$, $\hat{S}$,

$$J(\hat{\pi}, x) = \hat{v}(x).$$

Let $\hat{t}_1, \hat{t}_2, \ldots$ and $(\hat{x}_1, \hat{w}_1), (\hat{x}_2, \hat{w}_2), \ldots$ be elements of the stationary strategy. Then for arbitrary finite $\hat{t}_m$

$$\hat{v}(x) = \int_0^{\hat{t}_m} e^{-\alpha s} g\left(y^{\hat{\pi}, x}(s)\right) ds + \sum_{i=1}^m e^{-\alpha \hat{t}_i} c(\hat{x}_i, \hat{w}_i) + e^{-\alpha \hat{t}_m} \hat{v}(\hat{w}_m). \quad (2.17)$$

To check that (2.17) holds assume first that $m = 1$. If $\hat{t}_1 = 0$, then $x = \hat{x}_1 \in \hat{K}$ and $\hat{w}_1 = \hat{S}(\hat{x}_1)$. Hence

$$\hat{v}(x) = \hat{v}(\hat{x}_1) = c(\hat{x}_1, \hat{w}_1) + \hat{v}(\hat{w}_1).$$

If $0 < \hat{t}_1 < +\infty$, then $x \notin \hat{K}$ and $\hat{t}_1 = \inf\{t \geq 0; z^x(t) \in \hat{K}\}$. It follows from equation (2.11) and Lemma 2.1 that

$$\hat{v}(x) = \int_0^{\hat{t}_1} e^{-\alpha s} g(z^x(s)) \, ds + e^{-\alpha \hat{t}_0} M\hat{v}\left(z^x(\hat{t}_1)\right).$$

However $\hat{x}_1 = z^x(\hat{t}_1)$, $\hat{w}_1 = \hat{S}(\hat{x}_1)$ and

$$\hat{v}(\hat{x}_1) = M\hat{v}(\hat{x}_1) = c(\hat{x}_1, \hat{w}_1) + \hat{v}(\hat{w}_1);$$

consequently,

$$\hat{v}(x) = \int_0^{\hat{t}_1} e^{-\alpha s} g(z^x(s))\, ds + e^{-\alpha \hat{t}_1} c(\hat{x}, \hat{z}) + e^{-\alpha \hat{t}_1} \hat{v}(\hat{z}_1).$$

Similarly, if (2.17) holds for some m and $\hat{t}_{m+1} < +\infty$ then

$$\hat{t}_{m+1} - \hat{t}_m = \inf\left\{t \geq 0;\, z^{\hat{w}_m}(t) \in \hat{K}\right\} = \hat{t}(\hat{w}_m)$$

and, by the Bellman equation (2.11),

$$\hat{v}(\hat{w}_m) = \int_0^{\hat{t}(\hat{w}_m)} e^{-\alpha s} g(z^{\hat{w}_m}(s))\, ds + e^{-\alpha(\hat{t}(\hat{w}_m))} M\hat{v}(z^{\hat{w}_m}(\hat{t}(\hat{w}_m))).$$

Since $y^{\hat{w}_m}(\hat{t}(\hat{w}_m)) = \hat{x}_{m+1}$ and

$$\hat{v}(\hat{x}_{m+1}) = M\hat{v}(\hat{x}_{m+1}) = c(\hat{x}_{m+1}, \hat{w}_{m+1}) + \hat{v}(\hat{w}_{m+1}),$$

therefore

$$\begin{aligned}\hat{v}(x) = &\int_0^{\hat{t}_m} e^{-\alpha s} g(y^{\hat{\pi},x}(s))\, ds + \sum_{i=0}^m e^{-\alpha \hat{t}_i} c(\hat{x}_i, \hat{w}_i) \\ &+ e^{-\alpha \hat{t}_m}\Bigg\{\int_0^{\hat{t}_{m+1}-\hat{t}_m} e^{-\alpha s} g(z^{\hat{w}_m}(s))\, ds \\ &+ e^{-\alpha(\hat{t}_{m+1}-\hat{t}_m)}(c(\hat{x}_{m+1}, \hat{w}_{m+1}) + \hat{v}(\hat{w}_{m+1}))\Bigg\}.\end{aligned}$$

By the very definition of the strategy $\hat{\pi}$ we see that (2.17) holds for $m+1$ and thus for all $m = 1, 2, \ldots$.

In exactly the same way we show that for arbitrary strategy π and finite t_m,

$$\hat{v}(x) \leq \int_0^{t_m} e^{-\alpha s} g(y^{\pi,x}(s))\, ds + \sum_{i=0}^m e^{-\alpha t_i} c(x_i, w_i) + e^{-\alpha t_m}\hat{v}(w_m).$$

To prove that
$$\hat{v}(x) = J(\hat{\pi}, x),$$

assume for instance that $\hat{t}_m < +\infty$ for $m = 1, 2, \ldots$. Since

$$+\infty > \hat{v}(x) \geq \sum_{i=0}^{m} e^{-\alpha \hat{t}_i} c(\hat{x}_i, \hat{w}_i) \geq \gamma \sum_{i=0}^{m} e^{-\alpha \hat{t}_i},$$

$\hat{t}_m \uparrow +\infty$. Letting m tend to $+\infty$ in (2.17), we obtain

$$\hat{v}(x) = \int_0^{+\infty} e^{-\alpha s} g\left(y^{\hat{\pi},x}(s)\right) ds + \sum_{i=0}^{+\infty} e^{-\alpha \hat{t}_i} c(\hat{x}_i, \hat{w}_i).$$

Similarly, if $\hat{t}_1 = +\infty$ then $z^x(t) \notin \hat{K}$ for arbitrary $t \geq 0$. Lemma 2.1 and equation (2.11) imply

$$\hat{v}(x) = \int_0^{+\infty} e^{-\alpha s} g\left(z^x(s)\right) ds = J(\hat{\pi}, x).$$

If, finally, $\hat{t}_1 < \hat{t}_2 < \ldots < \hat{t}_m < \hat{t}_{m+1} = +\infty$, then $z^{\hat{w}_m}(t) \notin \hat{K}$ for all $t \geq 0$, and

$$\hat{v}(\hat{w}_m) = \int_0^{+\infty} e^{-\alpha s} g(z^{\hat{w}_m}(s)) ds.$$

By the identity (2.17) we obtain

$$\hat{v}(x) = \int_0^{\hat{t}_m} e^{-\alpha s} g(y^{\hat{\pi},x}(s)) ds + \sum_{i=0}^{m} e^{-\alpha \hat{t}_i} c(\hat{x}_i, \hat{w}_i) +$$
$$+ e^{-\alpha \hat{t}_m} \int_0^{+\infty} e^{-\alpha s} g(z^{\hat{w}_m}(s)) ds = J(\hat{\pi}, x).$$

Arguments of a similar type allow one to show that, for arbitrary strategy π,

$$J(\pi, x) \geq \hat{v}(x).$$

This way we have shown that the strategy $\hat{\pi}$ is optimal and at the same time that $\hat{v}$ is the unique solution of (2.11). □

Bibliographical notes

There is a vast literature devoted to impulse control problems. We refer the reader to the monograph by A. Bensoussan and J.L. Lions [6] and to the survey [69] for more information and references. Theorem 2.1 is a special case of a result due to M. Robin and Lemma 2.2 to B. Hanouzet and J.L. Joly, see, e.g., [69]. The abstract approach to dynamic programming equations based on monotonicity and concavity is due to the author, see [62], [63], [67] and references there.

Chapter 3
The maximum principle

The maximum principle is formulated and proved first for control problems with fixed terminal time. As an illustration, a new derivation of the solution to the linear regulator problem is given. Next the maximum principle for impulse control problems and for time optimal problems are discussed. In the latter case an essential role is played by the separation theorems for convex sets.

§ 3.1. Control problems with fixed terminal time

We start our discussion of the second approach to the optimality questions (see § 1.1) by considering a control system

$$\dot{y} = f(y, u), \quad y(0) = x \in \mathbf{R}^n, \tag{3.1}$$

and the cost functional

$$J_T(x, u(\,\cdot\,)) = \int_0^T g(y(t), u(t))\, dt + G(y(T)) \tag{3.2}$$

with $T > 0$, a fixed, finite number. The set of control parameters will be denoted, as before, by $U \subset \mathbf{R}^m$, the derivative of f, with respect to the state variable, by f_x and derivatives of g and G by g_x and G_x.

Here is a version of the maximum principle.

Theorem 3.1. *Assume that functions f, g, G and f_x, g_x, G_x are continuous and that a bounded control $\hat{u}(\,\cdot\,)$ and the corresponding absolutely continuous solution of (3.1) maximize the functional (3.2). For arbitrary $t \in (0, T)$, such that the left derivative $\frac{d^-}{dt}\hat{y}(t)$ exists and is equal to $f(\hat{y}(t), \hat{u}(t))$, the following inequality holds;*

$$\langle p(t), f(\hat{y}(t), \hat{u}(t))\rangle + g(\hat{y}(t), \hat{u}(t)) \\ \geq \max_{u \in U}(\langle p(t), f(\hat{y}(t), u)\rangle + g(\hat{y}(t), u)). \tag{3.3}$$

In (3.3), $p(t)$, $t \in [0, T]$, stands for the absolutely continuous solution of the equation

$$\dot{p} = -f_x^*(\hat{y}, \hat{u})p - g_x^*(\hat{y}, \hat{u}) \tag{3.4}$$

§ 3.1. Control problems with fixed terminal time

with the end condition
$$p(T) = G_x(\hat{y}(T)). \tag{3.5}$$

Remark. Since an arbitrary absolutely continuous function is differentiable almost everywhere, the inequality (3.3) holds for almost all $t \in [0,T]$.

Proof. Assume first that $g(x,y) = 0$ for $x \in E$, $u \in U$. Let, for some $t_0 \in (0,T)$,
$$\frac{d^-}{dt}\hat{y}(t_0) = f(\hat{y}(t_0), \hat{u}(t_0)).$$

For arbitrary control parameter $v \in U$ and sufficiently small $h \geq 0$ define the *needle variation* of $\hat{u}$ by the formula

$$u(t,h) = \begin{cases} \hat{u}(t), & \text{if } t \in [0, t_0 - h), \\ v, & \text{if } t \in [t_0 - h, t_0), \\ \hat{u}(t), & \text{if } t \in [t_0, T]. \end{cases}$$

Let $y(\cdot, h)$ be the output corresponding to $u(\cdot, h)$. It is clear that

$$\frac{d^+}{dh} G(y(T,0)) \leq 0, \tag{3.6}$$

provided that the right derivative in (3.6) exists. To prove that the derivative exists and to find its value, let us remark first that

$$y(t_0, h) = \hat{y}(t_0 - h) + \int_{t_0-h}^{t_0} f(y(s,h), v) \, ds,$$

$$\hat{y}(t_0) = \hat{y}(t_0 - h) + \int_{t_0-h}^{t_0} f(\hat{y}(s), \hat{u}(s)) \, ds.$$

Taking into account (3.5), we obtain

$$\frac{d^+}{dh} y(t_0, 0) = \lim_{h \downarrow 0} \frac{y(t_0, h) - \hat{y}(t_0)}{h} \tag{3.7}$$
$$= \lim_{h \downarrow 0} \frac{y(t_0, h) - \hat{y}(t_0 - h)}{h} - \lim_{h \downarrow 0} \frac{\hat{y}(t_0) - \hat{y}(t_0 - h)}{h}$$
$$= f(\hat{y}(t_0), v) - f(\hat{y}(t_0), \hat{u}(t_0)).$$

Moreover, $\frac{d}{dt} y(t,h) = f(y(t,h), \hat{u}(t))$ for almost all $t \in [t_0, T]$, so by Theorem II.1.4, Theorem I.1.1 and by (3.7)

$$\frac{d^+}{dh} y(T, 0) = S(T) S^{-1}(t_0)(f(\hat{y}(t_0), v) - f(\hat{y}(t_0), \hat{u}(t_0)),$$

where $S(\cdot)$ is the fundamental solution of

$$\dot q = f_x(\hat y, \hat u)q.$$

Therefore, the right derivative in (3.6) exists and

$$\begin{aligned}\frac{d^+}{dh}G(y(T,0)) &= \langle G_x(\hat y(T)), \frac{d^+}{dh}y(T,0)\rangle \\ &= \langle G_x(\hat y(T)), S(T)S^{-1}(t_0)(f(\hat y(t_0),v) - f(\hat y(t_0),\hat u(t_0)))\rangle \\ &= \langle S^*(t_0)^{-1}S^*(T)G_x(\hat y(T)), (f(\hat y(t_0),v) - f(\hat y(t_0),\hat u(t_0)))\rangle.\end{aligned}$$

Since the fundamental solution of the equation $\dot p = -f_x^*(\hat y, \hat u)p$ is equal to $(S^*(t))^{-1}$, $t \in [0,T]$, formula (I.1.7) and inequality (3.6) finish the proof of the theorem in the case when $g(x,u) = 0$, $x \in \mathbf{R}^n$, $u \in U$.

The case of the general function g can be reduced to the considered one. To do so we introduce an additional variable $w \in \mathbf{R}^1$ and define

$$\tilde f\left(\begin{bmatrix}x\\w\end{bmatrix}, u\right) = \begin{bmatrix}f(x,u)\\g(x,u)\end{bmatrix}, \quad \tilde G\begin{bmatrix}x\\w\end{bmatrix} = G(x) + w, \quad \begin{bmatrix}x\\w\end{bmatrix} \in \mathbf{R}^{n+1},\ u \in U.$$

It is easy to see that the problem of maximization of the function (3.2) for the system (3.1) is equivalent to the problem of maximization of

$$\tilde G\begin{bmatrix}y(T)\\w(T)\end{bmatrix}$$

for the following control system on $\mathbf{R}^{n+1}$,

$$\frac{d}{dt}\begin{bmatrix}y\\w\end{bmatrix} = \tilde f\left(\begin{bmatrix}y\\w\end{bmatrix}, u\right) = \begin{bmatrix}f(y,u)\\g(y,u)\end{bmatrix}$$

with the initial condition

$$\begin{bmatrix}y(0)\\w(0)\end{bmatrix} = \begin{bmatrix}x\\0\end{bmatrix} \in \mathbf{R}^{n+1}.$$

Applying the simpler version of the theorem, which has just been proved, we obtain the statement of the theorem in the general case. □

The obtained result can be reformulated in terms of the *Hamiltonian function*:

$$H(x,p,u) = \langle p, f(x,u)\rangle + g(x,u), \quad x \in \mathbf{R}^n,\ u \in U,\ p \in \mathbf{R}^n. \qquad (3.8)$$

Equations (3.1) and (3.4) are of the following form:

$$\dot{y} = \frac{\partial H}{\partial p}(y,p,u), \quad y(0) = x, \tag{3.9}$$

$$\dot{p} = -\frac{\partial H}{\partial x}(y,p,u), \quad p(T) = G_x(y(T)), \tag{3.10}$$

and condition (3.3) is equivalent to

$$\max_{u \in U} H(\hat{y}, \hat{p}, u) = H(\hat{y}, \hat{p}, \hat{u}) \tag{3.11}$$

at appropriate points $t \in (0, T)$.

§3.2. An application of the maximum principle

We will illustrate the maximum principle by applying it to the regulator problem of §1.3.

Assume that $\hat{u}$ is an optimal control for the linear regulator problem and $\hat{y}$ the corresponding output. Since $f(x,u) = Ax + Bu$, $g(x,u) = \langle Qx, x\rangle + \langle Ru, u\rangle$, $G(x) = \langle P_0 x, x\rangle$, $x \in \mathbf{R}^n$, $u \in \mathbf{R}^m$, equation (3.4) and the end condition (3.5) are of the form

$$\dot{p} = -A^*p - 2Q\hat{y} \quad \text{na } [0,T], \tag{3.12}$$

$$p(T) = 2P_0 \hat{y}(T). \tag{3.13}$$

In the considered case we minimize the cost functional (3.2) rather than, maximize, and, therefore, inequality (3.3) has to be reversed, and the maximization operation should be changed to the minimization one. Hence for almost all $t \in [0, T]$ we have

$$\min_{u \in \mathbf{R}^m} [\langle p(t), A\hat{y}(t) + Bu(t)\rangle + \langle Q\hat{y}(t), \hat{y}(t)\rangle + \langle Ru, u\rangle]$$

$$= \langle p(t), A\hat{y}(t) + B\hat{u}(t)\rangle + \langle Q\hat{y}(t), \hat{y}(t)\rangle + \langle R\hat{u}(t), \hat{u}(t)\rangle.$$

It follows from Lemma 1.2 that the minimum in the above expression is attained at exactly one point $-\frac{1}{2}R^{-1}B^*p(t)$. Consequently,

$$\hat{u}(t) = -\frac{1}{2}R^{-1}B^*p(t) \quad \text{for almost all } t \in [0,T].$$

Taking into account Theorem 3.1 we see that the optimal trajectory $\hat{y}$ and the function p satisfy the following system of equations:

$$\dot{\hat{y}} = A\hat{y} - \frac{1}{2}BR^{-1}B^*p, \quad \hat{y}(0) = x, \tag{3.14}$$

$$\dot{p} = -2Q\hat{y} - A^*p, \quad p(T) = 2P_0 \hat{y}(T). \tag{3.15}$$

System (3.14)-(3.15) has a unique solution and therefore determines the optimal trajectory. The exact form of the solution to (3.14)-(3.15) is given by

Lemma 3.1. *Let $P(t)$, $t \geq 0$, be the solution to the Riccati equation (3.18). Then*

$$p(t) = 2P(T-t)\hat{y}(t), \tag{3.16}$$

$$\dot{\hat{y}}(t) = (A - BR^{-1}B^*P(T-t))\hat{y}(t), \quad t \in [0,T]. \tag{3.17}$$

Proof. We show that (3.14) and (3.15) follow from (3.16) and (3.17). One obtains (3.14) by substituting (3.16) into (3.17). Moreover, taking account of the form of the Riccati equation we have

$$\dot{p}(t) = -2\dot{P}(T-t)\hat{y}(t) + 2P(T-t)\dot{\hat{y}}(t)$$
$$= -2(A^*P(T-t) + P(T-t)A + Q - P(T-t)BR^{-1}B^*P(T-t))\hat{y}(t)$$
$$+ 2P(T-t)(A - BR^{-1}B^*P(T-t))\hat{y}(t) = -A^*p(t) - 2Q\hat{y}(t)$$

for almost all $t \in [0,T]$. □

Remark. It follows from the maximum principle that $\hat{u} = -\frac{1}{2}R^{-1}B^*p$. But, by (3.16),

$$\hat{u}(t) = -R^{-1}B^*P(T-t)\hat{y}(t), \quad t \in [0,T].$$

This way we arrive at the optimal control in the feedback form derived earlier using the dynamic programming approach (see Theorem 1.3).

Exercise 3.1. We will solve system (3.14)-(3.15) in the case $n = m = 1$, $A = a \in \mathbf{R}$, $B = 1$, $Q = R = 1$, $P_0 = 0$. Then

$$\dot{\hat{y}} = a\hat{y} - \frac{1}{2}p, \quad x(0) = x \in \mathbf{R},$$
$$\dot{p} = -2\hat{y} - ap, \quad p(T) = 0.$$

Let us remark that the matrix $\begin{bmatrix} a & -\frac{1}{2} \\ -2 & -a \end{bmatrix}$ has two eigenvectors

$$\begin{bmatrix} 1 \\ 2(a - \sqrt{1+a^2}) \end{bmatrix}, \quad \begin{bmatrix} 1 \\ 2(a + \sqrt{1+a^2}) \end{bmatrix},$$

corresponding to the eigenvalues $\sqrt{1+a^2}$ and $-\sqrt{1+a^2}$. Therefore, for some constants ξ_1, ξ_2, $t \in [0,T]$

$$\begin{bmatrix} \hat{y}(t) \\ p(t) \end{bmatrix} = e^{t\sqrt{1+a^2}} \begin{bmatrix} 1 \\ 2(a - \sqrt{1+a^2}) \end{bmatrix} \xi_1 + e^{-t\sqrt{1+a^2}} \begin{bmatrix} 1 \\ 2(a + \sqrt{1+a^2}) \end{bmatrix} \xi_2.$$

Since $p(T) = 0$ and $\hat{y}(0) = x$,

$$\xi_1 = -e^{-2T\sqrt{1+a^2}} \frac{a + \sqrt{1+a^2}}{a - \sqrt{1+a^2}} \xi_2,$$

$$\xi_2 = \left(1 - e^{-2T\sqrt{1+a^2}} \frac{a + \sqrt{1+a^2}}{a - \sqrt{1+a^2}}\right)^{-1} x.$$

From the obtained formulae and the relation $\hat{u} = -\frac{1}{2}p$,

$$\hat{u}(t) = \frac{\hat{u}(t)}{\hat{y}(t)} \hat{y}(t) = -\frac{p(t)}{2\hat{y}(t)} \hat{y}(t)$$

$$= \frac{e^{2(T-t)\sqrt{1+a^2}} - 1}{(a + \sqrt{1+a^2}) e^{2(T-t)\sqrt{1+a^2}} + \sqrt{1+a^2} - a} \hat{y}(t), \quad t \in [0, T].$$

On the other hand, by Theorem 1.3, $\hat{u}(t) = -P(T-t)\hat{y}(t), t \in [0, T]$, where

$$\dot{P} = 1 + 2aP - P^2, \quad P(0) = 0.$$

Hence we arrive at the well known formula

$$P(t) = \frac{e^{2t\sqrt{1+a^2}} - 1}{(a + \sqrt{1+a^2}) e^{2t\sqrt{1+a^2}} + \sqrt{1+a^2} - a}, \quad t \geq 0,$$

for the solution of the above Riccati equation.

§ 3.3. The maximum principle for impulse control problems

We show now how the idea of the needle variation can be applied to impulse control problems. We assume, as in § 2.1, that the evolution of a control system between impulses is described by the equation

$$\dot{z} = f(z), \quad z(0) = x \in \mathbf{R}^n, \tag{3.18}$$

the solution of which will be denoted by $z(t, x)$, $t \in \mathbf{R}$.

It is convenient to assume that there exists a set U and a function $\varphi \colon \mathbf{R}^n \times U \longrightarrow \mathbf{R}^n$ such that $x + \varphi(x, U) = \Gamma(x)$, $x \in \mathbf{R}^n$, and to identify a shift (impulse) from x to $y \in \Gamma(x)$ with a choice of $u \in U$ such that $y = \varphi(x, u)$. An admissible impulse strategy on $I = [0, T]$ is then completely determined by specifying a sequence of moments $(t_1, \ldots, t_m)$, $0 \leq t_1 < t_2 < \ldots < t_m < T$ and a sequence $(u_1, \ldots, u_m)$ of elements in U, $m = 1, 2, \ldots$. We will denote it by $\pi(t_1, \ldots, t_m; u_1, \ldots, u_m)$ and the corresponding output trajectory starting from x by $y(t) = y^{\pi,x}(t)$, $t \in I$. Thus on each of the intervals $I_0 = [t_0, t_1]$, $I_k = (t_k, t_{k+1}]$, $k = 1, 2, \ldots, m$, $t_0 = 0$, $t_{m+1} = T$ the function $y(\cdot)$ satisfies equation (3.18) and boundary conditions

$$y(t_k^+) = y(t_k) + \varphi(y(t_k), u_k), \quad y(0) = x, \quad k = 1, 2, \ldots, m.$$

To simplify formulations, in contrast to the definition in § 2.1, we assume that output trajectories are left (not right) continuous.

The theorem below formulates necessary conditions which a strategy maximizing a *performance functional*

$$J_T(x, \pi) = G(y^{\pi,x}(T)) \tag{3.19}$$

has to satisfy. The case of a more general functional

$$J_T(x, \pi) = \int_0^T e^{-\alpha t} g(y^{\pi,x}(t))\, dt \tag{3.20}$$
$$+ \sum_{j=1}^m e^{-\alpha t_j} c(y^{\pi,x}(t_j), u_j) + e^{-\alpha T} G(y^{\pi,x}(T))$$

can be sometimes reduced to the one given by (3.19) (see comments after Theorem 3.2). To be in agreement with the setting of § 3.1 we do maximize rather than minimize the performance functional.

Theorem 3.2. *Assume that mappings $f(\cdot)$ and $\varphi(\cdot, u)$, $u \in U$, and the function $G(\cdot)$ are of class C^1 and that for arbitrary $x \in \mathbf{R}^n$ the set $\varphi(x, U)$ is convex. Let $\hat{\pi} = \hat{\pi}((\hat{t}_1, \ldots, \hat{t}_m), (\hat{u}_1, \ldots, \hat{u}_m))$ be a strategy maximizing (3.19) and $\hat{y}(\cdot)$ the corresponding output trajectory. Then, for arbitrary $k = 1, 2, \ldots, m$,*

$$\langle p(\hat{t}_k^+), \varphi(\hat{y}(\hat{t}_k), \hat{u}_k)\rangle \geq \max_{u \in U}\langle p(\hat{t}_k^+), \varphi(\hat{y}(\hat{t}_k), u)\rangle, \tag{3.21}$$

$$\langle p(\hat{t}_k^+), f(\hat{y}(\hat{t}_k^+))\rangle = \langle p(\hat{t}_k), f(\hat{y}(\hat{t}_k))\rangle, \quad \text{if } t_k > 0. \tag{3.22}$$

If $\hat{t}_k = 0$ then $=$ in (3.22) has to be replaced by $\geq$.

The function $p(\cdot): I \longrightarrow \mathbf{R}^n$ in (3.21) and (3.22) is left continuous and satisfies the equation

$$\dot{p} = -f_x^*(\hat{y})p, \tag{3.23}$$

in all intervals $\hat{I}_0 = [0, \hat{t}_1]$, $\hat{I}_k = (\hat{t}_k, \hat{t}_{k+1}]$, $k = 1, 2, \ldots, m$, with the end conditions

$$p(T) = G_x(\hat{y}(T)), \tag{3.24}$$

$$p(\hat{t}_k) = p(\hat{t}_k^+) + \varphi_x^*(\hat{y}(t_k), \hat{u}_k)p(\hat{t}_k^+), \quad k = 1, 2, \ldots, m. \tag{3.25}$$

Proof. We start from the following lemma which is a direct consequence of Theorem II.1.4 and Theorem I.1.1. The latter also imply that the derivative $z_x(t, x)$ exists for arbitrary $t \in \mathbf{R}$ and $x \in \mathbf{R}^n$.

§ 3.3. The maximum principle for impulse control problems

Lemma 3.2. Let $\Phi(t,s) = z_x(t-s, \hat{y}(s))$, $t,s \in [0,T]$. Then

$$p(s) = \Phi^*(\hat{t}_{k+1}, s)p(\hat{t}_{k+1}), \quad s \in \hat{I}_k,$$
$$p(\hat{t}_k^+) = \Phi^*(\hat{t}_{k+1}, \hat{t}_k^+)p(\hat{t}_{k+1}), \quad k = 0, 1, \ldots, m.$$

To prove the theorem we fix $u \in U$, a natural number $k \le m$ and a number $\varepsilon > 0$ such that $\hat{t}_{k-1} + \varepsilon < \hat{t}_k < \hat{t}_{k+1} - \varepsilon$. It follows from the convexity of the set $\varphi(\hat{y}(\hat{t}_k), U)$ that for an arbitrary number $\beta \in [0,1]$ there exists $u(\beta) \in U$ such that

$$\varphi(\hat{y}(\hat{t}_k), \hat{u}_k) + \beta(\varphi(\hat{y}(\hat{t}_k), u) - \varphi(\hat{y}(\hat{t}_k), \hat{u}_k)) = \varphi(\hat{y}(\hat{t}_k), u(\beta)).$$

We can assume that $u(0) = \hat{u}_k$ and $u(1) = u$. For arbitrary $\alpha \in (\hat{t}_k - \varepsilon, \hat{t}_k + \varepsilon)$ and $\beta \in [0,1]$ we define new strategies

$$\pi^{k-1}(\alpha, \beta), \pi^k(\alpha, \beta), \ldots, \pi^m(\alpha, \beta)$$

as follows:

$$\pi^{k-1}(\alpha, \beta) = \pi((\hat{t}_1, \ldots, \hat{t}_{k-1}), (\hat{u}_1, \ldots, \hat{u}_{k-1})),$$
$$\pi^k(\alpha, \beta) = \pi((\hat{t}_1, \ldots, \hat{t}_{k-1}, \alpha), (\hat{u}_1, \ldots, \hat{u}_{k-1}, u(\beta))),$$
$$\pi^j(\alpha, \beta) = \pi((\hat{t}_1, \ldots, \hat{t}_{k-1}, \alpha, \hat{t}_{k+1}, \ldots, \hat{t}_j),$$
$$(\hat{u}_1, \ldots, \hat{u}_{k-1}, u(\beta), \hat{u}_{k+1}, \ldots, \hat{u}_j)), \quad j = k+1, \ldots, m.$$

Denote

$$y^j(t; \alpha, \beta) = y^{\pi^j(\alpha, \beta), x}(t), \quad t \in [0,T], \ j = k-1, \ldots, m.$$

Then

$$\hat{y}(t) = y^m(t; \hat{t}_k, 0), \quad t \in [0,T].$$

For arbitrary $\alpha \in (\hat{t}_k - \varepsilon, \hat{t}_k + \varepsilon)$ and $\beta \in [0,1]$

$$\omega(\alpha, \beta)) = G(y^m(T; \alpha, \beta)) \le G(y^m(T; \hat{t}_k, 0)) \le \omega(\hat{t}_k, 0).$$

Hence

$$\frac{\partial \omega}{\partial \alpha}(\hat{t}_k, 0) = 0, \quad \text{if } \hat{t}_k > 0,$$
$$\frac{\partial \omega}{\partial \alpha}(\hat{t}_k, 0) \le 0, \quad \text{if } \hat{t}_k = 0, \tag{3.26}$$

$$\frac{\partial \omega}{\partial \beta}(\hat{t}_k, 0) \le 0, \tag{3.27}$$

under the condition that the function $\omega(\cdot, \cdot)$ has partial derivatives at $(\hat{t}_k, 0)$.

3. The maximum principle

The essence of the proof consists in showing that the partial derivatives exist and in calculating their values; it turns out that (3.26) and (3.27) imply (3.22) and (3.21) respectively.

To prove (3.21) define

$$\chi(\beta) = \omega(\hat{t}_k, \beta) = G(y^m(T; \hat{t}_k, \beta)), \quad \beta \in [0,1],$$

and note that for $\beta \in [0,1]$

$$\frac{\partial \chi}{\partial \beta}(\beta) = G_x(y^m(T; \hat{t}_k, \beta))y_\beta^m(T; \hat{t}_k, \beta)$$

$$= G_x(y^m(T; \hat{t}_k, \beta))z_x(T - \hat{t}_m, \delta_1(\beta))\frac{\partial}{\partial \beta}\delta_1(\beta),$$

where $\delta_1(\beta) = y^{m-1}(\hat{t}_m; \hat{t}_k, \beta) + \varphi(y^{m-1}(\hat{t}_m; \hat{t}_k, \beta), \hat{u}_m)$. Thus

$$\frac{\partial \omega}{\partial \beta}(\hat{t}_k, 0) = \chi_\beta(0) = \langle \Phi^*(T, \hat{t}_m^+)p(T), y_\beta^{m-1}(\hat{t}_m; \hat{t}_k, 0)$$
$$+ \varphi_x(y^{m-1}(\hat{t}_m; \hat{t}_k, 0), \hat{u}_m)y_\beta^{m-1}(\hat{t}_m; \hat{t}_k, 0)\rangle.$$

It follows from Lemma 3.2 and condition (3.25) that

$$\Phi^*(T, \hat{t}_m^+)p(T) = p(\hat{t}_m^+),$$

$$\varphi_x(y^{m-1}(\hat{t}_m; \hat{t}_k, 0), \hat{u}_m)p(\hat{t}_m^+) = p(\hat{t}_m) - p(\hat{t}_m^+).$$

Therefore

$$\chi_\beta(0) = \langle p(\hat{t}_m), y_\beta^{m-1}(\hat{t}_m; \hat{t}_k, 0)\rangle.$$

By induction, we obtain

$$\chi_\beta(0) = \langle p(\hat{t}_{k+1}), y_\beta^k(\hat{t}_{k+1}; \hat{t}_k, 0)\rangle. \tag{3.28}$$

Moreover

$$y^k(\hat{t}_{k+1}; \hat{t}_k, \beta) = z(\hat{t}_{k+1} - \hat{t}_k; \delta_k(\beta)), \tag{3.29}$$

where

$$\delta_k(\beta) = y^{k-1}(\hat{t}_k; \hat{t}_k, \beta) + \varphi(y^{k-1}(\hat{t}_k; \hat{t}_k, \beta), u(\beta)) \tag{3.30}$$
$$= \hat{y}(\hat{t}_k) + \varphi(\hat{y}(\hat{t}_k), \hat{u}_k) + \beta(\varphi(\hat{y}(\hat{t}_k), u) - \varphi(\hat{y}(\hat{t}_k), \hat{u}_k)).$$

Taking into account (3.28), (3.29) and (3.30) we finally see that

$$\chi_\beta(0) = \langle \Phi^*(\hat{t}_{k+1}, \hat{t}_k^+)p(\hat{t}_{k+1}), \varphi(\hat{y}(\hat{t}_k), u) - \varphi(\hat{y}(\hat{t}_k), \hat{u}_k)\rangle$$
$$= \langle p(\hat{t}_k^+), \varphi(\hat{y}(\hat{t}_k), u) - \varphi(\hat{y}(\hat{t}_k), \hat{u}_k)\rangle.$$

§3.3. The maximum principle for impulse control problems

Hence (3.21) follows.

To prove (3.22) define
$$\psi(\alpha) = \omega(\alpha, 0) = G(y^m(T; \alpha, 0)), \quad \alpha \in (\hat{t}_k - \varepsilon, \hat{t}_k + \varepsilon).$$

Then
$$\frac{\partial \psi}{\partial \alpha}(\alpha) = G_x(y^1(T; \alpha, 0)) \frac{\partial y^m}{\partial \alpha}(T; \alpha, 0)$$
$$= G_x(y^1(T; \alpha, 0)) z_x(T - \hat{t}_m; \gamma_1(\alpha)) \frac{\partial \gamma_1}{\partial \alpha}(\alpha),$$

where
$$\gamma_1(\alpha) = y^{m-1}(\hat{t}_m; \alpha, 0) + \varphi(y^{m-1}(\hat{t}_m; \alpha, 0)).$$

Consequently, by Lemma 3.2 and formula (3.25),
$$\frac{\partial \psi}{\partial \alpha} = \langle p(\hat{t}_m^+), y_\alpha^{m-1}(\hat{t}_m; \hat{t}_k, 0)\rangle + \langle p(\hat{t}_m) - p(\hat{t}_m^+), y_\alpha^{m-1}(\hat{t}_m; \hat{t}_k, 0)\rangle$$
$$= \langle p(\hat{t}_m), y_\alpha^{m-1}(\hat{t}_m; \hat{t}_k, 0)\rangle,$$

and, again by induction,
$$\frac{\partial \psi}{\partial \alpha}(t_k) = \langle p(\hat{t}_{k+1}), y_\alpha^k(\hat{t}_{k+1}; \hat{t}_k, 0)\rangle.$$

Moreover
$$y^k(\hat{t}_{k+1}; \alpha, 0) = z(\hat{t}_{k+1} - \alpha; \gamma_k(\alpha)),$$

where
$$\gamma_k(\alpha) = y^{k-1}(\alpha; \alpha, 0) + \varphi(y^{k-1}(\alpha; \alpha, 0), \hat{u}_k).$$

Hence
$$\frac{\partial \psi}{\partial \alpha}(t_k) = -\langle p(\hat{t}_{k+1}), f(\hat{y}(\hat{t}_{k+1}))\rangle + \langle p(\hat{t}_k^+), f(\hat{y}(\hat{t}_k))\rangle$$
$$+ \langle p(\hat{t}_k^+), \varphi_x(\hat{y}(\hat{t}_k), \hat{u}_k) f(\hat{y}(\hat{t}_k))\rangle.$$

Taking into account equation (3.25), we arrive at
$$\frac{\partial \psi}{\partial \alpha}(t_k) = -\langle p(\hat{t}_{k+1}), f(\hat{y}(\hat{t}_{k+1}))\rangle + \langle p(\hat{t}_k), f(\hat{y}(\hat{t}_k))\rangle.$$

On the other hand, the function $\langle p(\cdot), f(\hat{y}(\cdot))\rangle$ is constant on $(\hat{t}_k, \hat{t}_{k+1}]$ and therefore
$$\langle p(\hat{t}_k^+), f(y(\hat{t}_k^+))\rangle = \langle p(\hat{t}_{k+1}), f(\hat{y}(\hat{t}_{k+1}))\rangle.$$

In addition,
$$\frac{\partial \psi}{\partial \alpha}(t_k) = \langle p(\hat{t}_k), f(\hat{y}(\hat{t}_k))\rangle - \langle p(\hat{t}_k^+), f(\hat{y}(\hat{t}_k^+))\rangle,$$

and by (3.26) we finally obtain (3.22). $\square$

Remark. Theorem 3.2 can be reformulated in terms of appropriately defined *Hamiltonians*

$$H_1(x,p) = \langle p, f(x) \rangle, \quad H_2(x,p,u) = \langle p, \varphi(x,u) \rangle, \quad x,p \in \mathbf{R}^n, \ u \in U.$$

The conjugate equation (3.23) is of the form

$$\dot{p} = -H_{1x}^*(p,\hat{y}),$$

and condition (3.21) expresses the fact that the function $H_2(p(\hat{t}_k^+), \hat{y}(\hat{t}_k), \cdot)$ attains its maximal value at $\hat{u}_k$.

Remark. To cover the case of general functional (3.20) one has to replace the stated space $\mathbf{R}^n$ by $\mathbf{R}^{n+2}$ and define

$$\tilde{f}\begin{bmatrix} x \\ x_{n+1} \\ x_{n+2} \end{bmatrix} = \begin{bmatrix} f(x) \\ 1 \\ e^{-\alpha x_{n+1}} g(x) \end{bmatrix}, \quad \tilde{\varphi}\left(\begin{bmatrix} x \\ x_{n+1} \\ x_{n+2} \end{bmatrix}, u \right) = \begin{bmatrix} \varphi(x,u) \\ 0 \\ c(x,u) \end{bmatrix},$$

$$\tilde{G}\begin{bmatrix} x \\ x_{n+1} \\ x_{n+2} \end{bmatrix} = x_{n+2} + e^{-\alpha x_{n+1}} G(x), \quad x \in \mathbf{R}^n, \ x_{n+1}, x_{n+2} \in \mathbf{R}. \quad (3.31)$$

Formulating conditions from Theorem 3.2 for new functions $\tilde{f}$, $\tilde{\varphi}$ and $\tilde{G}$ we obtain the required maximum principle. The most stringent assumption here is the convexity of $\tilde{\varphi}(x,U)$. It can, however, be omitted, see [48].

§3.4. Separation theorems

An important role in the following section and in Part IV will be played by separation theorems for convex sets.

Assume that E is a Banach space with the norm $\|\cdot\|$. The space of all linear and continuous functionals φ on E with the norm $\|\varphi\|_* = \sup\{|\varphi(x)|: \|x\| \leq 1\}$ will be denoted by E^*.

Theorem 3.3. *Let K and L be disjoint convex subsets of E. If the interior of K is nonempty then there exists a functional $\varphi \in E^*$ different from zero such that*

$$\varphi(x) \leq \varphi(y) \quad \text{for } x \in K, \ y \in L. \quad (3.32)$$

Proof. Let x_0 be an interior point of K and y_0 an arbitrary point of L. Then the set $M = K - L + y_0 - x_0 = \{x - y + y_0 - x_0; \ x \in K, \ y \in L\}$ is convex and 0 is its interior point. Define

$$p(x) = \inf\{t > 0; \ \frac{x}{t} \in M\}. \quad (3.33)$$

§ 3.4. Separation theorems

It is easy to check that

$$p(x+y) \leq p(x) + p(y), \quad p(\alpha x) = \alpha p(x), \quad x,y \in E, \ \alpha \geq 0, \quad (3.34)$$
$$p(x) \leq c\|x\| \quad \text{for some } c > 0 \text{ and all } x \in E. \quad (3.35)$$

Define on the one dimensional linear space $\{\alpha z_0; \ \alpha \in \mathbf{R}\}$, where $z_0 = y_0 - x_0$, a functional φ_0 by setting $\varphi(\alpha z_0) = \alpha$, $\alpha \in \mathbf{R}$. Since $z_0 \notin M$, for all $\alpha > 0$

$$p(\alpha z_0) = \inf\{t > 0; \ \frac{\alpha}{t} z_0 \in M\} \geq \alpha = \varphi_0(\alpha z_0).$$

If $\alpha \leq 0$ then

$$\varphi_0(\alpha z_0) \leq 0 \leq p(\alpha z_0).$$

By Theorem A.4 (the Hahn–Banach theorem), there exists a linear functional φ such that

$$\varphi(x) \leq p(x), \quad x \in E. \quad (3.36)$$

It follows from (3.35) and (3.36) that $\varphi \in E^*$. Moreover, for $x \in M$

$$\varphi(x) \leq p(x) = \inf\{t > 0; \ \frac{x}{t} \in M\} \leq 1. \quad (3.37)$$

Hence from (3.37) and the identity $\varphi(z_0) = 1$, for $x \in K$ and $y \in L$,

$$\varphi(x - y + y_0 - x_0) = \varphi(x-y) + \varphi(z_0) \leq 1 \leq \varphi(z_0),$$

(3.32) follows. $\square$

If (3.32) holds we say that the functional $\varphi \in E^*$ separates the sets K and L.

Theorem 3.4. *Let $M \subset E$ be a convex set.*

(i) If the interior $\operatorname{Int} M$ of M is nonempty and $x_0 \notin M$ then there exists a functional $\varphi \in E^*$, $\varphi \neq 0$, such that

$$\varphi(x) \leq \varphi(x_0), \quad x \in M.$$

(ii) If a point $x_0 \in E$ does not belong to the closure of M then there exist a functional $\varphi \in E^*$, $\varphi \neq 0$, and a number $\delta > 0$ such that

$$\varphi(x) + \delta \leq \varphi(x_0), \quad x \in M.$$

Proof. (i) Define $K = \operatorname{Int} M$ and $L = \{x_0\}$. By Theorem 3.3 there exists $\varphi \in E^*$, $\varphi \neq 0$, such that

$$\varphi(x) \leq \varphi(x_0) \quad \text{for } x \in \operatorname{Int} M. \quad (3.38)$$

Since the set M is contained in the closure of Int M, inequality (3.38) holds for all $x \in M$.

(ii) For some $r > 0$ the closed ball $K = \{y \in E; \|y-x_0\| \leq r\}$ is disjoint from M. Setting $L = M$ we have from Theorem 3.3 that for some $\varphi \in E^*$, $\varphi \neq 0$, (3.32) holds. Since $\varphi \neq 0$ and $m = \inf\{\varphi(y); \|y-x_0\| \leq r\} < \varphi(x_0)$, it is sufficient to define δ as an arbitrary positive number smaller than $\varphi(x_0) - m$. □

Theorem 3.5. *Let M be a convex subset of $E = \mathbf{R}^n$. If a point $x_0 \in \mathbf{R}^n$ is not in the interior of M then there exists a vector $p \in \mathbf{R}^n$, $p \neq 0$, such that*

$$\langle p, x \rangle \leq \langle p, x_0 \rangle, \quad x \in M.$$

Proof. If M has a nonempty interior then the result follows from Theorem 3.4(i). If the interior of M is empty then the smallest linear space E_0 containing all vectors $x - x_0$, $x \in M$, is different from E. Therefore there exists a linear functional φ, different from zero and such that $\varphi(y) = 0$ for $y \in E_0$. In particular, $\varphi(x) = \varphi(x_0)$ for $x \in M$. The functional φ can be represented in the form $\varphi(z) = \langle p, z \rangle$, $z \in \mathbf{R}^n$ for some $p \in \mathbf{R}^n$. □

§3.5. Time-optimal problems

There are important optimization problems in control theory which cannot be solved using the idea of the needle variation as these variations are not admissible controls. In such cases it is often helpful to reformulate the problem as a geometric question of finding a supporting hyperplane to a properly defined convex set. We will apply this approach to prove the maximum principle for the so-called time optimal problem.

Let us consider again a linear system

$$\dot{y} = Ay + Bu, \quad y(0) = x \tag{3.39}$$

and assume that the set $U \in \mathbf{R}^m$ is convex and compact. Let $\hat{x}$ be a given element of $\mathbf{R}^n$ different from x. We say that a control $u(\,\cdot\,) \colon [0, +\infty) \longrightarrow U$ transfers x to $\hat{x}$ at time $T > 0$ if for the corresponding solution $y^{x,u}(\,\cdot\,)$ of (3.39), $y^{x,u}(T) = \hat{x}$.

The *time-optimal problem* consists of finding a control which transfers x to $\hat{x}$ in the minimal time. The following theorem is the original formulation of the maximum principle and is due to L. Pontriagin and his collaborators.

Theorem 3.6. (i) *If there exists a control transferring x to $\hat{x}$ then the time-optimal problem has a solution.*

(ii) *If $\hat{u}(\,\cdot\,)$ is a control transferring x to $\hat{x}$ in the minimal time $\hat{T} > 0$ then there exists a nonzero vector $\lambda \in \mathbf{R}^n$ such that for the solution $p(\,\cdot\,)$ of the equation*

$$\dot{p} = -A^*p, \quad p(\hat{T}) = \lambda, \tag{3.40}$$

§3.5. Time-optimal problems

the identity

$$\langle B^*p(t), \hat{u}(t)\rangle = \max_{u \in U}\langle B^*p(t), u\rangle \qquad (3.41)$$

holds for almost all $t \in [0,\hat{T}]$.

Remark. For a vector $b \in \mathbf{R}^m$, $b \neq 0$, there exists exactly one $(n-1)$-dimensional hyperplane $L(b)$ orthogonal to b, supporting U and such that the set U is situated below $L(b)$, the orientation being determined by b. The maximum principle (3.41) says that hyperplane $L(B^*p(t))$ rolling over U touches the set U at the values $\hat{u}(t)$, $t \in [0,\hat{T}]$, of the optimal control. The adjoint equation (3.40) indicates that the rolling hyperplanes are of a special form. It is therefore clear that the maximum principle provides important information about optimal control.

Proof. (i) Let $S(t) = e^{At}$, $t \in \mathbf{R}$. The solution to (3.9) is of the form

$$y(t) = S(t)\left[x + \int_0^t S^{-1}(r)Bu(r)\,dr\right], \quad t \geq 0.$$

For arbitrary $t \geq 0$ define a set $R(t) \subset \mathbf{R}^n$:

$$R(t) = \left\{\int_0^t S^{-1}(r)Bu(r)\,dr\colon u(r) \in U,\ r \in [0,t]\right\}.$$

We will need several properties of the multivalued function $R(t)$, $t \geq 0$, formulated in the following lemma.

Lemma 3.3. (i) *For arbitrary $t \geq 0$ the set $R(t)$ is convex and compact.*

(ii) *If $x_m \longrightarrow a$ and $t_m \downarrow t$ as $m \uparrow +\infty$ and $x_m \in R(t_m)$ for $m = 1,2,\ldots$, then $a \in R(t)$.*

(iii) *If $a \in \operatorname{Int} R(t)$ for some $t > 0$ then $a \in R(s)$ for all $s < t$ sufficiently close to t.*

Proof. In the proof we need a classic lemma about weak convergence. We recall that a sequence (h_m) of elements from a Hilbert space H is *weakly convergent* to $h \in H$ if for arbitrary $x \in H$, $\lim_m \langle h_m, x\rangle_H = \langle h, x\rangle_H$, where $\langle \cdot, \cdot\rangle_H$ denotes the scalar product in H.

Lemma 3.4. *An arbitrary bounded sequence (h_m) of elements of a separable Hilbert space H contains a subsequence weakly convergent to an element of H.*

Proof of Lemma 3.4. Let us consider linear functionals

$$\varphi_m(x) = \langle h_m, x\rangle_H, \quad x \in H,\ m = 1,2,\ldots.$$

It easily follows from the separability of H that there exists a subsequence (φ_{m_k}) and a linear subspace $H_0 \subset H$, dense in H, such that the sequence

$(\varphi_{m_k}(x))$ is convergent for arbitrary $x \in H_0$ to its limit $\varphi(x)$, $x \in H_0$. Since the sequence (h_m) is bounded, there exists a constant $c > 0$ such that

$$|\varphi(x)| \leq c|x|_H, \quad x \in H_0.$$

Therefore the linear functional φ has a unique extension to a linear, continuous functional $\tilde{\varphi}$ on H. By Theorem A.7 (the Riesz representation theorem) there exists $h \in H$ such that $\tilde{\varphi} = \langle h, x \rangle_H$ $x \in H$. The element h is the weak limit of (h_{m_k}). □

Proof of Lemma 3.3. Denote by $U(t)$ the set of all inputs defined on $[0,t]$ with values in U. It is a bounded, convex and closed subset of the Hilbert space $H = L^2(0,t;\mathbf{R}^n)$. The linear transformation from H into $\mathbf{R}^n$

$$u(\,\cdot\,) \longrightarrow \int_0^t S^{-1}(r)Bu(r)\,dr$$

maps the convex set $U(t)$ onto a convex set. Hence the set $R(t)$ is convex. It also maps weakly convergent sequences onto weakly convergent ones, and, by Lemma 3.4, the set $R(t)$ is closed. Since $R(t)$ is obviously a bounded set, the proof of (i) is complete.

To prove (ii), remark that

$$x_m = \int_0^t S^{-1}(r)Bu_m(r)\,dr + \int_t^{t_m} S^{-1}(r)Bu_m(r)\,dr = x_m^1 + x_m^2$$

for some $u_m(\,\cdot\,) \in U(t_m)$, $m = 1,2,\ldots$. Since $x_m^1 \in R(t)$, $x_m^2 \longrightarrow 0$ and $R(t)$ is a closed set, $a \in R(t)$.

To show (iii), assume, to the contrary, that $a \in \mathrm{Int}\, R(t)$ but for sequences $t_m \uparrow t$ and $x_m \longrightarrow a$, $x_m \notin R(t_m)$. Since the set $R(t_m)$ is convex, by Theorem 3.5 there exists a vector $p_m \in \mathbf{R}$, $|p_m| = 1$ such that

$$\langle x - a_m, p_m \rangle \leq 0 \quad \text{for all } x \in R(t_m). \tag{3.42}$$

Without any loss of generality we can assume that $p_m \longrightarrow p$, $|p| = 1$. Letting m tend to infinity in (3.42), we obtain that $\langle x - a, p \rangle \leq 0$ for arbitrary x from the closure of $\bigcup_m R(t_m)$ and thus for all $x \in R(t)$. Since $a \in R(t)$, p has to be zero, a contradiction. □

We go back to the proof of Theorem 3.6. Let

$$z(t) = S^{-1}(t)\hat{x} - x, \quad t \geq 0,$$

and let $\hat{T}$ be the infimum of all $t > 0$ such that $z(t) \in R(t)$. There exists a decreasing sequence $t_m \downarrow \hat{T}$ such that $x_m = z(t_m) \in R(t_m)$, $m = 1,2,\ldots$.

§ 3.5. Time-optimal problems

By Lemma 3.3(ii), $z(\hat{T}) \in R(\hat{T})$, and therefore there exists an optimal solution to the problem.

To show (ii) define $a = z(\hat{T})$. Then $a \notin \text{Int } R(\hat{T})$. For, if $a \in \text{Int } R(\hat{T})$, then, by Lemma 3.3(iii), there exists a number $t < \hat{T}$ such that $z(t) \in R(t)$, a contradiction with the definition of $\hat{T}$. Consequently, by Theorem 3.5, there exists $\lambda \in \mathbf{R}^n$, $\lambda \neq 0$, such that

$$\langle x - z(\hat{T}), \lambda \rangle \leq 0, \quad x \in R(\hat{T}). \tag{3.43}$$

Let $\hat{u}(\cdot)$ be an optimal strategy and $u(\cdot)$ an arbitrary control taking values in U. It follows from (3.43) that

$$\langle \int_0^{\hat{T}} S^{-1}(r)(Bu(r) - B\hat{u}(r))\, dr, \lambda \rangle \leq 0,$$

or, equivalently,

$$\int_0^{\hat{T}} \langle B^*(S^*(r))^{-1}(r)\lambda, \hat{u}(r) - u(r) \rangle\, dr \geq 0. \tag{3.44}$$

Since the function $p(t) = (S^*(t))^{-1}\lambda$, $t \in [0,T]$, is a solution to (3.40), see § I.1.1, and (3.44) holds for arbitrary admissible strategy $u(\cdot)$, relation (3.41) holds. □

Example 3.1. Consider the system

$$\ddot{z} = u \tag{3.45}$$

with the initial conditions

$$z(0) = x_1, \quad \dot{z}(0) = x_2. \tag{3.46}$$

We will find a control $u(\cdot)$ satisfying the constrains $-1 \leq u(t) \leq 1$, $t \geq 0$, and transferring, in the minimal time, initial state (3.46) to

$$z(\hat{T}) = 0, \quad \dot{z}(\hat{T}) = 0.$$

Transforming (3.45)-(3.46) into form (3.39), we obtain

$$A = \begin{bmatrix} 0 & 1 \\ 0 & 0 \end{bmatrix}, \quad B = \begin{bmatrix} 0 \\ 1 \end{bmatrix}.$$

The initial and the final states are, respectively,

$$x = \begin{bmatrix} x_1 \\ x_2 \end{bmatrix}, \quad \hat{x} = \begin{bmatrix} 0 \\ 0 \end{bmatrix}$$

and the set of control parameters $U = [-1, 1]$.

Since the pair (A, B) is controllable, an arbitrary initial state x close to $\hat{x}$ can be transferred to $\hat{x}$ by controls with values in $[-1, 1]$, see Theorem I.4.1. In fact, as we will see soon, all states x can be transferred to $\hat{x}$ this way.

An arbitrary solution of the adjoint equation

$$\dot{p} = \begin{bmatrix} 0 & 0 \\ -1 & 0 \end{bmatrix} p$$

is of the form $p(t) = \begin{bmatrix} -a \\ at+b \end{bmatrix}$, $t \geq 0$, where a and b are some constants. Assume that $x \neq \hat{x} = 0$ and $\hat{u}$ is an optimal control transferring x to 0 in the minimal time $\hat{T}$. Then, by Theorem 3.6, there exist constants a and b, at least one different from zero, such that for almost all $t \in [0, \hat{T}]$:

$$\hat{u}(t) = \operatorname{sgn} B^* p(t) = \operatorname{sgn}(at + b) \qquad (3.47)$$

where sgn r is equal 1, 0 or -1 according to whether r is positive, zero or negative. Without any loss of generality we can assume that (3.47) holds for all $t \in [0, \hat{T}]$.

It follows from (3.47) that the optimal control changes its value at most once and therefore $\hat{u}$ should be of one of the following forms:

(i) $\hat{u}(t) = 1, \quad t \in [0, \hat{T}]$,
(ii) $\hat{u}(t) = -1, \quad t \in [0, \hat{T}]$,
(iii) $\hat{u}(t) = 1, \quad t \in [0, s], \quad \hat{u}(t) = -1, \quad t \in (-s, \hat{T}]$,
(iv) $\hat{u}(t) = -1, \quad t \in [0, s], \quad \hat{u}(t) = 1, \quad t \in (s, \hat{T}]$,

where s is a number from $(0, \hat{T})$.

Denote by Γ_+ and Γ_- sets of these states $x \in \mathbf{R}^2$ which can be transferred to $\hat{x}$ by the constant controls taking values 1 and -1 respectively. Solutions to the equations

$$\dot{y}^+ = \begin{bmatrix} 0 & 1 \\ 0 & 0 \end{bmatrix} y^+ + \begin{bmatrix} 0 \\ 1 \end{bmatrix}, \quad \dot{y}^- = \begin{bmatrix} 0 & 1 \\ 0 & 0 \end{bmatrix} y^- + \begin{bmatrix} 0 \\ -1 \end{bmatrix}, \quad y(0) = \begin{bmatrix} x_1 \\ x_2 \end{bmatrix}$$

are given by the formulae

$$y_1^+(t) = x_1 + x_2 t + \frac{t^2}{2}, \qquad y_2^+(t) = x_2 + t,$$

$$y_1^-(t) = x_1 + x_2 t - \frac{t^2}{2}, \qquad y_2^-(t) = x_2 - t, \qquad t \geq 0.$$

Hence

$$\Gamma_+ = \left\{ \begin{bmatrix} x_1 \\ x_2 \end{bmatrix}; \quad x_1 = \frac{x_2^2}{2}, \; x_2 \leq 0 \right\},$$

$$\Gamma_- = \left\{ \begin{bmatrix} x_1 \\ x_2 \end{bmatrix}; \quad x_1 = -\frac{x_2^2}{2}, \; x_2 \geq 0 \right\}.$$

Sets Γ_+ and Γ_- contain exactly those states which can be transferred to $\hat{x}$ by optimal strategies (i) and (ii), and the minimal time to reach 0 is exactly x_2. Taking into account the shapes of the solutions y^+ and y^-, it is easy to see that if the initial state x is situated below the curve $\Gamma_+ \cup \Gamma_-$, then the optimal strategy is of type (iii) with s being equal to the moment of hitting Γ_-. If the initial state is situated above $\Gamma_+ \cup \Gamma_-$, then the optimal strategy is of the type (iv). We see also that in the described way an arbitrary state x can be transferred to $x \in \mathbf{R}^2$.

Bibliographical notes

Numerous applications of the maximum principle can be found in [27], [36] and [45]. The proof of the maximum principle for more general control problems is given, for instance, in W.H. Fleming and R.W. Rishel [27]. A functional analytic approach is presented in I.W. Girsanov [29]. Nonlinear optimization problems can be discussed in the framework of nonsmooth analysis, see F.H. Clarke [14].

Theorem 3.2 is borrowed from A. Blaquiere [9], and its proof follows the paper by R. Rempala and the author [49]. Applications can be found in [9].

Chapter 4
The existence of optimal strategies

This chapter starts with an example of a simple control problem without an optimal solution showing a need for existence results. A proof of the classic Fillipov theorem on existence is also presented.

§4.1. A control problem without an optimal solution

An important role in control theory is played by results on existence of optimal strategies. In particular, all the formulations of the maximum principle in Chapter 3 required the existence of an optimal strategy. The need for existence theorems is also more practical as natural control problems may not have optimal solutions.

Let us consider a control system on $\mathbf{R}^1$

$$\dot{y} = u, \quad y(0) = 0 \qquad (4.1)$$

with the (two-element) control set $U = \{-1, 1\}$. It is easy to see that a strategy minimizing the quadratic cost functional

$$J_T^1(0, u) = \int_0^T (y^2 + u^2)\, dt \qquad (4.2)$$

exists and is either of the form $\hat{u}(t) = 1$, $t \in [0, T]$, or $\hat{u}(t) = -1$, $t \in [0, T]$.

It turns out however that for a similar functional

$$J_T^2(0, u) = \int_0^T (y^2 - u^2)\, dt,$$

the optimal strategy does not exist. To see that this is really the case, let us remark first that for an arbitrary admissible strategy $u(\,\cdot\,)$, $y^2 - u^2 \geq -1$, hence $J_T^2(0, u) \geq -T$. On the other hand, if we define strategies $u_m(\,\cdot\,)$, $m = 1, 2, \ldots$

$$u_m(t) = \begin{cases} 1, & \text{if } \frac{1}{T}\frac{i-1}{m} \leq t < \frac{1}{T}\frac{2j-1}{2m}, \\ -1, & \text{if } \frac{1}{T}\frac{2j-1}{2m} \leq t < \frac{1}{T}\frac{j}{m},\ j = 1, \ldots, m, \end{cases}$$

then for the corresponding solutions $y_m(\,\cdot\,)$ of (4.1),

$$0 \leq y_m(t) \leq \frac{1}{2Tm}, \quad t \in [0, T],$$

and therefore $J_T^2(0, u_m) \longrightarrow -T$ as $m \uparrow +\infty$. So the infimum of the functional $J_T^2(0, \cdot)$ is $-T$. If, on the other hand, for a control $u(\,\cdot\,)$: $J_T^2(0, u(\,\cdot\,)) = -T$, then, for almost all $t \in [0, T]$, $y^2(t) - u^2(t) = y^2(t) - 1 = -1$, and thus $y(t) = 0$, and, consequently, $u(t) = 0$, for almost all $t \in [0, T]$. But this strategy is not an admissible one, a contradiction.

Before going to the main result of the chapter let us remark that some existence results were formulated in connection with the dynamic programming in § 1.2 and § 1.3, with the impulse control in § 2.1 and for the time optimal problem in § 3.5.

§ 4.2. Fillipov's theorem

Let us consider a control system

$$\dot{y} = f(y, u), \quad y(0) = x \in \mathbf{R}^n \tag{4.3}$$

with the cost functional

$$J(x, u) = G(T, y(T)). \tag{4.4}$$

Let us assume in addition that given are a set $U \subset \mathbf{R}^m$ of control parameters and a set $K \subset \mathbf{R}^{n+1}$ of the final constraints. Instead of fixing $T > 0$ we will require that

$$(T, y(T)) \in K. \tag{4.5}$$

Let $C > 0$ be a positive number such that

$$|f(x, u)| \leq C(1 + |x| + |u|), \tag{4.6}$$

$$|f(x, u) - f(z, u)| \leq C|x - z|(1 + |u|) \tag{4.7}$$

for arbitrary $x, z \in \mathbf{R}^n$ and $u \in U$, see § II.1.1.

An *admissible control* is an arbitrary Borel measurable function $u(\,\cdot\,)$ with values in U defined on an interval $[0, T]$ depending on $u(\,\cdot\,)$ and such that the solution $y(\,\cdot\,)$ of (4.3) satisfies condition (4.5).

The following theorem holds.

Theorem 4.1. *Assume that sets U and K are compact and the set $f(x, U)$ is convex for arbitrary $x \in \mathbf{R}^n$. Let moreover the function f satisfy (4.6)-(4.7) and G be a continuous function on K. If there exists an admissible control for (4.3) and (4.5) then there also exists an admissible control minimizing (4.4).*

Proof. It follows from the boundedness of the set K that there exists a number $\widetilde{T} > 0$ such that admissible controls $u(\,\cdot\,)$ are defined on intervals $[0, T] \subset [0, \widetilde{T}]$. Let us extend an admissible control $u(\,\cdot\,)$ onto $[0, \widetilde{T}]$ by setting $u(t) = u^+$ for $t \in [T, \widetilde{T}]$, where u^+ is an arbitrary selected element from U. Denote by $\mathcal{K}$ the set of all solutions of (4.3), defined on $[0, \widetilde{T}]$ and corresponding to the extended controls. By (4.6), for $y \in \mathcal{K}$ and $t \in [0, \widetilde{T}]$,

$$\gamma(t) = |y(t)| = \left| x + \int_0^t f(y(s), u(s))\, ds \right| \leq |x| + C \int_0^t (1 + |\gamma(s)| + |u(s)|)\, ds.$$

Hence, taking into account Gronwall's Lemma II.2.1,

$$|y(t)| \leq e^{Ct} \left[|x| + Ct + \int_0^t |u(s)|\, ds \right], \quad t \in [0, \widetilde{T}].$$

Since the set U is bounded, for some constant $C_1 > 0$

$$|y(t)| \leq C_1 \quad \text{for } y \in \mathcal{K}, \ t \in [0, \widetilde{T}]. \tag{4.8}$$

Similarly, for $y(\,\cdot\,) \in \mathcal{K}$ and $[t, s] \subset [0, \widetilde{T}]$,

$$|y(t) - y(s)| \leq \int_t^s |f(y(r), u(r))|\, dr \leq C \int_t^s (1 + |y(r)| + |u(r)|)\, dr.$$

By the boundedness of U by (4.8), for an appropriate constant $C_2 > 0$

$$|y(t) - y(s)| \leq C_2 |t - s| \quad \text{for } y(\,\cdot\,) \in \mathcal{K},\ t, s \in [0, \widetilde{T}]. \tag{4.9}$$

Conditions (4.8) and (4.9) imply that $\mathcal{K}$ is a bounded subset of $C(0, \widetilde{T}; \mathbf{R}^n)$ and that its elements are equicontinuous. By Theorem A.8 (the Ascoli theorem), the closure of $\mathcal{K}$ is compact in $C(0, \widetilde{T}; \mathbf{R}^n)$.

Let $\hat{J}$ denote the minimal value of (4.4) and let (y_m) be a sequence of elements from $\mathcal{K}$ corresponding to controls (u_m) such that

$$J(x, u_m) = G(T_m, y_m(T_m)) \longrightarrow \hat{J}. \tag{4.10}$$

Since K and $\overline{\mathcal{K}}$ are compact and the function G is continuous, we can assume, without any loss of generality, that the sequence (y_m) is uniformly convergent on $[0, \widetilde{T}]$ to a continuous function $\hat{y}(\,\cdot\,)$ and $T_m \longrightarrow \hat{T}$. We will show that the function $\hat{y}(\,\cdot\,)$ is an optimal trajectory.

Since $\hat{y}(\,\cdot\,)$ satisfies the Lipschitz condition (4.9), it has finite derivatives $\frac{d\hat{y}}{dt}(t)$ for almost all $t \in [0, \widetilde{T}]$. Let t be a fixed number from $(0, \widetilde{T})$ for which the derivative exists. The continuity of $f(\cdot, \cdot)$ implies that for arbitrary $\varepsilon > 0$ there exists $\delta > 0$ such that if $|z - \hat{y}(t)| < \delta$ then the set $f(z, U)$ is contained in the δ-neighbourhood $f_\delta(\hat{y}(t), U)$ of the set $f(\hat{y}(t), U)$. By uniform convergence of the sequence (y_m), there exist $\eta > 0$ and $N > 0$

such that $|y_m(s) - \hat{y}(t)| < \delta$, for $s \in (t, t+\eta)$ and $m > N$. Let us remark that

$$\frac{y_m(s) - y_m(t)}{s - t} = \frac{1}{s-t} \int_t^s f(y_m(r), u_m(r))\, dr \qquad (4.11)$$

and that the integral $\frac{1}{s-t}\int_s^t f(y_m(r), u_m(r))\, dr$ is in the closure $\overline{f_\delta(\hat{y}(t), U)}$ of the convex set $f_\delta(\hat{y}(t), U)$. Letting m tend to $+\infty$ in (4.10) we obtain

$$\frac{\hat{y}(s) - \hat{y}(t)}{s-t} \in \overline{f_\delta(\hat{y}(t), U)},$$

and consequently

$$\frac{d\hat{y}}{dt}(t) \in \overline{f_\delta(\hat{y}(t), U)}.$$

Since $\delta > 0$ was arbitrary and the set $f(\hat{y}(t), U)$ is convex and closed,

$$\frac{d\hat{y}}{dt}(t) \in f(\hat{y}(t), U). \qquad (4.12)$$

This way we have shown that if for some $t \in (0, \widetilde{T})$ the derivative $\frac{d\hat{y}}{dt}(t)$ exists, then (4.12) holds. From this and the lemma below on measurable selectors, we will deduce that there exists a control $\hat{u}(\cdot)$ with values in U such that

$$\frac{d\hat{y}}{dt}(t) = f(\hat{y}(t), \hat{u}(t)) \quad \text{for almost all } t \in [0, \widetilde{T}]. \qquad (4.13)$$

The control $\hat{u}$ is an optimal strategy and $\hat{y}$ is an optimal output. For, from the convergence $T_m \longrightarrow \hat{T}$, $y_m \longrightarrow \hat{y}$ in $C(0, \widetilde{T}; \mathbf{R}^n))$ as $m \to +\infty$, the compactness of K and the continuity of G, it follows, see (4.10), that $\hat{J} = G(\hat{T}, \hat{y}(\hat{T}))$ and $(\hat{T}, \hat{y}(T)) \in K$.

We proceed now to the already mentioned lemma on measurable selectors.

Lemma 4.1. *Assume that D is a compact subset of $\mathbf{R}^p \times \mathbf{R}^m$ and let Δ be the projection of D on $\mathbf{R}^p$. There exists a Borel function $v: \Delta \longrightarrow \mathbf{R}^m$ such that*

$$(w, v(w)) \in D \quad \text{for arbitrary } w \in \Delta.$$

Proof. The set D is compact as an image of a compact set by a continuous transformation, and thus also Borel.

If $m = 1$ then we define

$$v(w) = \min\{z : (w, z) \in D\}, \quad w \in \Delta. \qquad (4.14)$$

The function v is lower-semicontinuous and thus measurable. Hence the lemma is true for $m = 1$.

If m is arbitrary then we can regard the set D is a subset of $\mathbf{R}^{p+m-1} \times \mathbf{R}^1$ and use the fact that the lemma is true if $m = 1$. This way we obtain that there exists a measurable function v_1 defined on the projection $D_1 \subset \mathbf{R}^{p+m-1}$ of the set D such that $(w_1, v_1(w_1)) \in D$ for arbitrary $w_1 \in \Delta_1$. So if we assume that the lemma is true for $m - 1$ and regard the set Δ_1 as a subset $\mathbf{R}^p \times \mathbf{R}^{m-1}$ we obtain that there exists a measurable function $v_2 \colon \Delta \longrightarrow \mathbf{R}^{m-1}$ such that

$$(w, v_2(w)) \in \Delta_1, \quad \text{for all } w \in \Delta.$$

The function

$$v(w) = (v_2(w), v_1(w, v_2(w))), \quad w \in \Delta,$$

has the required property. Hence the lemma follows by induction. $\square$

We will now complete the proof of Theorem 4.1. It follows from Theorem A.10 (Luzin's theorem) that there exists an increasing sequence (Δ_k) of compact subsets of $[0, \widetilde{T}]$ such that the Lebesgue measure of $\bigcup_k \Delta_k$ is $\widetilde{T}$ and, on each Δ_k, the function $d\hat{y}/dt$ is continuous. We check easily that sets

$$D_k = \left\{ (t, u) \in \Delta_k \times U \colon \frac{d\hat{y}(t)}{dt} = f(\hat{y}(t), u) \right\} \subseteq \hat{\mathbf{R}} \times \mathbf{R}^m \quad k = 1, 2, \ldots \quad (4.15)$$

are compact, and by Lemma 4.1 there exist Borel measurable functions $\hat{u}_k \colon \Delta_k \longrightarrow U$ such that

$$\frac{d\hat{y}}{dt}(t) = f(\hat{y}(t), \hat{u}_k(t)), \quad t \in \Delta_k, \ k = 1, 2, \ldots.$$

The required control $\hat{u}$ satisfying (4.13) can finally be defined by the formula

$$\hat{u}(t) = \begin{cases} \hat{u}_1(t) & \text{for } t \in \Delta_1, \\ \hat{u}_k(t) & \text{for } t \in \Delta_k \setminus \Delta_{k-1}, \ k = 2, \ldots. \end{cases}$$

The proof of Theorem 4.1 is complete. $\square$

We deduce from Theorem 4.1 several corollaries.

Theorem 4.2. *Assume that a function f and a set U satisfy the conditions of Theorem 4.1 and that for system (4.3) there exists a control transferring x to $\hat{x}$ in a finite time. Then there exists a control transferring x to $\hat{x}$ at the minimal time.*

Proof. Let us assume that there exists a control transferring x to $\hat{x}$ at time $\widetilde{T} > 0$ and define $K = [0, \widetilde{T}] \times \{\hat{x}\}$, $G(T, \hat{x}) = T$, $T \in [0, \widetilde{T}]$. The set K is compact and the function G continuous. Therefore all the assumptions of Theorem 4.1 are satisfied and there exists a strategy minimizing functional (4.4). It is clear that this is a strategy which transfers x to $\hat{x}$ at the minimal time. $\square$

In particular, assumptions of Theorem 4.2 are satisfied if system (4.3) is linear and the set U compact and convex. Hence the theorem implies the existence of an optimal solution to the time-optimal problem for linear systems, see Theorem 3.6.

As a corollary of Theorem 4.1 we also have the following result.

Theorem 4.3. *Assume that a function f and a set U satisfy the assumptions of Theorem 4.1 and that G is a continuous function on $\mathbf{R}^n$. Then, for arbitrary $T > 0$, there exists a strategy minimizing the functional*

$$J_T(x, u) = G(y(T)). \tag{4.16}$$

Proof. As in the proof of Theorem 4.1 we show that there exists $C > 0$ such that $|y(T)| \leq C$ for arbitrary output trajectory $y(\,\cdot\,)$. It is enough now to apply Theorem 4.1 with $K = \{T\} \times \{z \colon |z| \leq C\}$ and $G(T, z) = G(z)$, $|z| \leq C$. □

Remark. Theorem 4.3 is not true for more general functionals

$$J(x, u) = \int_0^T g(y(s), u(s))\,ds + G(y(T)), \tag{4.17}$$

as an example from § 4.1 shows. The method of introducing an additional variable, used at the end of the proof of Theorem II.3.1, cannot be applied here. This is because the new system is of the form

$$\dot{y} = f(y, u), \quad y(0) = x,$$
$$\dot{w} = g(y, u), \quad W(0) = 0,$$

and even if the sets $f(x, U)$, $x \in \mathbf{R}^n$, are convex subsets of $\mathbf{R}^n$, the set

$$\left\{ \begin{bmatrix} f(x, u) \\ g(x, u) \end{bmatrix} ; u \in U \right\}$$

is not, in general, a convex subset of $\mathbf{R}^{n+1}$.

Theorems on the existence of optimal controls for functionals (4.17) require more sophisticated methods than those used in the proof of Theorem 4.1.

Bibliographical notes

Theorem 4.1 is due to Fillipov, see [27]. In its proof we followed [27]. The proof contains essential ingredients – in a simple form – of stronger existence results which can be found in the book by I. Cesari [13] or in the paper by C. Olech [41].

PART IV

INFINITE DIMENSIONAL LINEAR SYSTEMS

Chapter 1
Linear control systems

This chapter starts with basic results on semigroups of linear operators on Banach spaces. Characterizations of the generators of the semigroups due to Hille and Yosida and to Lions are given. Abstract material is illustrated by self-adjoint and differential operators. The final section is devoted to nonhomogenous differential equations in Banach spaces which are the mathematical models of infinite dimensional control systems.

§ 1.1. Introduction

The control systems considered in the preceding chapters were described by ordinary differential equations and the stated space was $\mathbb{R}^n$. There is, however, a large number of systems which cannot be represented by a finite number of parameters. If we identify, for instance, the state of a heated bar with its temperature distribution, then the state is an element of an infinite dimensional function space. Parametrizing points of the bar by numbers from the interval $[0, L]$ and denoting by $y(t, \xi)$ the temperature at moment $t \geq 0$ and at point $\xi \in [0, L]$, we arrive at the parabolic equation (0.11) with boundary conditions (0.12)-(0.13), see Example 0.7.

The variety of situations described by partial differential equations is enormous and quite often control questions, similar to those considered in Parts I and III appear.

Control theory of infinite dimensional systems requires more sophisticated methods than those of finite dimension. The difficulties increase to the same extent as passing from ordinary differential equations to equations of parabolic and hyperbolic types. A complete theory exists only for a linear system. We will limit our discussion to several fundamental questions of linear theory and use the so-called semigroup approach.

§1.2. Semigroups of operators

Assume for a while that E and U are finite dimensional linear spaces and $A: E \to E$ and $B: U \to E$ are linear operators. Solutions to the linear differential equation

$$\dot{y} = Ay(t) + Bu(t), \quad y(0) = x \in E, \; t \geq 0, \tag{1.1}$$

are given, see § I.1.1, by the formula

$$y(t) = S(t)x + \int_0^t S(t-s)Bu(s)\,ds, \quad t \geq 0, \tag{1.2}$$

where

$$S(t) = e^{At}, \quad t \geq 0, \tag{1.3}$$

is the fundamental solution of the equation

$$\dot{z} = Az, \quad z(0) = x \in E. \tag{1.4}$$

The theory of linear semigroups extends the concept of fundamental solutions to arbitrary Banach spaces E and gives a precise meaning to (1.2) and (1.3) in more general situation. It turns out that a number of control systems modelled by partial differential equations can be treated as special cases of the general theory.

Uncontrolled systems (1.4) on Banach spaces are discussed in §§ 1.2-1.6 and basic properties of controlled systems are the object of § 1.7. The following considerations are valid for both real and complex Banach spaces. The norm on the Banach space E will be denoted by $\|\cdot\|$, and, if the state space is Hilbert, usually by $|\cdot|$.

Let us remark that a function $S(t)$, $t \geq 0$, with values in $\mathbf{M}(n,n)$ is the fundamental solution of the equation (1.4) if and only if it is a continuous solution of the operator equation

$$S(t+s) = S(t)S(s), \quad t, s \geq 0, S(0) = I. \tag{1.5}$$

This leads us to the following generalization.

Let E be a Banach space. A *semigroup of operators* is an arbitrary family of bounded linear operators $S(t): E \longrightarrow E$, $t \geq 0$, satisfying (1.5) and such that

$$\lim_{t \downarrow 0} S(t)x = x \quad \text{for arbitrary } x \in E. \tag{1.6}$$

Because of the continuity condition (1.6), the full name of the concept is C_0-*semigroups of operators*; we will however say, in short, semigroups of operators or even semigroups.

178   1. Linear control systems

In the case of finite dimensional spaces E the operator A in (1.3) and (1.4) is identical with the derivative of $S(\cdot)$ at 0:

$$\frac{dS}{dt}(0) = \lim_{h\downarrow 0} \frac{S(h) - I}{h} = A. \tag{1.7}$$

In the general case, the counterpart of A is called the *infinitesimal operator* or the *generator* of $S(t)$, $t \geq 0$. It is not, in general, an operator defined on the whole of E, but its domain $D(A)$ consists of all those $x \in E$ for which there exists the limit

$$\lim_{h\downarrow 0} \frac{S(h)x - x}{h}, \tag{1.8}$$

and the operator A itself is given by the formula

$$Ax = \lim_{h\downarrow 0} \frac{S(h)x - x}{h}, \quad x \in D(A). \tag{1.9}$$

Example 1.1. If A is a bounded linear operator on E then the family

$$S(t) = e^{tA} = \sum_{m=0}^{+\infty} \frac{t^m}{m!} A^m, \quad t \geq 0,$$

is a semigroup with generator A.

Example 1.2. Let H be a Hilbert space with a complete and orthonormal basis (e_m) and let (λ_m) be a sequence of real numbers diverging to $-\infty$. Define

$$S(t)x = \sum_{m=1}^{+\infty} e^{\lambda_m t} \langle x, e_m \rangle e_m, \quad x \in H,\ t \geq 0. \tag{1.10}$$

It is not difficult to prove that the family given by (1.10) is a semigroup. We will show later, see page 198, that the domain of its generator A is given by

$$D(A) = \left\{ x \in H;\ \sum_{m=1}^{+\infty} (\lambda_m \langle x, e_m \rangle)^2 < +\infty \right\},$$

and the generator itself by

$$Ax = \sum_{m=1}^{+\infty} \lambda_m \langle x, e_m \rangle, \quad x \in D(A).$$

Example 1.3. Let $H = L^2[0, \pi]$, $Ax = d^2x/d\xi^2$ and the domain $D(A)$ of the operator A consist of absolutely continuous functions $x(\cdot)$ defined on $[0, \pi]$ equal zero at 0 and π and such that the first derivative $dx/d\xi$ is absolutely continuous on $[0,\pi]$ and the second derivative $d^2x/d\xi^2$ belongs to H.

We will show later, see page 199, that the operator A defined this way is the generator of a semigroup given by (1.10) with $e_m(\xi) = \sqrt{\frac{2}{\pi}}\sin m\xi$, $\xi \in [0,\pi]$, $\lambda_m = -m^2$, $m = 1, 2, \ldots$. This semigroup is sometimes denoted by
$$S(t) = e^{td^2/d\xi^2}, \quad t \geq 0.$$
It is not difficult to guess that the formula
$$y(t) = S(t)x + \int_0^t S(t-s)bu(s)\,ds, \quad t \geq 0,$$
defines the solution to the parabolic equation (0.11)-(0.13) of Example 0.7 in which $\sigma^2 = 1$ and $L = \pi$.

Example 1.4. Let us consider the wave equation
$$\frac{\partial^2 y}{\partial t^2}(t,\xi) = \frac{\partial^2 y}{\partial \xi^2}(t,\xi), \quad t \in \mathbf{R},\ \xi \in [0,\pi],$$
with initial and boundary conditions
$$y(t,0) = y(t,\pi) = 0, \quad t \in \mathbf{R},$$
$$y(0,\xi) = a(\xi), \quad \frac{\partial y}{\partial t}(0,\xi) = b(\xi), \quad \xi \in (0,\pi).$$
We identify functions $a(\,\cdot\,)$ and $b(\,\cdot\,)$ with their Fourier expansions
$$a(\xi) = \sum_{m=1}^{+\infty} \alpha_m \sin m\xi, \tag{1.11}$$
$$b(\xi) = \sum_{m=1}^{+\infty} \beta_m \sin m\xi, \quad \xi \in (0,\pi). \tag{1.12}$$
If one assumes that $\sum_{m=1}^{+\infty}(m^2|\alpha_m| + m|\beta_m|) < +\infty$ then is it easy to check that the Fourier expansion of the solution $y(\cdot,\cdot)$ of the wave equation is of class C^2 with respect to both variables and
$$y(t,\xi) = \sum_{m=1}^{+\infty}\left(\alpha_m \cos mt + \frac{\beta_m}{m}\sin mt\right)\sin m\xi. \tag{1.13}$$
We have also
$$\frac{\partial y}{\partial t}(t,\xi) = \sum_{m=1}^{+\infty}(-m\alpha_m \sin mt + \beta_m \cos mt)\sin m\xi, \quad \xi \in (0,\pi). \tag{1.14}$$

Define E to be the set of all pairs $\begin{bmatrix} a \\ b \end{bmatrix}$ of functions with expansions (1.11) and (1.12) such that

$$\sum_{m=1}^{+\infty} m^2|\alpha_m|^2 + |\beta_m|^2 < +\infty.$$

This is a Hilbert space with the following scalar product:

$$\langle\!\langle \begin{bmatrix} a \\ b \end{bmatrix}, \begin{bmatrix} \tilde{a} \\ \tilde{b} \end{bmatrix} \rangle\!\rangle = \sum_{m=1}^{+\infty} \left(m^2 \alpha_m \tilde{\alpha}_m + \beta_m \tilde{\beta}_m \right).$$

Formulae (1.13) and (1.14) suggest that the semigroup of solutions to the wave equation should be defined as follows:

$$S(t) \begin{bmatrix} a \\ b \end{bmatrix} = \sum_{m=1}^{+\infty} \begin{bmatrix} \cos mt & \frac{\sin mt}{m} \\ -m \sin mt & \cos mt \end{bmatrix} \begin{bmatrix} \alpha_m \\ \beta_m \end{bmatrix} \sin m(\,\cdot\,), \quad t \geq 0.$$

We check directly that the above formula defines a semigroup of operators. The formula is meaningful for all $t \in \mathbf{R}$ and

$$S^*(t) = S^{-1}(t) = S(-t), \quad t \in \mathbf{R}.$$

Hence solutions of the wave equations define a continuous group of linear, unitary transformations of E into E (see page 246).

It follows from Examples 1.2 and 1.3 that infinitesimal generators are, in general, noncontinuous and not everywhere defined operators. Therefore a complete description of the class of all generators, due to Hille and Yosida, is rather complicated. We devote §1.3 to it and the rest of the present section will be on elementary properties of the semigroups.

In our considerations we will often integrate E-valued functions F defined on $[\alpha, \beta] \subset [0, +\infty)$. The integral $\int_\alpha^\beta F(s)\, ds$ should be understood in the Bochner sense, see §A.4; in particular, an arbitrary continuous function is Bochner integrable and its integral is equal to the limit of Riemann's sums

$$\int_\alpha^\beta F(s)\, ds = \lim_n \sum_{0 \leq k < m_n} F(s_{nk})(t_{n,k+1} - t_{nk}),$$

where $\alpha = t_{n0} < t_{n1} < \ldots < t_{nm_n} = \beta$, $s_{nk} \in [t_{nk}, t_{n,k+1}]$, $k = 0, 1, \ldots, m_n - 1$, $n = 0, 1, \ldots$, $\max(t_{n,k+1} - t_{nk};\ k = 0, 1, \ldots, m_n - 1) \longrightarrow 0$ if $n \uparrow +\infty$. Moreover, for an arbitrary Bochner integrable function f,

$$\|\int_\alpha^\beta F(s)\, ds\| \leq \int_a^b \|F(s)\|\, ds < +\infty.$$

§ 1.2. Semigroups of operators

Theorem 1.1. *Let $S(t)$, $t \geq 0$, be a semigroup of operators on a Banach space E. Then*

(i) *there exist constants $M > 0$ and ω such that*

$$\|S(t)\| \leq M e^{\omega t}, \quad t \geq 0;$$

(ii) *for arbitrary $x \in E$, the function $S(\cdot)x$, $x \in E$, is continuous on $[0, +\infty)$.*

Proof. (i) By Theorem A.5 (the Banach–Steinhaus theorem) and condition (1.6), for some numbers $M > 0$, $m > 0$ and arbitrary $t \in [0, m]$,

$$\|S(t)\| \leq M.$$

Let $t \geq 0$ be arbitrary and $k = 0, 1, \ldots$ and $s \in [0, m)$ such that $t = km + s$. Then

$$\|S(t)\| = \|S(km)S(s)\| \leq M M^k \leq M M^{t/m} \leq M e^{\omega t},$$

where $\omega = \frac{1}{m} \log M$.

(ii) For arbitrary $t \geq 0$, $h \geq 0$ and $x \in E$

$$\|S(t+h)x - S(t)x\| \leq M e^{\omega t} \|S(h)x - x\|.$$

If, in addition, $t - h \geq 0$, then

$$\|S(t-h)x - S(t)x\| \leq M e^{\omega(t-h)} \|S(h)x - x\|.$$

Hence the continuity of $S(\cdot)x$ at $t \geq 0$ follows from (1.6). □

Theorem 1.2. *Assume that an operator A with the domain $D(A)$ generates a semigroup $S(t)$, $t \geq 0$. Then for arbitrary $x \in D(A)$ and $t \geq 0$,*

(i) $\qquad\qquad\qquad S(t)x \in D(A),$

(ii) $\qquad\qquad\qquad \dfrac{d}{dt} S(t)x = A S(t)x = S(t)Ax,$

(iii) $\qquad\qquad\qquad S(t)x - x = \displaystyle\int_0^t S(r)Ax\, dr.$

Moreover,

(iv) *the domain $D(A)$ is dense in E and the operator A is closed (i.e., its graph $\{(x, A(x)); x \in D(A)\}$ is a closed subset of $E \times E$).*

Proof. For $h > 0$, $t > 0$

$$\frac{S(t+h)x - S(t)x}{h} = S(t)\frac{S(h)x - x}{h} = \frac{S(h) - I}{h} S(t)x. \tag{1.15}$$

If $x \in D(A)$, then
$$\lim_{h \downarrow 0} S(t) \frac{S(h)x - x}{h} = S(t)Ax.$$

Hence, $S(t)x \in D(A)$ and
$$\frac{d^+}{dt} S(t)x = S(t)Ax = AS(t)x.$$

If $h > 0$, $t > 0$ and $t - h > 0$ then
$$\frac{S(t-h)x - S(t)x}{h} = -S(t-h)\frac{S(h)x - x}{h}$$

and therefore
$$\frac{d^-}{dt} S(t)x = S(t)Ax.$$

This way the proofs of (i) and (ii) are complete.

To prove (iii) note that the right hand side of (iii) is well defined as continuous functions are Bochner integrable. Since the function $S(\,\cdot\,)x$ is continuously differentiable and (ii) holds, we have for all $\varphi \in E^*$

$$\varphi(S(t)x - x) = \int_0^t \frac{d}{dr} \varphi(S(r)x)\, dr$$
$$= \varphi\left(\int_0^t S(r)Ax\, dr\right),$$

and (iii) follows by Theorem A.4 (the Hahn–Banach theorem).

For $h > 0$, $t > 0$ and $x \in E$
$$\frac{S(h) - I}{h}\left(\int_0^t S(r)x\, dr\right) = \frac{1}{h}\int_0^h S(r)(S(t)x - x)\, dt, \qquad (1.16)$$

hence, letting h in (1.16) tend to 0 we obtain $\int_0^t S(r)x\, dr \in D(A)$. But $\lim_{t \downarrow 0} \frac{1}{t}\int_0^t S(r)x\, dr = x$ and thus the set $D(A)$ is dense in E.

To prove that the operator A is closed assume that $(x_m) \in D(A)$, $m = 1, 2, \ldots$, $x_m \to x$ and $Ax_m \to y$ as $m \uparrow +\infty$. Since, for arbitrary $r \geq 0$, $m = 1, 2, \ldots$,

$$\|S(r)Ax_m - S(r)y\| \leq Me^{\omega r}\|Ax_m - y\|,$$

$S(\,\cdot\,)Ax_m \longrightarrow S(\,\cdot\,)y$ uniformly on an arbitrary interval $[0, t]$, $t \geq 0$. On the other hand,
$$S(t)x_m - x_m = \int_0^t S(r)Ax_m\, dr. \qquad (1.17)$$

§ 1.2. Semigroups of operators

Letting m in (1.17) tend to $+\infty$ we obtain
$$S(t)x - x = \int_0^t S(r)y\, dr,$$
and consequently
$$\lim_{t\downarrow 0} \frac{S(t)x - x}{t} = y.$$
Hence $x \in D(A)$ and $Ax = y$. The proof of Theorem 1.2 is complete. □

As a corollary we have the following important proposition.

Proposition 1.1. *A given operator A can be a generator of at most one semigroup.*

Proof. Let $S(t)$, $\widetilde{S}(t)$, $t \geq 0$, be semigroups with the generator A. Fix $x \in D(A)$, $t \geq 0$, and define a function $z(s) = S(t-s)\widetilde{S}(s)x$, $s \in [0, t]$. The function $z(\cdot)$ has a continuous first derivative and
$$\frac{d}{ds} z(s) = -AS(t-s)\widetilde{S}(s)x + S(t-s)A\widetilde{S}(s)x$$
$$= -S(t-s)A\widetilde{S}(s)x + S(t-s)A\widetilde{S}(s)x = 0.$$
Hence
$$S(t)x - \widetilde{S}(t)x = z(0) - z(t) = \int_0^t \frac{d}{ds} z(s)\, ds = 0.$$

□

We will need a result on the family $S^*(t)$, $t \geq 0$, of the adjoint operators on E^*, see §III.3.4 and §A.2. In the theorem below, E is a Hibert space and we identify E with E^*.

Theorem 1.3. *Assume that E is a Hilbert space and $S(t)$, $t \geq 0$, a semigroup on E. Then the adjoint operators $S^*(t)$, $t \geq 0$, form a semigroup on E as well.*

Proof. By the definition of the adjoint operator, $(S(t+s))^* = (S(t)S(s))^* = S^*(s)S^*(t)$, $t, s \geq 0$, and therefore (1.5) holds in the present situation.

To show that
$$\lim_{t\downarrow 0} S^*(t)x = x \quad \text{for } x \in E^*,$$
we need the following lemma.

Lemma 1.1 *If f is a function with values in a Banach space E, defined and Bochner integrable on an interval $(\alpha, \beta + h_0)$, $h_0 > 0$, $\beta > \alpha$, then*
$$\lim_{h\downarrow 0} \int_\alpha^\beta \|f(t+h) - f(t)\|\, dt = 0.$$

Proof. The lemma is true if E is one dimensional, by Theorem A.4, and hence it is true for simple functions f which take on only a finite number of values. By the very definition of Bochner integrable functions, see § A.4, there exists a sequence (f_n) of simple functions such that

$$\lim_{n\to\infty} \int_\alpha^{\beta+h_0} \|f_n(t) - f(t)\| \, dt = 0.$$

It follows from the estimate

$$\int_\alpha^\beta \|f(t+h) - f(t)\| \, dt$$

$$\leq \int_\alpha^\beta \|f_n(t+h) - f(t+h)\| \, dt$$

$$+ \int_\alpha^\beta \|f_n(t+h) - f_n(t)\| \, dt + \int_\alpha^\beta \|f_n(t) - f(t)\| \, dt$$

$$\leq 2 \int_\alpha^{\beta+h_0} \|f_n(t) - f(t)\| \, dt + \int_\alpha^\beta \|f_n(t+h) - f_n(t)\| \, dt,$$

valid for $h \in (0, h_0)$ and $n = 1, 2, \ldots$, that the lemma is true in general. □

For arbitrary $s \geq 0$, $\|S(s)\| = \|S^*(s)\|$. Without any loss of generality we can assume that for some $c > 0$ and all $s \geq 0$, $\|S^*(s)\| \leq c$. Moreover, for arbitrary $z, y \in E$ the function $\langle S^*(s)z, y\rangle = \langle z, S(s)y\rangle$, $s \geq 0$, is continuous. This easily implies that $S^*(\cdot)z$ is a bounded, Borel measurable function with values in a separable Hilbert space, and, therefore, it is Bochner integrable on arbitrary finite interval.

To complete the proof of Theorem 1.3 we first fix $t > 0$ and an element $x \in E$ and show that

$$\lim_{h \downarrow 0} S^*(t+h)x = S^*(t)x. \tag{1.18}$$

If $\gamma \in (0, t)$ then

$$\|S^*(t+h)x - S^*(t)x\| = \|\gamma^{-1} \int_0^\gamma (S^*(t+h)x - S^*(t)x) \, du\|$$

$$= \|\gamma^{-1} \int_0^\gamma S^*(u)(S^*(t+h-u)x - S^*(t-u)x) \, du\|$$

$$\leq \gamma^{-1} M \int_0^\gamma \|S^*(t+h-u)x - S^*(t-u)x\| \, du,$$

and, by Lemma 1.1, formula (1.18) holds.

§ 1.3. The Hille–Yosida theorem

Let us finally consider an arbitrary sequence (t_l) of positive numbers convergent to zero and an arbitrary element x_0 of E and denote by M the closure of the convex combinations of the elements $S^*(t_l)x_0$, $l = 1, 2, \ldots$. Since $S^*(t_l)x_0 \longrightarrow x_0$ weakly as $l \to +\infty$, so, by Theorem III.3.4(ii), $x_0 \in M$. Therefore, the space $E_0 \subset E$ of all linear combinations of $S^*(t_l)x$, $l = 1, 2, \ldots$, $x \in E$, is dense in E. By (1.18),

$$\lim_l S^*(t_l)x = x, \quad \text{for all } x \in E_0.$$

Moreover, $\sup_l \|S^*(t_l)\| < +\infty$ and therefore

$$\lim_l S^*(t_l)x = x \quad \text{for all } x \in E.$$

□

Exercise 1.1. Give an example of a separable Banach space E and of a semigroup $S^*(t)$, $t \geq 0$, on E such that for some $x \in E^*$ the function $S^*(t)x$, $t \geq 0$, is not continuous at 0.

§ 1.3. The Hille–Yosida theorem

Of great importance in applications are theorems which give sufficient conditions for an operator A to be the infinitesimal generator of a semigroup. A fundamental result in this respect is due to Hille and Yosida. Conditions formulated in their theorem are at the same time sufficient and necessary. The proof of the necessity part (not needed in the book) is rather easy and is left as an exercise for the interested reader.

Theorem 1.4. *Let A be a closed linear operator defined on a dense set $D(A)$ contained in a Banach space E. If there exist $\omega \in \mathbf{R}$ and $M \geq 1$ such that, for arbitrary $\lambda > \omega$, the operator $\lambda I - A$ has an inverse $R(\lambda) = (\lambda I - A)^{-1}$ satisfying*

$$\|R^m(\lambda)\| \leq \frac{M}{(\lambda - \omega)^m} \quad \text{for } m = 1, 2, \ldots, \tag{1.19}$$

then A is the infinitesimal generator of a semigroup $S(t)$, $t \geq 0$, on E such that

$$\|S(t)\| \leq M e^{\omega t}, \quad t > 0. \tag{1.20}$$

The family of the operators $(\lambda I - A)^{-1}$, $\lambda > \omega$ is called the *resolvent* of A.

Proof. For $\lambda > \omega$, define linear bounded operators $A_\lambda = \lambda(\lambda R(\lambda) - I)$ and semigroups

$$S^\lambda(t) = e^{tA_\lambda} = e^{-\lambda t} \sum_{m=0}^{+\infty} \frac{(\lambda^2 t)^m}{m!} R^m(\lambda).$$

We will prove that the limit of the semigroups $S^\lambda(\cdot)$ as $\lambda \to +\infty$ is the required semigroup.

Let us show first that
$$A_\lambda x \longrightarrow Ax \quad \text{for all } x \in D(A). \tag{1.21}$$

If $x \in D(A)$, then $R(\lambda)(\lambda x - Ax) = x$ and therefore
$$\|\lambda R(\lambda)x - x\| = \|R(\lambda)Ax\| \le \frac{M}{\lambda - \omega}\|Ax\|, \quad \lambda > \omega.$$

Hence
$$\lim_{\lambda \uparrow +\infty} \lambda R(\lambda)x = x \quad \text{for } x \in D(A). \tag{1.22}$$

Since $\|\lambda R(\lambda)\| \le \frac{|\lambda|M}{\lambda - \omega}$, $\lambda > \omega$ and the set $D(A)$ is dense in E, the formula (1.22) is true for all $x \in E$ by Theorem A.5(ii) (the Banach–Steinhaus theorem). Taking into account that
$$A_\lambda x = \lambda R(\lambda)Ax, \quad \lambda > \omega,$$

we arrive at (1.21).

Let us remark that
$$\|S^\lambda(t)\| \le e^{-\lambda t} \sum_{m=0}^{+\infty} \frac{(\lambda^2 t)^m}{m!} \frac{M}{(\lambda - \omega)^m} \le M e^{\frac{\omega \lambda}{\lambda - \omega} t}, \quad \lambda > \omega, \ t > 0. \tag{1.23}$$

Since $A_\lambda A_\mu = A_\mu A_\lambda$, $S^\lambda(t)A_\mu = A_\mu S^\lambda(t)$, $t \ge 0$, $\lambda, \mu > \omega$, so, for $x \in D(A)$,

$$S^\lambda(t)x - S^\mu(t)x = \int_0^t \frac{d}{ds} S^\mu(t-s) S^\lambda(s) x \, ds$$
$$= \int_0^t S^\mu(t-s)(A_\lambda - A_\mu) S^\lambda(s) x \, ds$$
$$= \int_0^t S^\mu(t-s) S^\lambda(s)(A_\lambda - A_\mu) x \, ds.$$

Therefore, by (1.23),
$$\|S^\lambda(t)x - S^\mu(t)x\| \le \|A_\lambda x - A_\mu x\| M^2 \int_0^t e^{\frac{\omega\mu}{\mu-\omega}(t-s) + \frac{\omega\lambda}{\lambda-\omega}s} \, ds$$
$$\le M^2 t \|A_\lambda x - A_\mu x\| e^{\frac{\omega\mu}{\mu-\omega}t}.$$

Consequently, by (1.21),
$$\|S^\lambda(t)x - S^\mu(t)x\| \longrightarrow 0, \quad \text{for } \lambda, \mu \uparrow +\infty \text{ and } x \in D(A), \tag{1.24}$$

§1.3. The Hille–Yosida theorem

uniformly in $t \geq 0$ from an arbitrary bounded set. It follows from Theorem A.5(ii) (the Banach–Steinhaus theorem) and from (1.23) that the convergence in (1.24) holds for all $x \in E$.

Define
$$S(t)x = \lim_{\lambda \uparrow +\infty} S^\lambda(t)x, \quad x \in E. \tag{1.25}$$

We check easily that $S(t)$, $t \geq 0$, is a semigroup of operators and that for $x \in D(A)$,
$$S(t)x - x = \int_0^t S(r)Ax\, dr, \quad t \geq 0. \tag{1.26}$$

Let B be the infinitesimal generator of $S(t)$, $t \geq 0$. By (1.26), $D(A) \subset D(B)$ and $Ax = Bx$ for $x \in D(A)$. It is therefore enough to show that $D(A) = D(B)$. Since
$$\|S(t)\| \leq Me^{\omega t},$$
the operator
$$\widetilde{R}(\lambda)x = \int_0^{+\infty} e^{-\lambda t} S(t)x\, dt, \quad x \in E \ (\lambda > \omega),$$
is well defined and continuous. For arbitrary $\lambda > \omega$, $y \in E$ and $h > 0$
$$\frac{1}{h}[S(h)\widetilde{R}(\lambda)x - \widetilde{R}(\lambda)x]$$
$$= e^{\lambda h}\frac{1}{h}\int_h^{+\infty} e^{-\lambda(t+h)}S(t+h)x\, dt - \frac{1}{h}\int_0^{+\infty} e^{-\lambda t}S(t)x\, dt$$
$$= \frac{e^{\lambda h} - 1}{h}\widetilde{R}(\lambda)x - e^{\lambda h}\frac{1}{h}\int_0^h e^{-\lambda t}S(t)x\, dt.$$

Hence $\widetilde{R}(\lambda)y \in D(B)$ and
$$(\lambda - B)\widetilde{R}(\lambda)y = y, \quad y \in E. \tag{1.27}$$

If $x \in D(B)$ and $\lambda > \omega$, then
$$\widetilde{R}(\lambda)Bx = \int_0^{+\infty} e^{-\lambda t}S(t)Bx\, dt = \int_0^{+\infty} e^{-\lambda t}\frac{d}{dt}S(t)x\, dt = -x + \lambda\widetilde{R}(\lambda)x,$$
and therefore
$$\widetilde{R}(\lambda)(\lambda - B)x = x, \quad x \in D(B). \tag{1.28}$$

From (1.27) and (1.28)
$$\widetilde{R}(\lambda) = (\lambda - B)^{-1}, \quad \lambda > \omega. \tag{1.29}$$

In particular, for arbitrary $y \in E$, the equation

$$\lambda x - Bx = y, \quad x \in D(B), \qquad (1.30)$$

has exactly one solution in $D(B)$. Since $x = R(\lambda)y \in D(A)$ is a solution of this equation, $\widetilde{R}(\lambda) = R(\lambda)$. In particular, $D(B) = \widetilde{R}(\lambda)E = R(\lambda)E = D(A)$. □

It follows from the proof of Theorem 1.4 that

Proposition 1.2. *If for the generator A of a semigroup $S(t)$, $t \geq 0$, condition (1.19) is satisfied or if a semigroup $S(t)$ satisfies (1.20), then*

$$(\lambda - A)^{-1}x = R(\lambda)x = \int_0^{+\infty} e^{-\lambda t}S(t)x\,dt, \quad x \in E, \ \lambda > \omega. \qquad (1.31)$$

§1.4. Phillips' theorem

For control theory, and for the stabilizability question in particular, of great importance is a result saying that the sum of a generator and a bounded linear operator is a generator. It is due to Phillips.

Theorem 1.5. *If an operator A generates a semigroup on a Banach space E and $K: E \longrightarrow E$ is a bounded linear operator then the operator $A + K$ with the domain identical to $D(A)$ is also a generator.*

Proof. Let $S(t)$, $t \geq 0$, be the semigroup generated by A. Then for some constants $M \geq 1$ and $\omega \in \mathbf{R}$

$$\|S(t)\| \leq Me^{\omega t}, \quad t \geq 0.$$

It follows from Proposition 1.2 that the operator $R(\lambda) = (\lambda I - A)^{-1}$ is given by

$$R(\lambda)x = \int_0^{+\infty} e^{-\lambda t}S(t)x\,dt, \quad x \in E, \ \lambda > \omega.$$

Since the semigroup $e^{-\omega t}S(t)$, $t \geq 0$, is generated by $A - \omega I$, without any loss of generality, we can assume that

$$\|S(t)\| \leq M, \quad t \geq 0.$$

Let us introduce a new norm $\|\cdot\|_0$ on E,

$$\|x\|_0 = \sup_{t \geq 0} \|S(t)x\|.$$

Then

$$\|x\| \leq \|x\|_0 \leq M\|x\|, \quad x \in E.$$

§1.4. Phillips' theorem

Hence the norms $\|\cdot\|_0$ and $\|\cdot\|$ are equivalent, and we check easily that

$$\|S(t)\|_0 \leq 1, \quad t \geq 0,$$
$$\|R(\lambda)\|_0 \leq \frac{1}{\lambda}, \quad \lambda > 0.$$

Assume that $\lambda > \|K\|_0$. Since

$$\|KR(\lambda)\|_0 \leq \|K\|_0 \|R(\lambda)\|_0 < 1,$$

the operator $I - KR(\lambda)$ is invertible. Define

$$\widetilde{R}(\lambda) = R(\lambda)(I - KR(\lambda))^{-1} = \sum_{m=0}^{+\infty} R(\lambda)(KR(\lambda))^m, \quad \lambda > \|K\|_0. \quad (1.32)$$

We will show that

(i) $$\|\widetilde{R}(\lambda)\|_0 \leq \frac{1}{\lambda - \|K\|_0},$$

(ii) $$\widetilde{R}(\lambda) = (\lambda - A - K)^{-1}.$$

From (1.33)

$$\|\widetilde{R}(\lambda)\|_0 \leq \sum_{m=0}^{+\infty} \|R(\lambda)\|_0 \|KR(\lambda)\|_0^m$$
$$\leq \frac{1}{\lambda} \frac{1}{1 - \|KR(\lambda)\|_0} \leq \frac{1}{\lambda - \|K\|_0}, \quad \lambda > \|K\|_0,$$

so the inequality (i) holds.

To prove (ii) notice that

$$(\lambda - A - K)\widetilde{R}(\lambda) = (\lambda - A)\widetilde{R}(\lambda) - K\widetilde{R}(\lambda)$$
$$= (I - KR(\lambda))^{-1} - KR(\lambda)(I - KR(\lambda))^{-1} = I.$$

Since

$$\widetilde{R}(\lambda)(\lambda - A - K) = R(\lambda)(\lambda - A - K) + \sum_{m=1}^{+\infty} R(\lambda)(KR(\lambda))^m(\lambda - A - K)$$
$$= I - R(\lambda) + \sum_{m=1}^{+\infty} (R(\lambda)K)^m - \sum_{m=2}^{+\infty} (R(\lambda)K)^m = I,$$

(ii) takes place as well.

It follows from (i) that

$$\|(\widetilde{R}(\lambda))^m\|_0 \leq \frac{1}{(\lambda - \|K\|_0)^m}, \quad m = 1, 2, \ldots.$$

Moreover, the operator $A + K$ is closed and the theorem follows from Theorem 1.4. □

Remark. It follows from the proof of the theorem that if $\widetilde{S}(t)$, $t \geq 0$, is the semigroup generated by $A + K$ then

$$\|\widetilde{S}(t)\| \leq M e^{(\omega + \|K\|)t}, \quad t \geq 0.$$

§ 1.5. Important classes of generators and Lions' theorem

We will now deduce some consequences of Hille–Yosida's theorem which help to prove that a given operator is the generator of a semigroup.

If A is a closed, linear operator with a dense domain $D(A)$ then the *domain $D(A^*)$* of the *adjoint to operator A* consists of all functionals $f \in E^*$ for which the transformation $x \longrightarrow f(Ax)$ has a continuous extension from $D(A)$ to E. If the extension exists then it is unique, belongs to E^* and by the very definition it is the value $A^* f$ of the adjoint operator A^* on f.

The following theorem holds.

Theorem 1.6. *Let A be a closed operator with a dense domain $D(A)$. If for $\omega \in \mathbf{R}$ and all $\lambda > \omega$*

$$\|(\lambda I - A)x\| \geq (\lambda - \omega)\|x\| \quad \textit{for all } x \in D(A), \tag{1.33}$$
$$\|(\lambda I - A^*)f\| \geq (\lambda - \omega)\|f\| \quad \textit{for all } f \in D(A^*), \tag{1.34}$$

then the operator A generates a semigroup of operators $S(t)$, $t \geq 0$, such that

$$\|S(t)\| \leq e^{\omega t}, \quad t \geq 0.$$

Proof. It follows from (1.33) that $\operatorname{Ker}(\lambda I - A) = \{0\}$, for $\lambda > \omega$ and that the image of $\lambda I - A$ is a closed, linear space $E_0 \subset E$. If $E_0 \neq E$, then, by Theorem A.4 (the Hahn–Banach theorem), there exists a functional $f \in E^*$, $f \neq 0$, such that

$$f(\lambda x - Ax) = 0, \quad x \in D(A). \tag{1.35}$$

Hence $f \in D(A^*)$ and $A^* f = \lambda f$. By (1.34), $f = 0$, so, necessarily, $E_0 = E$. The operator $\lambda - A$ is invertible and, taking into account (1.33),

$$\|R(\lambda)\| \leq \frac{1}{\lambda - \omega}. \tag{1.36}$$

Therefore

$$\|R^m(\lambda)\| \leq \frac{1}{(\lambda - \omega)^m}, \quad \lambda > \omega, \; m = 1, 2, \ldots, \tag{1.37}$$

and the result follows by the Hille–Yosida theorem. $\square$

§ 1.5. Important classes of generators and Lions' theorem

Conditions (1.33) and (1.34) have particularly simple form if the space E is Hilbert (over real or complex numbers).

Theorem 1.7. *Assume that a closed operator A with a dense domain $D(A)$ is defined on a Hilbert space H. If there exists $\omega \in \mathbf{R}$ such that*

$$\operatorname{Re}\langle Ax, x\rangle \leq \omega \|x\|^2 \quad \text{for } x \in D(A), \tag{1.38}$$
$$\operatorname{Re}\langle A^*x, x\rangle \leq \omega \|x\|^2 \quad \text{for } x \in D(A^*), \tag{1.39}$$

then the operator A generates a semigroup $S(t)$, $t \geq 0$, such that

$$\|S(t)\| \leq e^{\omega t}, \quad t \geq 0. \tag{1.40}$$

Proof. We apply Theorem 1.6 and check that (1.33) holds; condition (1.34) can be proved in the same way. Let $\lambda > \omega$, $x \in D(A)$, then

$$\|\lambda x - Ax\|^2 = \|(\lambda - \omega)x + (\omega x - Ax)\|^2$$
$$= (\lambda - \omega)^2 \|x\|^2 + \|\omega x - Ax\|^2 + 2(\lambda - \omega)\operatorname{Re}\langle \omega x - Ax, x\rangle.$$

From (1.38)

$$\operatorname{Re}\langle \omega x - Ax, x\rangle = \omega \|x\|^2 - \operatorname{Re}\langle Ax, x\rangle \geq 0,$$

and (1.33) follows. □

A linear operator A on a Hilbert space H, densely defined, is said to be *selfadjoint* if $D(A) = D(A^*)$ and $A = A^*$.

We have the following corollary of Theorem 1.7

Corollary 1.1. *A selfadjoint operator A such that*

$$\operatorname{Re}\langle Ax, x\rangle \leq \omega \|x\|^2 \quad \text{for } x \in D(A), \tag{1.41}$$

generates a semigroup $S(t)$, $t \geq 0$. Since the semigroups $S^\lambda(\cdot)$, $\lambda > \omega$ generated by $A_\lambda = \lambda(\lambda R(\lambda) - I)$, are selfadjoint, the semigroup $S(t) = \lim_{\lambda \uparrow +\infty} S^\lambda(t)$, $t \geq 0$, consists of selfadjoint operators as well.

The following way of defining semigroups is due to J.L. Lions, see, e.g., [33].

Let us consider a pair of Hilbert spaces $V \subset H$, with scalar products and norms $((\cdot,\cdot))$, $\langle\cdot,\cdot\rangle$, $\|\cdot\|$, $|\cdot|$ such that V is a dense subset of H and the embedding of V into H is continuous. Let $a(\cdot,\cdot)$ be a continuous, bilinear function defined on V. In the case of complex Hilbert spaces V and H instead of linearity with respect to the second variable we require from $a(\cdot,\cdot)$ antilinearity

$$a(u, \lambda v) = \overline{\lambda} a(u, v), \quad u, v \in V, \quad \lambda \in \mathbf{C}.$$

Define
$$D(A) = \{u \in V;\ a(u,\cdot) \text{ is continuous in norm } |\cdot|\}. \qquad (1.42)$$

Since V is dense in H, for $u \in D(A)$ there exists exactly one element $Au \in H$ such that
$$\langle Au, v\rangle = a(u,v), \quad v \in V. \qquad (1.43)$$

The following theorem is due to J.L. Lions.

Theorem 1.8. *If for some constants $\omega \in \mathbf{R}$, $\alpha > 0$,*
$$\operatorname{Re} a(u,u) + \alpha\|u\|^2 \le \omega|u|^2, \quad u \in V, \qquad (1.44)$$

then the operator A defined by (1.42), (1.43) generates a semigroup $S(t)$, $t \ge 0$, on H and
$$|S(t)| \le e^{\omega t}, \quad t \ge 0.$$

Proof. Let us fix $\lambda > \omega$ and define a new bilinear functional
$$a_\lambda(u,v) = \lambda\langle u,v\rangle - a(u,v), \quad u,v \in V.$$

From (1.44),
$$\begin{aligned}
\alpha\|u\|^2 &\le \lambda\langle u,u\rangle - \operatorname{Re} a(u,u) \\
&\le \operatorname{Re}(\lambda\langle u,u\rangle - a(u,u)) \\
&\le \operatorname{Re} a_\lambda(u,u) \\
&\le |a_\lambda(u,u)|, \quad u \in V.
\end{aligned}$$

We will need the following abstract result.

Proposition 1.3. (the Lax–Milgram theorem). *Let $a(\cdot,\cdot)$ be a continuous bilinear functional defined on a Hilbert space H such that, for some $\alpha > 0$,*
$$|a(u,u)| \ge \alpha|u|^2, \quad u \in H.$$

Then there exists a continuous, invertible, linear operator F from H onto H such that
$$\langle Fu, v\rangle = a(u,v), \quad u,v \in H.$$

Proof. For arbitrary u, the functional $v \longrightarrow \overline{a(u,v)}$ is linear and continuous, and by Theorem A.7 (the Riesz representation theorem) there exists exactly one element in H, which we denote by Fu, such that
$$\overline{a(u,v)} = \langle v, Fu\rangle, \quad v \in V.$$

Hence
$$\langle Fu, v\rangle = a(u,v),$$

§ 1.5. Important classes of generators and Lions' theorem

and it is easy to check that F is a linear, continuous transformation. Moreover,
$$\begin{aligned}|Fu| &= \sup\{|\langle Fu,v\rangle|;\ |v|\le 1\}\\ &= \sup\{|a(u,v)|;\ |v|\le 1\}\\ &\ge |a(u,\tfrac{u}{|u|})|\\ &\ge \alpha|u|,\quad u\in H.\end{aligned}$$

Consequently the image $F(H)$ of the transformation F is a closed linear subspace of H. If $F(H)\ne H$, then by the Hahn-Banach theorem there exists $v\in H$, $v\ne 0$, such that
$$\langle Fu,v\rangle = 0 \text{ for all } u\in H.$$

In particular,
$$\langle Fv,v\rangle = 0.$$

But
$$\langle Fv,v\rangle = a(v,v) \ne 0,$$

a contradiction. □

It follows from Proposition 1.3 that there exists a linear, invertible operator $\widetilde{A}_\lambda\colon V\longrightarrow V$ such that
$$a_\lambda(u,v) = ((\widetilde{A}_\lambda u,v))\quad \text{for } u,v\in V.$$

Let A_λ be defined by (1.42)-(1.43) with $a(\cdot,\cdot)$ replaced by $a_\lambda(\cdot,\cdot)$. Then $A_\lambda = \lambda I - A$, $D(A_\lambda) = D(A)$. Let, moreover, an operator $J\colon H\longrightarrow V$ be defined by the relation
$$\langle h,v\rangle = ((Jh,v)),\quad h\in H,\ v\in V.$$

We directly check that the image $J(H)$ is dense in V. For $h\in H$ we set
$$u = \widetilde{A}_\lambda^{-1} J(h).$$

Then $\widetilde{A}_\lambda u = Jh$,
$$a_\lambda(u,v) = ((\widetilde{A}_\lambda u,v)) = ((Jh,v)) = \langle h,v\rangle,$$

hence $u\in D(A_\lambda)$ and $A_\lambda u = h$. On the other hand, if $u\in D(A_\lambda)$ and $A_\lambda u = h$ then for $v\in V$
$$a_\lambda(u,v) = \langle A_\lambda u,v\rangle = ((JA_\lambda u,v)) = ((\widetilde{A}_\lambda u,v)).$$

194   1. Linear control systems

Consequently $u = \tilde{A}_\lambda^{-1} Jh$. This way we have shown that

$$D(A_\lambda) = \tilde{A}_\lambda^{-1} J(H), \qquad (1.45)$$
$$A_\lambda^{-1} = \tilde{A}_\lambda^{-1} J.$$

For arbitrary $\lambda > \omega$ the operator

$$R(\lambda) = (\lambda I - A)^{-1} = A_\lambda^{-1}$$

is a well defined linear operator. In particular, operators A_λ and A are closed. It follows from (1.45) and the denseness of $J(H)$ in V that $D(A)$ is dense in H and, moreover, for $u \in D(A)$

$$|(\lambda I - A)u|^2 = |(\lambda - \omega)u + (\omega - A)u|^2$$
$$= (\lambda - \omega)^2 |u|^2 + 2(\lambda - \omega)\left[\omega |u|^2 - \operatorname{Re}\langle Au, u\rangle\right] + |(\omega - A)u|^2$$
$$\geq (\lambda - \omega)^2 |u|^2.$$

Therefore, for $\lambda > \omega$, we have $|R(\lambda)| \leq \frac{1}{\lambda - \omega}$ and thus $|R^m(\lambda)| \leq \frac{1}{(\lambda - \omega)^m}$, $m = 1, 2, \ldots$. The conditions of the Hille–Yosida theorem are satisfied. The proof of the theorem is complete. □

§1.6. Specific examples of generators

We will illustrate the general theory by showing that specific differential operators are generators. We will also complete examples introduced earlier.

Example 1.5. Let $(a, b) \subset \mathbf{R}$ be a bounded interval and $H = L^2(a, b)$. Let us consider an operator

$$A = \frac{d}{d\xi}$$

with the following domain:

$$D(A) = \left\{x \in H;\ x \text{ is absolutely continuous on } [a, b],\ \frac{dx}{d\xi} \in H,\ x(b) = 0\right\}.$$

We will show that the operator A is the generator of the so-called left-shift semigroup $S(t)$, $t \geq 0$:

$$S(t)x(\xi) = \begin{cases} x(t + \xi), & \text{if } t + \xi \in (a, b), \\ 0, & \text{if } t + \xi \notin (a, b). \end{cases} \qquad (1.46)$$

We will give two proofs, first a direct one and then a longer one based on Theorem 1.7.

§ 1.6. Specific examples of generators

Let $\lambda > 0$ and $y \in H$. The equation
$$\lambda x - Ax = y, \quad x \in D(A),$$
is equivalent to a differential equation
$$\frac{dx}{d\xi}(\xi) = \lambda x(\xi) - y(\xi), \quad \xi \in (a, b),$$
$$x(b) = 0,$$
and has a solution $x = R(\lambda)y$ of the form
$$x(\xi) = \int_\xi^b e^{\lambda(\xi-s)} y(s)\, ds, \quad \xi \in [a,b].$$

On the other hand, for the semigroup $S(t)$, $t \geq 0$, given by (1.46)
$$\left(\int_0^{+\infty} e^{-\lambda t} S(t)\, dt \right) y(\xi) = \int_0^{b-\xi} e^{-\lambda t} y(\xi + t)\, dt$$
$$= \int_\xi^b e^{\lambda(\xi-s)} y(s)\, ds, \quad \xi \in [a,b],\ \lambda > 0.$$

So the resolvent of the generator $S(t)$, $t \geq 0$, is identical with the resolvent of the operator A and therefore A is the generator of $S(t)$, $t \geq 0$ (see Proposition 1.1 and Proposition 1.2).

To give a proof based on Theorem 1.7, let us remark that for $x \in D(A)$
$$\langle Ax, x \rangle = \int_a^b \frac{dx}{ds}(\xi) x(\xi)\, d\xi = -x^2(a) - \int_a^b \frac{dx}{d\xi}(\xi) x(\xi)\, d\xi,$$
hence
$$\langle Ax, x \rangle = -\frac{1}{2} x^2(a) \leq 0.$$

We find now the adjoint operator A^*. If $y \in D(A^*)$, then there exists $z \in H$ such that
$$\langle Ax, y \rangle = \langle x, z \rangle \quad \text{for all } x \in D(A).$$

Therefore
$$\int_a^b \frac{dx}{d\xi} y\, d\xi = \int_a^b x(\xi) z(\xi)\, d\xi \tag{1.47}$$
$$= -\int_a^b z(\xi) \left(\int_\xi^b \frac{dx}{ds}\, ds \right) d\xi = \int_a^b \frac{dx}{ds} \left(\int_a^s z(\xi)\, d\xi \right) ds.$$

To proceed further we need the following lemma, which will also be used in some other examples.

Lemma 1.2. *Assume that ψ is an integrable function on an interval $[a,b]$ such that*

$$\int_a^b \frac{d^{(m)}\varphi}{d\xi^{(m)}}(\xi)\psi(\xi)\,d\xi = 0 \qquad (1.48)$$

for some $m = 0, 1, \ldots$ and arbitrary $\varphi \in C_0^\infty(a,b)$. Then ψ is identical, almost everywhere, with a polynomial of the order $m-1$.

Proof. If $m = 0$ the identity (1.48) is of the form

$$\int_a^b \varphi(\xi)\psi(\xi)\,d\xi = 0. \qquad (1.49)$$

By a standard limit argument we show first that (1.49) holds for all continuous and bounded functions and therefore also for all bounded and measurable functions φ. Let

$$\varphi_n = (\operatorname{sgn}\psi)\chi_{(-n,n)}(\psi), \quad n = 1, 2, \ldots .$$

Then

$$0 = \int_a^b \varphi_n(\xi)\psi(\xi)\,d\xi = \int_{|\psi|\geq n} \psi(\xi)\,d\xi + \int_{|\psi|<n} |\psi(\xi)|\,d\xi.$$

Since ψ is integrable, $\int_{|\psi|\geq n} \psi(\xi)\,d\xi \longrightarrow 0$ as $n \uparrow +\infty$. Hence

$$\int_a^b |\psi(\xi)|\,d\xi = 0,$$

and therefore $\psi(\xi) = 0$ for almost all $\xi \in (a,b)$. The proof is complete for $m = 0$.

Assume that $m = 1$ and let φ_0 be a function from $C_0^\infty(a,b)$ for which $\int_a^b \varphi_0(\xi)\,d\xi = 1$. If $\varphi \in C_0^\infty(a,b)$ then the function

$$\widetilde{\varphi}(\xi) = \int_a^\xi [\varphi(\eta) - (\int_a^b \varphi(s)\,ds)\varphi_0(\eta)]\,d\eta, \quad \xi \in (a,b),$$

also belongs to $C_0^\infty(a,b)$ and therefore

$$\int_a^b \frac{d}{d\xi}\widetilde{\varphi}(\xi)\psi(\xi)\,d\xi = 0.$$

§ 1.6. Specific examples of generators

Hence
$$\int_a^b \varphi(\xi)\psi(\xi)\,d\xi = \int_a^b \varphi(\xi)\,d\xi \cdot \int_a^b \varphi_0(s)\psi(s)\,ds.$$

Consequently
$$\int_a^b \varphi(\xi) \left[\psi(\xi) - \int_a^b \varphi_0(s)\psi(s)\,ds\right] d\xi = 0.$$

Since the lemma is true for $m = 0$, we obtain that
$$\psi(\xi) = \int_a^b \varphi_0(s)\psi(s)\,ds \quad \text{for almost all } \xi \in (a,b).$$

This way we have shown that the lemma is true for $m = 1$. The case of general m can by proved by induction and is left to the reader as an exercise. □

Returning to Example 1.5, let us remark that identity (1.47) and Lemma 1.2 for $m = 1$ imply
$$y(s) = -\int_a^s z(\xi)\,d\xi + \gamma, \quad s \in (a,b), \tag{1.50}$$

for a constant γ. Hence the function y is necessarily absolutely continuous and its derivative $dy/d\xi = -z$ belongs to H.

We will calculate γ. From (1.47) and (1.50) we have
$$\int_a^b \frac{dx}{d\xi} y\,d\xi = -x(a)y(a) - \int_a^b x\frac{dy}{d\xi}\,ds = -x(a)\gamma + \langle x,z\rangle.$$

Hence $x(a)\gamma = 0$ for arbitrary $x \in D(A)$. Therefore $\gamma = y(a) = 0$. This way we have proved that
$$D(A^*) \subseteq \left\{y \in H;\, y \text{ absolutely continuous, } \frac{dy}{d\xi} \in H, y(a) = 0\right\} \tag{1.51}$$

and
$$A^* y = -\frac{dy}{d\xi}, \quad y \in D(A^*).$$

It is not difficult to see that in (1.51) the equality holds, and, since
$$\langle A^* y, y\rangle = -\frac{1}{2}y^2(b) \leq 0,$$

the operator A generates a semigroup $S(t)$, $t \geq 0$, satisfying
$$\|S(t)\| \leq 1, \quad t \geq 0.$$

Example 1.2. (Continuation, see page 178.) The operator A has a dense domain. Assume that $y \in D(A^*)$. Then there exists $z \in H$ such that

$$\langle Ax, y \rangle = \sum_{m=1}^{+\infty} \lambda_m \langle x, e_m \rangle \langle y, e_m \rangle = \sum_{m=1}^{+\infty} \langle x, e_m \rangle \langle z, y_m \rangle \quad \text{for } x \in D(A).$$

Letting $x = e_m$ in the formula, we obtain $\langle z, y_m \rangle = \lambda_m \langle y, e_m \rangle$, $m = 1, \ldots$, and, since $z \in H$,

$$\sum_m (\langle z, y_m \rangle)^2 = \sum_m (\lambda_m \langle y, e_m \rangle)^2 < +\infty.$$

Consequently, $y \in D(A)$ and $A^* y = Ay$. We easily check that $D(A) \subseteq D(A^*)$. We see that the operator A is selfadjoint and

$$\langle Ax, x \rangle = \sum_{m=1}^{+\infty} \lambda_m (\langle x, e_m \rangle)^2 \leq (\sup \lambda_m) \|x\|^2.$$

By Corollary 1.1, the operator A generates a semigroup $S(t)$, $t \geq 0$. That the semigroup $S(t)$, $t \geq 0$, is given by (1.10) can be shown in a similar way to the first part of Example 1.5.

Example 1.3. (Continuation, see page 178.) We prove that A is selfadjoint. If $y \in D(A^*)$, then, for some $z \in H$,

$$\langle Ax, y \rangle = \langle x, z \rangle \quad \text{for all } x \in D(A).$$

Since $x(\xi) = \int_0^\xi \frac{dx}{ds}(s)\, ds$, $\frac{dx}{d\xi}(\xi) = \frac{dx}{d\xi}(0) + \int_0^\xi \frac{d^2 x}{ds^2}\, ds$, $\xi \in (0, \pi)$,

$$\langle x, z \rangle = \int_0^\pi \frac{dx}{d\xi}(\xi) \left(\int_\xi^\pi z(s)\, ds \right) d\xi$$

$$= \frac{dx}{d\xi}(0) \int_0^\pi \left(\int_r^\pi z(s)\, ds \right) dr + \int_0^\pi \frac{d^2 x}{d\xi^2}(\xi) \left[\int_\xi^\pi \left(\int_r^\pi z(s)\, ds \right) dr \right] d\xi.$$

Hence

$$0 = \langle Ax, y \rangle - \langle x, z \rangle \tag{1.52}$$

$$= \int_0^\pi \frac{d^2 x}{d\xi^2}(\xi) \left[y(\xi) - \int_\xi^\pi \left(\int_r^\pi z(s)\, ds \right) dr \right] d\xi - \frac{dx}{d\xi}(0) \int_0^\pi \left(\int_r^\pi z(s)\, ds \right) dr.$$

In particular, the above identity holds for arbitrary $x \in C_0^\infty(0, \pi)$. By Lemma 1.2, for some constants γ, δ

$$y(\xi) = \int_\xi^\pi \left(\int_r^\pi z(s)\, ds \right) dr + \gamma + \delta \xi, \quad \xi \in [0, \pi]. \tag{1.53}$$

§ 1.6. Specific examples of generators

Inserting expression (1.53) into (1.52) we obtain

$$\int_0^\pi \frac{d^2x}{d\xi^2}(\gamma + \delta\xi)\, d\xi = \eta \frac{dx}{d\xi}(0),$$

where

$$\eta = \int_0^\pi \left(\int_r^\pi z(s)\, ds \right) dr,$$

and therefore

$$\frac{dx}{d\xi}(\pi)(\gamma + \delta\pi) = \frac{dx}{d\xi}(0)(\gamma + \eta).$$

Since the values $\frac{dx}{d\xi}(\pi)$, $\frac{dx}{d\xi}(0)$ can be chosen arbitrarily,

$$\gamma = -\eta, \quad \delta = \frac{\eta}{\pi}.$$

But then $y(0) = y(\pi) = 0$. We have shown that $D(A^*) \subset D(A)$ and $A^* = Ay$ for $y \in D(A^*)$. In a similar way one shows that $D(A) \subset D(A^*)$. Therefore the operator A is selfadjoint, and, since

$$\langle Ax, x \rangle = \int_0^\pi \frac{d^2x}{d\xi^2} x\, d\xi = -\int_0^\pi \left(\frac{dx}{d\xi}\right)^2 d\xi \leq 0,$$

it is a generator by Corollary 1.1. Denote by $\mathcal{A}$ the generator of the semigroup from Example 1.2 with

$$e_m(\xi) = \sqrt{\frac{2}{\pi}} \sin m\xi, \ \xi \in (0,\pi), \ \lambda_m = -m^2, \ m = 1,2,\ldots.$$

Let us remark that for arbitrary numbers $\alpha_1, \alpha_2, \ldots, \alpha_m$, $m = 1, 2, \ldots$

$$A\left(\sum_{j=1}^m \alpha_j e_j\right) = \mathcal{A}\left(\sum_{j=1}^m \alpha_j e_j\right).$$

If $x \in D(A)$ and $x_m = \sum_{j=1}^m \langle x, e_j \rangle e_j$, $m = 1, 2, \ldots$, then $x_m \to x$ and $A(x_m) \to A(x)$. Therefore also $\mathcal{A}(x_m) \to \mathcal{A}(x)$. Since the operator A is closed, we obtain that $x \in D(A)$ and $Ax = \mathcal{A}x$. Therefore the generator A is an extension of the generator $\mathcal{A}$. By Proposition 1.1 and Proposition 1.2 the generators are identical. As a consequence we have that the semigroup $S(t)$, $t \geq 0$, generated by the differential operator A has the representation

$$S(t)x(\xi) = \frac{2}{\pi} \sum_{m=1}^{+\infty} e^{-tm^2} \sin m\xi \left(\int_0^\pi x(s) \sin ms\, ds\right), \quad \xi \in [0,\pi], \ t \geq 0.$$

Exercise 1.2. Applying the Lions theorem, prove that the operator A from Example 1.3 is a generator.

Hint. Define

$$V = H_0^1(0,\pi) = \left\{ x \in H; \frac{dx}{d\xi} \in H, \ x(0) = x(\pi) = 0 \right\},$$

$$((x,y)) = \int_0^\pi \frac{dx}{d\xi} \frac{dy}{d\xi} \, d\xi,$$

$$a(x,y) = -((x,y)), \quad x, y \in V.$$

Here is a different illustration of the Lions theorem.

Example 1.6. Let $H = L^2(a,b)$, $V = H^1(a,b) = \{x \in H; \frac{dx}{d\xi} \in H\}$,

$$((x,y)) = \int_a^b \frac{dx}{d\xi} \frac{dy}{d\xi} \, d\xi + \int_a^b x(\xi) y(\xi) \, d\xi$$

and

$$a(x,y) = -((x,y)).$$

It is clear that the bilinear functional $a(\cdot,\cdot)$ satisfies the conditions of Theorem 1.8.

Let A be the generator defined by (1.43) and (1.42). We show that

$$D(A) = \left\{ x \in H; \frac{dx}{d\xi}, \frac{d^2x}{d\xi^2} \in H, \ \frac{dx}{d\xi}(a) = \frac{dx}{d\xi}(b) = 0 \right\}, \tag{1.54}$$

$$Ax = \frac{d^2x}{d\xi^2} - x. \tag{1.55}$$

If $x \in D(A)$, then, for some $z \in H$ and arbitrary $y \in V$,

$$a(x,y) = \langle z, y \rangle,$$

or, equivalently,

$$\int_a^b \frac{dx}{d\xi} \frac{dy}{d\xi} + \int_a^b x(\xi) y(\xi) \, d\xi = -\int_a^b z(\xi) y(\xi) \, d\xi. \tag{1.56}$$

From (1.56) and the representation $y(\xi) = y(a) + \int_a^\xi \frac{dy}{ds} \, ds$, $\xi \in (a,b)$,

$$\int_a^b \frac{dy}{d\xi}(\xi) \left[\frac{dx}{d\xi}(\xi) + \int_\xi^b x(s) \, ds + \int_\xi^b z(s) \, dz \right] d\xi \tag{1.57}$$

$$= -y(a) \int_a^b (x(\xi) + z(\xi)) \, d\xi.$$

§ 1.6. Specific examples of generators

Since $C_0^\infty(a,b) \subset V$, by Lemma 1.2, for a constant γ

$$\frac{dx}{d\xi}(\xi) = -\int_a^b (x(s) + z(s))\, ds + \gamma, \quad \xi \in (a,b). \tag{1.58}$$

This way we see that $dx/d\xi$ is absolutely continuous and

$$\frac{d^2x}{d\xi^2} = x + z.$$

In particular,

$$Ax = z = \frac{d^2x}{d\xi^2} - x.$$

To show that $\frac{dx}{d\xi}(a) = \frac{dx}{d\xi}(b) = 0$, we remark first that

$$\frac{dx}{d\xi}(b) = \gamma, \quad \frac{dx}{d\xi}(a) = \gamma - \int_a^b (x(s) + z(s))\, ds.$$

By (1.58) and (1.57)

$$\gamma \int_a^b \frac{dy}{d\xi}\, d\xi = \gamma(y(b) - y(a)) = -y(a) \int_a^b (x(s) + z(s))\, ds,$$

or, equivalently,

$$\gamma y(b) = \left(\gamma - \int_a^b (x(s) + z(s))\, ds \right) y(a).$$

Since y is an arbitrary element of $V = H^1(a,b)$,

$$\gamma = 0 = \int_a^b (x(s) + z(s))\, ds.$$

Hence $\frac{dx}{d\xi}(a) = \frac{dx}{d\xi}(b) = 0$, and the required result easily follows. □

Example 1.7. On the domain $D(A)$ given by (1.54) define $A+B$ where B is a bounded operator and A given by (1.55). The operator $A+B$ defines a semigroup by Theorem 1.5. In particular, if B is the identity operator, then we obtain that the operator $d^2/d\xi^2$ with the Neuman boundary conditions generates a semigroup.

§1.7. The integral representation of linear systems

Let us consider a linear control system

$$\dot{y} = Ay + Bu, \quad y(0) = x, \tag{1.59}$$

on a Banach space E. We will assume that the operator A generates a semigroup of operators $S(t)$, $t \geq 0$, on E and that B is a linear, bounded operator acting from a Banach space U into E. We will study first the case when $U = E$ and the equation (1.59) is replaced by

$$\dot{y} = Ay + f, \quad y(0) = x, \tag{1.60}$$

with f an E-valued function.

Any continuous function $y: [0,T] \longrightarrow E$ such that
(i) $y(0) = x$, $y(t) \in D(A)$, $t \in [0,T]$,
(ii) y is differentiable at any $t \in [0,T]$ and

$$\frac{dy}{dt}(t) = Ay(t) + f(t), \quad t \in [0,T]$$

is called a *strong* solution of (1.60) on $[0,T]$.

Theorem 1.9. *Assume that $x \in D(A)$, $f(\cdot)$ is a continuous function on $[0,T]$ and $y(\cdot)$ is a strong solution of (1.60) on $[0,T]$. Then*

$$y(t) = S(t)x + \int_0^t S(t-s)f(s)\,ds, \quad t \in [0,T]. \tag{1.61}$$

Proof. Let us fix $t > 0$ and $s \in (0,t)$. It follows from the elementary properties of generators that

$$\frac{d}{ds}S(t-s)y(s) = -AS(t-s)y(s) + S(t-s)\frac{dy(s)}{ds}$$
$$= -S(t-s)Ay(s) + S(t-s)[Ay(s) + f(s)] = S(t-s)f(s).$$

Integrating the above identity from 0 to t we obtain

$$y(t) - S(t)y(0) = \int_0^t S(t-s)f(s)\,ds.$$

$\square$

The formula does not always define a strong solution to (1.60). Additional conditions either on f or on the semigroup $S(t)$, $t \geq 0$, are needed. Here is a typical result in this direction.

§ 1.7. The integral representation of linear systems

Theorem 1.10. *Assume that $f(\cdot)$ is a function with continuous first derivatives on $[0,+\infty]$ and $x \in D(A)$. Then equation (1.60) has a strong solution.*

Proof. The function $S(\cdot)x$ is, by Theorem 1.2, a strong solution of equation (1.60) with $f = 0$. Therefore we can assume that $x = 0$. For all $t \geq 0$,

$$y(t) = \int_0^t S(t-s)f(s)\,ds = \int_0^t S(t-s)\left(f(0) + \int_0^s \frac{df}{dr}(r)\,dr\right)ds$$

$$= \int_0^t S(t-s)f(0)\,ds + \int_0^t \left(\int_r^t S(t-s)\frac{df}{dr}(r)\,ds\right)dr. \tag{1.62}$$

We will show that for $x \in E$ and $r \geq 0$

$$\int_0^r S(s)x\,ds \in D(A) \quad \text{and} \quad S(r)x - x = A\left(\int_0^r S(s)x\,ds\right). \tag{1.63}$$

It follows from Theorem 1.2 that (1.63) holds for all $x \in D(A)$. For an arbitrary $x \in E$ let (x_n) be a sequence of elements from $D(A)$ converging to x. Then

$$S(r)x_n - x_n = A\left(\int_0^r S(s)x_n\,ds\right) \longrightarrow S(r)x - x$$

and

$$\int_0^r S(s)x_n\,ds \longrightarrow \int_0^r S(r)x\,ds.$$

Since the operator A is closed, (1.63) holds for all x.

For arbitrary $t \geq r \geq 0$, we have, by (1.63),

$$\int_0^t S(t-s)f(0)\,ds \in D(A)$$
$$S(t)f(0) - f(0) = A\left(\int_0^t S(t-s)f(0)\,ds\right), \tag{1.64}$$

$$\int_0^t S(t-s)\frac{df}{dr}(r)\,ds \in D(A)$$
$$(S(t-r) - I)\frac{df}{dr}(r) = A\left(\int_r^t S(t-s)\frac{df}{dr}(r)\,ds\right). \tag{1.65}$$

Let us remark that for an arbitrary closed operator A, if functions $\psi(\cdot)$ and $A\psi(\cdot)$ are continuous, then for arbitrary $r \geq 0$

$$\int_0^r \psi(s)\,ds \in D(A) \quad \text{and} \quad A\left(\int_0^r \psi(s)\,ds\right) = \int_0^r A\psi(s)\,ds.$$

Therefore (1.62), (1.64) and (1.65) imply that $y(t) \in D(A)$ and

$$Ay(t) = (S(t) - I)f(0) + \int_0^t (S(t-r) - I)\frac{df}{dr}(r)\,dr, \quad t \geq 0.$$

Since

$$y(t) = \int_0^t S(s)f(t-s)\,ds,$$

$$\frac{dy}{dt}(t) = S(t)f(0) + \int_0^t S(r)\frac{df}{dr}(t-r)\,dr = S(t)f(0) + \int_0^t S(t-r)\frac{df}{dr}(r)\,dr.$$

Summarizing,

$$\frac{dy}{dt}(t) - Ay(t) = S(t)f(0) - f(0) + \int_0^t (S(t-r) - I)\frac{df}{dr}(r)\,dr$$

$$- S(t)f(0) + \int_0^t S(t-r)\frac{df}{dr}(r)\,dr = 0.$$

The proof of the theorem is complete. $\square$

Corollary 1.2. *If a control function $u(\cdot)$ is of class C^1 and $x \in D(A)$, then equation (1.59) has exactly one strong solution, and the solution is given by*

$$y(t) = S(t)x + \int_0^t S(t-s)Bu(s)\,ds, \quad t \geq 0. \tag{1.66}$$

The function $y(t)$, $t \geq 0$, given by the formula (1.66) is well defined for an arbitrary Bochner integrable function $u(\cdot)$. It will be called a *weak solution* of the equation

$$\dot{y} = Ay + Bu. \tag{1.67}$$

It is easy to show that the weak solution of (1.67) is a continuous function.

The following proposition, the proof of which is left as an exercise, shows that weak solutions are uniform limits of strong solutions.

Proposition 1.4. *Assume that $u(\cdot)$ is a Bochner integrable function on $[0,T]$ and $x \in E$. If $(u_k(\cdot))$ and (x_k) are sequences such that*

(i) $x_k \in D(A)$, *for all* $k = 1, 2, \ldots$ *and* $\lim_k x_k = x$,

(ii) $u_k(\cdot)$ *are C^1 functions from $[0,T] \longrightarrow E$, $n = 1, 2, \ldots$ and*

$$\lim_k \int_0^T \|u_k(s) - u(s)\|\,ds = 0,$$

then the sequence of functions

$$y_k(t) = S(t)x_k + \int_0^t S(t-s)Bu_k(s)\,ds, \quad k = 1, 2, \ldots,\ t \in [0,T],$$

converges uniformly to the weak solution $y(\cdot)$ *of* (1.66).

Strong, and therefore weak, solutions of (1.67) are not in general identical with the classical solutions of partial differential equations of which they are abstract models. To see the difference, let us go back to heat equation (1.10). The physical arguments which led to its derivation did not determine which function space should be taken as the stated space E. There are always several possible choices. Let us assume for instance that the stated space is the Hilbert space $H = L^2(0, L)$ and that the generator A is given by

$$Ax = \sigma^2 \frac{d^2 x}{d\xi^2},$$

with the domain $D(A)$ described in Example 1.3. Assume that $b(\cdot) \in H$ and let $Bu = ub$, $u \in \mathbf{R}$. Equation (1.67) is then a version of (0.11), (0.12) and (0.13).

If a function $u(\cdot)$ is of class C^1 and $x \in D(A)$ then, by Theorem 1.10,

$$y(t) = S(t)x + \int_0^t S(t-s)bu(s)\,ds \in D(A), \qquad (1.68)$$

$$\frac{dy}{dt}(t) = Ay(t) + bu(t) \quad \text{for } t \geq 0.$$

A solution $y(\cdot)$ given by (1.68) is nonclassical because, for each $t \geq 0$, $y(t)$ is a class of functions and the inclusion $y(t) \in D(A)$ implies only that there exists a representative of the class which is sufficiently smooth and satisfies the Dirichlet boundary conditions. Whether one could find a representative $y(t, \xi)$, $\xi \in [0, L]$, which is a smooth function of $(t, \xi) \in [0, +\infty) \times [0, L]$ satisfying (0.11)-(0.13), requires an additional analysis. Moreover, the continuity of the strong solution and its differentiability are in the sense of the space $L^2(0, L)$, which is again different form the classical one.

However, despite the mentioned limitations, the abstract theory is a powerful instrument for studying evolutionary problems. It gives a first orientation of the problem. Moreover, the strong solution often has a classical interpretation, and the classical solution aways is the strong one.

Bibliographical notes

Additional material on semigroups and their applications can be found in Pazy [44] and Kisyński [34]. The semigroup approach to control theory of infinite dimensional systems is the subject of the monograph by Curtain and Pritchard [17] and the author's paper of survey character [68]. This approach was initiated by Balakrishnan [3] and Fattorini [26].

Chapter 2
Controllability

This chapter is devoted to the controllability of linear systems. The analysis is based on characterizations of images of linear operators in terms of their adjoint operators. The abstract results lead to specific descriptions of approximately controllable and exactly controllable systems which are applicable to parabolic and hyperbolic equations. Formulae for controls which transfer one state to another are given as well.

§ 2.1. Images and kernels of linear operators

Assume that U is a separable Hilbert space. For arbitrary numbers $b > a \geq 0$ denote by $L^2(a,b;U)$ the space of functions, more precisely, equivalence classes of functions, $u(\,\cdot\,)\colon [a,b] \longrightarrow U$, Bochner integrable (see § A.4 and § A.6), on $[a,b]$ and such that

$$\int_a^b |u(s)|^2 ds < +\infty.$$

It is also a Hilbert space with the scalar product

$$\langle\langle u(\,\cdot\,), v(\,\cdot\,)\rangle\rangle = \int_a^b \langle u(s), v(s)\rangle\, ds, \quad u(\,\cdot\,), v(\,\cdot\,) \in L^2(a,b;U),$$

where $\langle \cdot, \cdot \rangle$ is the scalar product in U. Norms in U and in $L^2(a,b;H)$ will be denoted by $|\cdot|$ and $\|\cdot\|$ respectively.

As in Part I, in the study of controllability of infinite dimensional systems

$$\dot{y} = Ay + Bu, \quad y(0) = x \in E, \tag{2.1}$$

an important role will be played by the operator

$$\mathcal{L}_T u = \int_0^T S(T-s) Bu(s)\, ds, \quad u(\,\cdot\,) \in L^2(0,T;U) \tag{2.2}$$

(see § 1.7), acting from $U_T = L^2(0,T;U)$ into E. Note that

$$y(T) = S(T)x + \mathcal{L}_T u, \quad u(\,\cdot\,) \in L^2(0,T;U),$$

where $y(\,\cdot\,)$ is the weak solution to (2.1).

§ 2.1. Images and kernels of linear operators

To analyse the operator $\mathcal{L}_T$ we will need several results from operator theory which we discuss now.

Let X, Y and Z be Banach spaces and F and G linear, bounded operators from X into Z and Y into Z. The adjoint spaces and adjoint operators will be denoted, as before, by X^*, Y^* and Z^* and by F^* and G^* respectively. The image $F(X)$ and the kernel $F^{-1}\{0\}$ of the transformation F will be denoted by $\operatorname{Im} F$ and $\operatorname{Ker} F$.

Our main aim in this section is to give characterizations of the inclusions

$$\operatorname{Im} F \subset \operatorname{Im} G, \qquad \overline{\operatorname{Im} F} \subset \overline{\operatorname{Im} G}$$

in terms of the adjoint operators F^* and G^*.

Theorem 2.1. *The following two conditions are equivalent*

$$\overline{\operatorname{Im} F} \subset \overline{\operatorname{Im} G}, \tag{2.3}$$

$$\operatorname{Ker} F^* \supset \operatorname{Ker} G^*. \tag{2.4}$$

Proof. Assume that (2.3) holds and that for same $f \in Z^*$, $G^* f = 0$ and $F^* f \neq 0$. Then for arbitrary $y \in Y$, $f(G(y)) = 0$ and $f = 0$ on $\operatorname{Im} G$. It follows from (2.3) that $f = 0$ on $\operatorname{Im} F$ and $f(F(x)) = 0$ for $x \in X$. Hence $F^* f = 0$, a contradiction. This way we have shown that (2.3) implies (2.4).

Assume now that (2.4) holds and that there exists $z \in \overline{\operatorname{Im} F} \setminus \overline{\operatorname{Im} G}$. There exists a functional $f \in Z^*$ such that $f(z) \neq 0$ and $f = 0$ on $\operatorname{Im} G$. Moreover, for a sequence of elements $x_m \in X$, $m = 1, 2, \ldots$, $F(x_m) \longrightarrow z$. For sufficiently large m, $F^* f(x_m) \neq 0$. Therefore $F^* f \neq 0$, and, at the same time, $G^* f = 0$, which contradicts (2.4). □

Under rather general conditions, the inclusion

$$\operatorname{Im} F \subset \operatorname{Im} G \tag{2.5}$$

takes place if the condition

$$\text{for some } c > 0 \text{ and all } f \in Z^* \quad \|F^* f\| \leq c \|G^* f\| \tag{2.6}$$

holds.

We will prove first

Lemma 2.1. *Inclusion (2.5) holds if and only if*

$$for \ some \ c > 0, \quad \{F(x); \|x\| \leq 1\} \subset \{Gy; \|y\| \leq c\}. \tag{2.7}$$

Moreover (2.6) is equivalent to the condition

$$for \ some \ c > 0, \quad \{F(x); \|x\| \leq 1\} \subset \overline{\{G(y); \|y\| \leq c\}}. \tag{2.8}$$

Proof. Assume that (2.5) holds, and, in addition, $\operatorname{Ker} G = \{0\}$. Then the operator $G^{-1}F$ is well defined, closed and, by Theorem A.2 (the closed graph theorem), continuous. Hence there exists a constant $c > 0$ such that

$$\|G^{-1}(F(x))\| \le c, \quad \text{provided } \|x\| \le 1,$$

and (2.7) holds. If $\operatorname{Ker} G \ne \{0\}$, consider the induced transformation $\hat{G}$ from the quotient space $\hat{Y} = Y/\operatorname{Ker} G$ into X. We recall that the norm of the equivalence class $[y] \in \hat{Y}$ of an element $y \in Y$ is given by

$$\|[y]\| = \inf\{\|y + \tilde{y}\|; G(\tilde{y}) = 0\}.$$

Since $\operatorname{Im} G = \operatorname{Im} \hat{G}$, inclusion (2.5) implies $\operatorname{Im} F \subset \operatorname{Im} \hat{G}$, and, by the proven part of the lemma,

$$\{F(x); \|x\| \le 1\} \subset \{\hat{G}([y]); \|[y]\| \le c\}$$
$$\subset \{G(y + \tilde{y}); \|y + \tilde{y}\| \le c + 1\}.$$

Hence (2.5) implies (2.7) for arbitrary G. It is obvious that (2.7) implies (2.5).

To prove the second equivalence, assume that (2.8) holds. Then, for arbitrary $f \in Z^*$,

$$\|F^*f\| = \sup_{\|x\| \le 1} |f(F(x))| \le c \sup_{\|y\| \le 1} |f(G(y))| \le c\|G^*f\|,$$

and (2.6) is true.

Finally assume that (2.6) holds and for some $x_0 \in X$, $\|x_0\| \le 1$, $F(x_0) \notin \{Gy; \|y\| \le c\}$. By Theorem III.3.4(ii) there exists $f \in Z^*$ for which $f(F(x_0)) > 1$ and for arbitrary $y \in Y$, $\|y\| \le 1$, $c|f(G(y))| \le 1$. Hence at the same time $\|F^*f\| > 1$ and $c\|G^*f\| \le 1$. The obtained contradiction with the condition (2.6) completes the proof of the lemma. □

Corollary 2.1. *If the set $\{Gy; \|y\| \le c\}$ is closed, then conditions (2.5) and (2.6) are equivalent.*

In particular we have the following theorem.

Theorem 2.2. *If Y is a Hilbert space then conditions (2.5) and (2.6) are equivalent.*

Proof. Assume that Y is a Hilbert space and $z \in \overline{\{Gy; \|y\| \le c\}}$. Then there exists a sequence of elements $y_m \in Y$, $m = 1, 2, \ldots$, such that $G(y_m) \longrightarrow z$. We can assume that (y_m) converges weakly to $\tilde{y} \in Y$, $\|\tilde{y}\| \le c$ (see Lemma III.3.4). For arbitrary $f \in Z^*$, $f(G(y_m)) \longrightarrow f(z)$ and $f(G(y_m)) = G^*f(y_m) \longrightarrow G^*f(y)$, hence $f(z) = f(G(y))$. Since $f \in Z^*$ is arbitrary, $z = G(\tilde{y}) \in \{G(y); \|y\| \le c\}$ and the set $\{G(y); \|y\| \le c\}$ is closed. Corollary 2.1 implies the result. □

Remark. It follows from the proof that Theorem 2.2 is valid for Banach spaces such that an arbitrary bounded sequence has a weakly convergent subsequence. As was shown by Banach, this property characterizes the *reflexive spaces* in which, by the very definition, an arbitrary linear, bounded functional f on Y^* is of the form $f \longrightarrow f(y)$, for some $y \in Y$.

Example 2.1. Let $Y = C[0,1]$, $X = Z = L^2[0,1]$ and let $Fx = x$, $Gy = y$, $x \in X$, $y \in Y$. Then condition (2.8) is satisfied but not (2.7). So, without additional assumptions on the space Y, Theorem 2.2 is not true.

We will often meet, in what follows, inverses of operators which are not one-to-one and onto. For simplicity, assume that X and Z are Hilbert spaces and $F: X \longrightarrow Z$ is a linear continuous map. Then $X_0 = \text{Ker } F$ is a closed subspace of X. Denote by X_1 the orthogonal complement of X_0:

$$X_1 = \{x \in X;\ \langle x, y \rangle = 0 \text{ for all } y \in X_0\}.$$

The subspace X_1 is also closed and the restriction F_1 of F to X_1 is one-to-one. Moreover

$$\text{Im } F_1 = \text{Im } F.$$

We define

$$F^{-1}(z) = F_1^{-1}(z), \quad z \in \text{Im } F.$$

The operator $F^{-1}: \text{Im } F \longrightarrow Y$ defined this way is called the *pseudoinverse* of F. It is linear and closed but in general noncontinuous. Note that for arbitrary $z \in \text{Im } F$

$$\inf\{\|x\|;\ F(x) = z\} = \|F^{-1}(z)\|.$$

§ 2.2. The controllability operator

We start our study of the controllability of (2.1) by extending Proposition I.1.1 and give an explicit formula for a control transferring state $a \in E$ to $b \in E$. Generalizing the definition of the controllability matrix we introduce the controllability operator by the formula

$$Q_T x = \int_0^T S(r) B B^* S^*(r) x\, dr, \quad x \in E. \tag{2.9}$$

It follows from Theorems 1.1 and 1.3 that for arbitrary $x \in E$ the function $S(r) B B^* S^*(r) x$, $r \in [0, T]$, is continuous, and the Bochner integral in (2.9) is well defined. Moreover, for a constant $c > 0$,

$$\int_0^T |S(r) B B^* S^*(r) x|\, dr \leq c |x|, \quad x \in E.$$

Hence the operator Q_T is linear and continuous. It is also self-adjoint and nonnegative definite:

$$\langle Q_T x, x \rangle = \int_0^T |B^*S^*(r)x|^2 \, dr \geq 0, \quad x \in E. \tag{2.10}$$

We denote in what follows by $Q_T^{1/2}$ the unique self-adjoint and nonnegative operator whose square is equal Q_T. There exists exactly one such operator. Well defined are operators Q_T^{-1} and $(Q_T^{1/2})^{-1}$ (see §2.1); the latter will also be denoted by $Q_T^{-1/2}$. In connection with these definitions we propose to solve the following

Exercise 2.1. Let (e_m) be an orthonormal, not necessarily complete, sequence in a Hilbert space E and (γ_m) a bounded sequence of positive numbers. Define

$$Qx = \sum_{m=1}^{+\infty} \gamma_m \langle x, e_m \rangle e_m, \quad x \in E.$$

Find formulae for $Q^{1/2}$, Q^{-1}, $(Q^{1/2})^{-1}$.

Theorem 2.3. (i) *There exists a strategy $u(\cdot) \in U_T$ transferring $a \in E$ to $b \in E$ in time T if and only if*

$$S(T)a - b \in \operatorname{Im} Q_T^{1/2}.$$

(ii) *Among the strategies transferring a to b in time T there exists exactly one strategy $\hat{u}$ which minimizes the functional $J_T(u) = \int_0^T |u(s)|^2 ds$. Moreover,*

$$J_T(\hat{u}) = |Q_T^{-1/2}(S(T)a - b)|^2.$$

(iii) *If $S(T)a - b \in \operatorname{Im} Q_T$, then the strategy $\hat{u}$ is given by*

$$\hat{u}(t) = -B^*S^*(T-t)Q_T^{-1}(S(T)a - b), \quad t \in [0, T].$$

Proof. (i) Let us remark that a control $u \in U_T$ transfers a to b in time T if and only if

$$b \in S(T)a + \mathcal{L}_T(u).$$

It is therefore sufficient to prove that

$$\operatorname{Im} \mathcal{L}_T = \operatorname{Im} Q_T^{1/2}. \tag{2.11}$$

Let $x \in E$ and $u(\cdot) \in U_T$. Then

$$\langle\langle u(\cdot), \mathcal{L}_T^* x \rangle\rangle = \langle \mathcal{L}_T u, x \rangle = \int_0^T \langle S(T-r)Bu(r), x \rangle \, dr$$

$$= \int_0^T \langle u(r), B^*S^*(T-r)x \rangle \, dr,$$

§2.2. The controllability operator

and, by the very definition of the scalar product in U_T,

$$\mathcal{L}_T^* x(r) = B^* S^*(T-r)x, \quad \text{for almost all } r \in (0,T).$$

Since

$$\begin{aligned}\|\mathcal{L}_T^* x\|^2 &= \int_0^T |B^* S^*(T-r)x|^2 dr \\ &= \langle Q_T x, x\rangle = |Q_T^{1/2} x|^2, \quad x \in E,\end{aligned} \qquad (2.12)$$

part (i) follows immediately from Theorem 2.2.

(ii) Let us remark that for arbitrary $x \in E$,

$$\mathcal{L}_T \mathcal{L}_T^* x = Q_T x. \qquad (2.13)$$

Let us fix $y \in \operatorname{Im} Q_T^{1/2}$ and let

$$\hat{u} = \mathcal{L}_T^{-1} y, \quad z = Q_T^{-1/2} y.$$

It follows from (2.13) that

$$\mathcal{L}_T \mathcal{L}_T^* Q_T^{-1/2} z = Q_T^{1/2} z = y. \qquad (2.14)$$

If $\mathcal{L}_T u = 0$, $u \in U_T$, then

$$\langle\langle \mathcal{L}_T^* Q_T^{-1/2} z, u\rangle\rangle = \langle Q_T^{-1/2} z, \mathcal{L}_T u\rangle = 0.$$

Taking into account the definition of $\mathcal{L}_T^{-1}$ and identity (2.14) we obtain

$$\hat{u} = \mathcal{L}_T^* Q_T^{-1/2} z, \qquad (2.15)$$

and, by (2.12),

$$\|\hat{u}\| = \|\mathcal{L}_T^* Q_T^{-1/2} z\| = |z| = |Q_T^{-1/2} y|.$$

Defining

$$y = S(T)a - b,$$

we arrive at (ii).

The final part of the theorem follows immediately from (2.15) and the definition of z. □

§2.3. Various concepts of controllability

The results of the two preceding sections lead to important analitical characterizations of various concepts of controllability.

Let $R_T(a)$ be the set of all states attainable from a in time $T \geq 0$. We have the obvious relation

$$R_T(a) = S(T)a + \operatorname{Im} \mathcal{L}_T.$$

We say that system (2.1) is *exactly controllable* from a in time T if

$$R_T(a) = E.$$

If

$$\overline{R_T(a)} = E,$$

then we say that system (2.1) is *approximately controllable* from a in time T.

We say that system (2.1) is *null controllable* in time T if an arbitrary state can be transferred to 0 in time T or, equivalently, if and only if

$$\operatorname{Im} S(T) \subset \operatorname{Im} \mathcal{L}_T. \tag{2.16}$$

We have the following characterizations.

Theorem 2.4. *The following conditions are equivalent.*
  (i) *System (2.1) is exactly controllable from an arbitrary state in time $T > 0$.*
  (ii) *There exists $c > 0$ such that for arbitrary $x \in E$*

$$\int_0^T |B^* S^*(t)x|^2 dt \geq c|x|^2. \tag{2.17}$$

  (iii) $\operatorname{Im} Q_T^{1/2} = E.$ \hfill (2.18)

Proof. If a system is exactly controllable from an arbitrary state, it is controllable in particular from 0. However, the exact controllability from 0 in time T is equivalent to

$$\operatorname{Im} \mathcal{L}_T \supset E.$$

Applying Theorem 2.2 to $G = \mathcal{L}_T$ and $F = I$, we obtain that condition (2.17) is equivalent to the exact controllability of (2.1) from 0. If, however, $\operatorname{Im} \mathcal{L}_T = E$, then, for arbitrary $a \in E$, $R_T(a) = E$, and (2.1) is exactly controllable from arbitrary state.

The final part of the theorem follows from Theorem 2.3(i). □

Theorem 2.5. *The following conditions are equivalent.*
  (i) *System* (2.1) *is approximately controllable in time* $T > 0$ *from an arbitrary state.*
  (ii) *If* $B^* S^*(r) x = 0$ *for almost all* $r \in [0, T]$, *then* $x = 0$. (2.19)
  (iii) $\operatorname{Im} Q_T^{1/2}$ *is dense in* E. (2.20)

Proof. It is clear that the set $R_T(a)$ is dense in E if and only if the set $\operatorname{Im} \mathcal{L}_T$ is dense in E. Hence (i) is equivalent to (ii) by Theorem 2.1. Moreover (i) is equivalent to (iii) by Theorem 2.3. □

Theorem 2.6. *The following conditions are equivalent.*
  (i) *System* (2.1) *is null controllable in time* $T > 0$.
  (ii) *There exists* $c > 0$ *such that for all* $x \in E$

$$\int_0^T |B^* S^*(r) x|^2 dr \geq c |S^*(T) x|^2. \tag{2.21}$$

  (iii) $\operatorname{Im} Q_T^{1/2} \supset \operatorname{Im} S(T)$. (2.22)

Proof. Since null controllability is equivalent to (2.16), characterizations (i) and (ii) and characterizations (i) and (iii) are equivalent by Theorem 2.2 and Theorem 2.3 respectively. □

§2.4. Systems with self-adjoint generators

Let us consider system (2.1) in which the generator A is a self-adjoint operator on a Hilbert space H, such that for an orthonormal and complete basis (e_m) and for a decreasing to $-\infty$ sequence (λ_m), $A e_m = \lambda_m e_m$, $m = 1, 2, \ldots$. (See Example 1.2 with $E = H$, pages 177 and 198).

We will consider two cases, B as the identity operator and B as one-dimensional. In the former case, we have the system

$$\dot{y} = Ay + u, \quad y(0) = a, \tag{2.23}$$

with the set U of control parameters identical to H. In the latter case,

$$\dot{y} = Ay + uh, \quad y(0) = a, \tag{2.24}$$

$U = \mathbf{R}^1$ and h is a fixed element in H.

Lemma 2.2. *Sets* $\operatorname{Im} \mathcal{L}_T$, *for system* (2.23), *are identical for all* $T > 0$. *Moreover, state b is reachable from 0 in a time $T > 0$ if and only if*

$$\sum_{m=1}^{+\infty} |\lambda_m| |\langle b, e_m \rangle|^2 < +\infty. \tag{2.25}$$

Proof. In the present situation the operator Q_T is of the form

$$Q_T x = \int_0^T S(2t)x\, dt = \sum_{m=1}^{+\infty} \left(\int_0^T e^{2\lambda_m t}\, dt\right) \langle x, e_m\rangle e_m$$

$$= \sum_{m=1}^{+\infty} \left(\frac{e^{2\lambda_m T} - 1}{2\lambda_m}\right) \langle x, e_m\rangle e_m.$$

Therefore

$$Q_T^{1/2} x = \sum_{m=1}^{+\infty} \left(\frac{|e^{2\lambda_m T} - 1|}{2|\lambda_m|}\right)^{1/2} \langle x, e_m\rangle e_m.$$

Let us remark that for arbitrary $T > 0$

$$|e^{2\lambda_m T} - 1| \longrightarrow 1, \quad \text{as } m \uparrow +\infty.$$

Hence $b = \sum_{m=1}^{+\infty} \langle b, e_m\rangle e_m \in \operatorname{Im} Q_T^{1/2}$ if and only if condition (2.25) holds. Since $\operatorname{Im} \mathcal{L}_T = \operatorname{Im} Q_T^{1/2}$ the result follows. $\square$

Lemma 2.3. *Under the conditions of Lemma 2.2, for arbitrary $T > 0$,*

$$\operatorname{Im} S(T) \subset \operatorname{Im} Q_T^{1/2}. \tag{2.26}$$

Proof. If $a \in H$, then

$$S(T)a = \sum_{m=1}^{+\infty} e^{\lambda_m T} \langle a, e_m\rangle e_m.$$

Since

$$\sum_m |\lambda_m| e^{2\lambda_m T} |\langle a, e_m\rangle|^2 \leq \left(\sup_m |\lambda_m| e^{2\lambda_m T}\right) |a|^2 < +\infty,$$

(2.26) holds. $\square$

As an immediate corollary of Lemma 2.2, Lemma 2.3 and results of §2.3, we obtain the following theorem.

Theorem 2.7. *System (2.23) has the following properties:*
   (i) *It is not exactly controllable from any state and at any moment $T > 0$.*
   (ii) *It is approximately controllable from arbitrary a and in arbitrary $T > 0$.*
   (iii) *Set $R_T(a)$ is characterized by (2.25).*

We proceed now to system (2.24).

§ 2.4. Systems with self-adjoint generators

It follows from Theorem 2.6 that (2.24) is not exactly controllable. We will now formulate a result giving a characterization of the set of all attainable points showing at the same time the applicability of theorems from § 2.1. An explicit and complete characterization of the set is not known (see [52]).

First of all we have the following

Theorem 2.8. *System (2.24) is approximately controllable from an arbitrary state at a time $T > 0$ if and only if*

$$\lambda_n \neq \lambda_m \quad for \ n \neq m, \tag{2.27}$$

$$\langle h, e_m \rangle \neq 0 \quad for \ m = 1, 2, \ldots. \tag{2.28}$$

Proof. We apply Theorem 2.5. For arbitrary $x \in H$

$$B^* S^*(t) x = \sum_{m=1}^{+\infty} e^{\lambda_m t} \langle h, e_m \rangle \langle x, e_m \rangle.$$

Since

$$\sum_{m=1}^{+\infty} |\langle h, e_m \rangle \langle x, e_m \rangle| \leq |h| |x| < +\infty$$

and $\lambda_m \longrightarrow -\infty$ as $m \uparrow +\infty$, the function

$$\varphi(t) = \sum_{m=1}^{+\infty} e^{\lambda_m t} \langle h, e_m \rangle \langle x, e_m \rangle, \quad t > 0,$$

is well defined and analytic. Hence $\varphi(t) = 0$ for almost all $t \in [0, T]$ if and only if $\varphi(t) = 0$ for all $t \geq 0$. It is clear that conditions (2.27) and (2.28) are necessary for the approximate controllability of (2.24). To prove that they also are sufficient we can assume, without any loss of generality, that the sequence (λ_m) is decreasing. Then

$$\lim_{t \uparrow +\infty} e^{-\lambda_1 t} \varphi(t) = \langle h, e_1 \rangle \langle x, e_1 \rangle = 0.$$

Since $\langle h, e_1 \rangle \neq 0$, $\langle x, e_1 \rangle = 0$. Moreover,

$$\varphi(t) = \sum_{m=2}^{+\infty} e^{\lambda_m t} \langle h, e_m \rangle \langle x, e_m \rangle, \quad t > 0.$$

In a similar way

$$\lim_{t \uparrow +\infty} e^{-\lambda_2 t} \varphi(t) = \langle h, e_2 \rangle \langle x, e_2 \rangle = 0,$$

and thus $\langle x, e_2 \rangle = 0$. By induction $\langle x, e_m \rangle = 0$ for all $m = 1, 2, \ldots$, and, since the basis (e_m) is complete, $x = 0$. This completes the proof of the theorem. □

Let (α_m) be a bounded sequence of positive numbers. Let H_0 be a subspace of H given by

$$H_0 = \left\{ x \in H; \sum_{m=1}^{+\infty} \frac{\langle x, e_m \rangle^2}{\alpha_m^2} < +\infty \right\}. \tag{2.29}$$

We say that the space H_0 is *reachable at a moment* $T > 0$ if 0 can be transferred to an arbitrary element of H_0 in time T by the proper choice of a control from $L^2(0,T)$.

It turns out that one can give necessary conditions, which are very close to sufficient conditions, for the space H_0 to be reachable at time $T > 0$. To formulate them, consider functions $f_m(t) = e^{\lambda_m t}$, $t \in [0, T]$, and closed subspaces of Z_m of $Z = L^2(0,T)$, $m = 1, 2, \ldots$, generated by all f_k, $k \neq m$. Let δ_m be the distance, in the sense of the space Z, from f_m to Z_m, $m = 1, 2, \ldots$.

Theorem 2.9. (i) *If H_0 is reachable at time T, then there exists $\gamma > 0$ such that*

$$0 < \alpha_m \le \gamma \delta_m |\langle h, e_m \rangle|, \quad m = 1, 2, \ldots.$$

(ii) *If, for some $\gamma > 0$,*

$$0 < \alpha_m \le \frac{\gamma}{m} \delta_m |\langle h, e_m \rangle|, \quad m = 1, 2, \ldots,$$

then H_0 is reachable at time T.

Proof. Let $F: H \longrightarrow H$ be an operator given by

$$Fx = \sum_{m=1}^{+\infty} \alpha_m \langle x, e_m \rangle e_m, \quad x \in H.$$

Then

$$\operatorname{Im} F = H_0.$$

Hence the space H_0 is reachable at $T > 0$ if and only if

$$\operatorname{Im} F \subset \operatorname{Im} \mathcal{L}_T. \tag{2.30}$$

It follows from Theorem 2.2 that (2.30) holds if and only if for a number $\gamma > 0$

$$\int_0^T |\langle S(t)h, x \rangle|^2 dt \ge \frac{1}{\gamma^2} |Fx|^2, \quad x \in H. \tag{2.31}$$

Denoting the norm in Z by $\|\cdot\|$ we can reformulate (2.31) equivalently as

$$\left\| \sum_{j=1}^{k} \frac{\beta_j}{\alpha_j} f_j \xi_j \right\| \ge \frac{1}{\gamma} \quad \text{provided} \quad \sum_{j=1}^{k} \xi_j^2 = 1, \quad k = 1, 2, \ldots. \tag{2.32}$$

§ 2.4. Systems with self-adjoint generators

Let us fix m and $\varepsilon > 0$ and choose $k \geq m$ and $k \geq m$ and numbers η_j, $j \leq k$, $j \neq m$ such that

$$\delta_m + \varepsilon \geq \|f_m + \sum_{\substack{j=1 \\ j \neq m}}^{k} \frac{\beta_j}{\alpha_j} \frac{\alpha_m}{\beta_m} f_j \eta_j\|. \tag{2.33}$$

Defining $\eta_m = 1$, $\delta = \left(\sum_{j=1}^{k} \eta_j^2\right)^{1/2} \geq 1$, we obtain from (2.32) and (2.33)

$$\frac{|\beta_m|}{\alpha_m}(\delta_m + \varepsilon) \geq \|\sum_{j=1}^{k} \frac{\beta_j}{\alpha_j} f_j \eta_j\| \geq \delta \|\sum_{j=1}^{k} \frac{\beta_j}{\alpha_j} f_j \frac{\eta_j}{\delta}\| \geq \delta \frac{1}{\gamma} \geq \frac{1}{\gamma}.$$

Since $\varepsilon > 0$ was arbitrary, (i) follows.

To prove (ii) denote by $\hat{f}_m$ the orthogonal projection of f_m onto Z_m and let

$$\tilde{f}_m = \frac{1}{\delta_m^2}(f_m - \hat{f}_m), \quad m = 1, 2, \ldots.$$

Let us remark that

$$\langle\langle f_m, \tilde{f}_m \rangle\rangle = \frac{1}{\delta_m^2}\langle\langle f_m, f_m - \hat{f}_m\rangle\rangle = 1,$$

$$\langle\langle f_m, \tilde{f}_k \rangle\rangle = 0, \quad \|\tilde{f}_m\| = 1, \quad k \neq m,$$

where $\langle\langle \cdot, \cdot \rangle\rangle$ denotes the scalar product on Z.
Therefore, if, for a sequence (ξ_m),

$$\sum_{m=1}^{+\infty} \frac{|\xi_m|}{|\beta_m|} \frac{1}{\delta_m} < +\infty,$$

then

$$u(\cdot) = \sum_{m=1}^{+\infty} \frac{\xi_m}{\beta_m} \tilde{f}_m(\cdot) \in Z$$

and

$$\int_0^T e^{\lambda_m t} u(t)\, dt = \langle\langle f_m, u \rangle\rangle = \frac{\xi_m}{\beta_m}, \quad m = 1, 2, \ldots.$$

Consequently,

$$\sum_{m=1}^{+\infty} \frac{|\langle x, e_m \rangle|}{|\beta_m| \delta_m} < +\infty$$

implies $x \in \operatorname{Im} \mathcal{L}_T$. On the other hand,

$$\sum_{m=1}^{+\infty} \frac{|\langle x, e_m \rangle|}{|\beta_m| \delta_m} = \sum_{m=1}^{+\infty} |\langle x, e_m \rangle| \frac{m}{|\beta_m| \delta_m} \frac{1}{m}$$

$$\leq \left(\sum_{m=1}^{+\infty} \frac{|\langle x, e_m \rangle|^2 m^2}{\beta_m^2 \delta_m^2} \right)^{1/2} \left(\sum_{m=1}^{+\infty} \frac{1}{m^2} \right)^{1/2}.$$

Hence, if

$$0 < \alpha_m \leq \frac{1}{m} |\beta_m| \delta_m, \quad m = 1, 2, \ldots,$$

then $H_0 \subset \operatorname{Im} \mathcal{L}_T$. The proof of the theorem is complete. $\square$

§2.5. Controllability of the wave equation

As another application of the abstract results, we discuss the approximate controllability of the wave equation, see Example 1.4, page 179.

Let h be a fixed function from $L^2(0, \pi)$ with the Fourier expansion

$$h(\xi) = \sum_{m=1}^{+\infty} \gamma_m \sin m\xi, \quad \xi \in (0, \pi), \quad \sum_{m=1}^{+\infty} \gamma_m^2 < +\infty.$$

Let us consider the equation

$$\frac{\partial^2 y}{\partial t^2} = \frac{\partial^2 y}{\partial \xi^2} + hu(t), \quad t \geq 0, \ \xi \in (0, \pi), \tag{2.34}$$

with initial and boundary conditions as in Example 1.4 and with real valued $u(\cdot)$. Equation (2.34) is equivalent to the system of equations

$$\frac{\partial y}{\partial t} = v, \tag{2.35}$$

$$\frac{\partial v}{\partial t} = \frac{\partial^2 y}{\partial \xi^2} + hu(t), \quad t \geq 0, \ \xi \in (0, \pi). \tag{2.36}$$

Let E be the Hilbert space and $S(t)$, $t \in \mathbf{R}$, the group of the unitary operators introduced in Example 1.4. In accordance with the considerations of §1.7, the solution to (2.35)-(2.36) with the initial conditions $u(0) = a$, $v(0) = b$ will be identified with the function $Y(t)$, $t \geq 0$, given by

$$Y(t) = \begin{bmatrix} y(t) \\ v(t) \end{bmatrix} = S(t) \begin{bmatrix} a \\ b \end{bmatrix} + \int_0^t S(t-s) \begin{bmatrix} 0 \\ h \end{bmatrix} u(s). \tag{2.37}$$

§ 2.5. Controllability of the wave equation

Theorem 2.10. *If*

$$T \geq 2\pi, \quad \gamma_m \neq 0 \quad \text{for } m = 1, 2, \ldots,$$

then system (2.35)-(2.36) is approximately controllable in time T from an arbitrary state.

Proof. In the present situation $U = \mathbf{R}$ and the operator $B: \mathbf{R} \longrightarrow E$ is given by

$$Bu = \begin{bmatrix} 0 \\ h \end{bmatrix} u, \quad u \in \mathbf{R}.$$

Since $S^*(t) = S(-t)$, $t \geq 0$,

$$B^* S^*(t) \begin{bmatrix} a \\ b \end{bmatrix} = \langle\!\langle \begin{bmatrix} 0 \\ h \end{bmatrix}, S(-t) \begin{bmatrix} a \\ b \end{bmatrix} \rangle\!\rangle$$

$$= \sum_{m=1}^{+\infty} \gamma_m (m\alpha_m \sin mt + \beta_m \cos mt), \quad t \geq 0.$$

Taking into account that

$$\sum_{m=1}^{+\infty} |m\gamma_m \alpha_m| < +\infty, \quad \sum_{m=1}^{+\infty} |\gamma_m \beta_m| < +\infty,$$

we see that the formula

$$\varphi(t) = \sum_{m=1}^{+\infty} \gamma_m (m\alpha_m \sin mt + \beta_m \cos mt), \quad t \geq 0,$$

defines a continuous, periodic function with the period 2π. Moreover,

$$m\gamma_m \alpha_m = \frac{1}{\pi} \int_0^{2\pi} \varphi(t) \cos mt \, dt,$$

$$\gamma_m \beta_m = \frac{1}{\pi} \int_0^{2\pi} \varphi(t) \sin mt \, dt, \quad m = 1, 2, \ldots.$$

Hence if $T \geq 2\pi$ and $\varphi(t) = 0$ for $t \in [0, T]$,

$$m\gamma_m \alpha_m = 0 \text{ and } \beta_m \gamma_m = 0, \quad m = 1, 2, \ldots.$$

Since $\gamma_m \neq 0$ for all $m = 1, 2, \ldots$, $\alpha_m = \beta_m = 0$, $m = 1, 2, \ldots$, and we obtain that $a = 0$, $b = 0$. By Theorem 2.6 the theorem follows. $\square$

Bibliographical notes

Theorems from §2.1 are taken from the paper by S. Dolecki and D.L. Russell [21] as well as from the author's paper [68]. They are generalizations of a classical result due to Douglas [22].

An explicit description of reachable spaces in the case of finite dimensional control operators B has not yet been obtained; see an extensive discussion of the problem in the survey paper by D.L. Russell [52].

Asymptotic properties of the sequences (f_m) and (δ_m) were the object of intensive studies in D.L. Russell's paper [52]. Theorem 2.9 is from the author's paper [68]. Part (i) is due to D.L. Russell [52].

Chapter 3
Stability and stabilizability

We will show first that the asymptotic stability of an infinite dimensional linear system does not imply its exponential stability and is not determined by the spectrum of the generator. Then we will prove that stable systems are characterized by their corresponding Liapunov equations. It is also proved that null controllability implies stabilizability and that under additional conditions a converse implication takes place.

§3.1. Various concepts of stability

Characterizations of stable or stabilizable infinite dimensional systems are much more complicated then the finite dimensional one. We will restrict our discussion to some typical results underlying specific features of the infinite dimensional situation.

Let A be the infinitesimal generator of a semigroup $S(t)$, $t \geq 0$, on a Banach space E. If $x \in D(A)$ then a strong solution, see §1.7, of the equation
$$\dot{z} = Az, \quad z(0) = x \in E, \tag{3.1}$$
is the function
$$z^x(t) = S(t)x, \quad t \geq 0.$$

For arbitrary $x \in E$, $z^x(\cdot)$ is the limit of strong solutions to (3.1) and is also called a weak solution to (3.1), see §1.7.

If the space E is finite dimensional then, by Theorem I.2.3, the following conditions are equivalent:

For some $N > 0$, $\nu > 0$ and all $t \geq 0$ $\|S(t)\| \leq Ne^{-\nu t}$. (3.2)

For arbitrary $x \in E$, $z^x(t) \longrightarrow 0$ exponentially as $t \to 0$. (3.3)

For arbitrary $x \in E$, $\displaystyle\int_0^{+\infty} \|z^x(t)\|^2 dt < +\infty.$ (3.4)

For arbitrary $x \in E$, $z^x(t) \longrightarrow 0$ as $t \to +\infty$ (3.5)

$$\sup\{\operatorname{Re} \lambda; \lambda \in \sigma(A)\} < 0. \tag{3.6}$$

In general, in the infinite dimensional situation the above conditions are not equivalent, as the theorem below shows.

Let us recall that for linear operators A on an infinite dimensional space E, $\lambda \in \sigma(A)$ if and only if for some $z \in E$ the equation $\lambda x - Ax = z$ either does not have a solution $x \in D(A)$ or has more than one solution.

Theorem 3.1. *Let E be an infinite dimensional Banach space. Then*
  (i) *Conditions (3.2), (3.3) and (3.4) are all equivalent.*
  (ii) *Conditions (3.2), (3.3) and (3.4) are essentially stronger than (3.5) and (3.6).*
  (iii) *Condition (3.5) does not imply, in general, condition (3.6), even if E is a Hilbert space.*
  (iv) *Condition (3.6) does not imply, in general, condition (3.5), even if E is a Hilbert space.*

Proof. (i) It is obvious that (3.2) implies (3.3) and (3.4). Assume that (3.4) holds. Then the transformation $x \longrightarrow S(\cdot)x$ from E into $L^2(0, +\infty; E)$ is everywhere defined and closed. Therefore, by Theorem A.2 (the closed graph theorem), it is continuous.

Consequently, there exists a constant $K > 0$ such that

$$\int_0^{+\infty} \|S(t)x\|^2 dt < K\|x\|^2, \quad \text{for all } x \in E.$$

Moreover, by Theorem 1.1, for some $M \geq 1$ and $\omega > 0$,

$$\|S(t)\| \leq M e^{\omega t} \quad \text{for all } t \geq 0.$$

For arbitrary $t > 0$, $x \in E$,

$$\frac{1 - e^{-2\omega t}}{2\omega} \|S(t)x\|^2 = \int_0^t e^{-2\omega r} \|S(t)x\|^2 dr$$

$$\leq \int_0^t e^{-2\omega r} \|S(r)\|^2 \|S(t-r)x\|^2 dr$$

$$\leq M^2 \int_0^t \|S(s)x\|^2 ds \leq M^2 K \|x\|^2.$$

Therefore, for a number $L > 0$,

$$\|S(t)\| \leq L, \quad t \geq 0.$$

Consequently,

$$t\|S(t)x\|^2 = \int_0^t \|S(t)x\|^2 ds \leq \int_0^t \|S(s)\|^2 \|S(t-s)x\|^2 ds$$

$$\leq L^2 K \|x\|^2, \quad t \geq 0, \, x \in E,$$

§ 3.1. Various concepts of stability

and
$$\|S(t)\| \leq L\sqrt{\frac{k}{t}}, \quad t > 0.$$

Hence, there exists $\bar{t} > 0$ such that
$$\|S(\bar{t})\| < 1.$$

For arbitrary $t \geq 0$ there exist $m = 0, 1, \ldots$ and $s \in [0, \bar{t})$ such that $t = m\bar{t} + s$. Therefore
$$\|S(t)\| \leq \|s(\bar{t})\|^m M e^{\omega s} \leq \|S(\bar{t})\|^{t/\bar{t}} M \|S(\bar{t})\|^{-1} e^{\omega \bar{t}},$$

and (3.2) follows.

(ii) It is clear that (3.2) implies (3.5). Moreover, if (3.2) holds and $\operatorname{Re}\lambda > -\nu$ then
$$R(\lambda) = (\lambda I - A)^{-1} = \int_0^{+\infty} e^{-\lambda t} S(t)\, dt,$$

see Proposition 1.2. Consequently
$$\sup\{\operatorname{Re}\lambda;\ \lambda \in \sigma(A)\} \leq -\nu < 0,$$

and (3.6) holds as well.

To see that (3.5) does not imply, in general, (3.2), we define in $E = l^2$ a semigroup
$$S(t)x = (e^{-\gamma_m t} \xi_m), \quad x = (\xi_m) \in l^2,$$

where (γ_m) is a sequence monotonically decreasing to zero. Then
$$\|S(t)x\|^2 = \sum_{m=1}^{+\infty} e^{-2\gamma_m t} |\xi_m|^2 \longrightarrow 0, \quad \text{as } t \uparrow +\infty.$$

However
$$\left\|S\left(\frac{1}{\gamma_m}\right)\right\| \geq e^{-1}, \quad m = 1, 2, \ldots,$$

and (3.2) is not satisfied.

An example showing that (3.6) does not imply (3.5) and (3.2) will be constructed in the proof of (iv).

(iii) Note that the semigroup $S(t)$, $t \geq 0$, constructed in the proof of (ii) satisfies (3.5). Since $-\gamma_m \in \sigma(A)$, $m = 1, 2, \ldots$, $\sup\{\operatorname{Re}\lambda;\ \lambda \in \sigma(A)\} \geq 0$ and (3.6) does not hold.

(iv) We will prove first a lemma.

Lemma 3.1. *There exists a semigroup $S(t)$, $t \geq 0$, on E, the complex Hilbert space $l_{\mathbf{C}}^2$, such that*

$$\|S(t)\| = e^t, \ t \geq 0 \ \text{and} \ \sigma(A) = \{i\lambda_m; m = 1, 2, \ldots\}$$

where (λ_m) is an arbitrary sequence of real numbers such that $|\lambda_m| \longrightarrow +\infty$.

Proof. Let us represent an arbitrary element $x \in l_{\mathbf{C}}^2$, regarded as an infinite column, in the form $x = (x^m)$, where $x^m \in \mathbf{C}^m$. The required semigroup $S(t)$, $t \geq 0$, is given by the formula

$$S(t)x = (e^{i\lambda_m t} e^{A_m t} x^m), \tag{3.7}$$

where matrices $A_m = (a_{i,j}^m) \in \mathbf{M}(n,n)$ are such that

$$a_{ij}^m = \begin{cases} 1 & \text{for } j = i+1, \ i = 1, 2, \ldots, m-1, \\ 0 & \text{for the remaining } i, j. \end{cases}$$

It is easy to show that $\lim_{t \downarrow 0} S(t)x = x$ for all $x \in E$, and we leave this as an exercise.

To calculate $\|S(t)\|$ remark that

$$\|S(t)\| = \sup_m \|e^{i\lambda_m t} e^{A_m t}\| \leq \sup_m \|e^{A_m t}\| \leq \sup_m e^{t\|A_m\|} \leq e^t, \quad t \geq 0,$$

because $\|A_m\| = 1$, $m = 1, 2, \ldots$. On the other hand, for the vector $e_m \in \mathbf{C}^m$ with all elements equal $\frac{1}{\sqrt{m}}$,

$$\|e^{A_m t} e_m\| = \left(\frac{1}{m} \left[1^2 + (1 + \frac{t}{1!})^2 + \ldots + (1 + \ldots + \frac{t^{m-1}}{(m-1)!})^2 \right] \right)^{1/2}$$
$$\longrightarrow e^t, \text{ as } m \uparrow +\infty.$$

Hence $\|S(t)\| = e^t$.

For arbitrary $\lambda \notin \{i\lambda_m; m = 1, 2, \ldots\}$ let $W(\lambda)$ be the following operator:

$$W(\lambda)x = ((\lambda I - (i\lambda_m + A_m))^{-1} x^m).$$

An elementary calculation shows that $W(\lambda)$ is a bounded operator such that

$$D = \operatorname{Im} W(\lambda) = \left\{ (x^m) \in l^2; \sum_{m=1}^{+\infty} |\lambda_m|^2 |x^m|^2 < +\infty \right\}.$$

Moreover, if D_0 is the set of all sequences $x = (x^m)$ such that $x^m \neq 0$ for a finite number of m, then for $x \in D_0$,

$$A(x^m) = ((\lambda_m i + A_m)x^m)$$

and
$$W(\lambda)x = R(\lambda)x,$$
where $R(\lambda)$, $\lambda \notin \sigma(A)$ is the resolvent of $S(t)$, $t \geq 0$. Consequently $D(A) = D$ and for $\lambda \notin \sigma(A)$, $R(\lambda) = W(\lambda)$.

One checks that for $\lambda \in \{i\lambda_m; m = 1, 2, \ldots\}$,
$$(\lambda I - A)W(\lambda) = I \quad \text{on } l^2,$$
$$W(\lambda)(\lambda I - A) = I \quad \text{on } D(A).$$

Hence $\sigma(A) = \{i\lambda_m; m = 1, 2, \ldots\}$. The proof of the lemma is complete. □

We go back to the proof of (iv). For arbitrary $\alpha \in \mathbf{R}^1$ and $\beta > 0$ define a new semigroup $\widetilde{S}(\cdot)$,
$$\widetilde{S}(t) = e^{\alpha t} S(\beta t), \quad t \geq 0.$$

The generator $\widetilde{A}$ of the semigroup $\widetilde{S}(\cdot)$ is given by
$$\widetilde{A} = \alpha I + \beta A, \quad D(\widetilde{A}) = D(A).$$

Moreover
$$\|\widetilde{S}(t)\| = e^{(\alpha+\beta)t} \quad \text{and} \quad \sup\{\operatorname{Re} \lambda; \lambda \in \sigma(\widetilde{A})\} = \alpha.$$

Taking in particular $\alpha = -\frac{1}{2}$ and $\beta = 1$ we have that
$$\|\widetilde{S}(t)\| = e^{\frac{1}{2}t}, \ t \geq 0 \ \text{and} \ \sup\{\operatorname{Re}\lambda; \lambda \in \sigma(\widetilde{A})\} = -\frac{1}{2}.$$

Since $\lim_{t \to +\infty} \|\widetilde{S}(t)\| = +\infty$, there exists, by Theorem A.5(i) (the Banach-Steinhaus theorem), $x \in E$ such that $\overline{\lim_{t \to +\infty}} \|\widetilde{S}(t)x\| = +\infty$.

Hence (3.6) does not imply (3.5) in the case of E being the complex Hilbert space $l_{\mathbf{C}}^2$. The case of E being a real Hilbert space follows from the following exercise. □

Exercise 3.1. Identify the complex Hilbert space $l_{\mathbf{C}}^2$ with the Cartesian product $l^2 \times l^2$ of two real l^2 spaces and the semigroup $S(t)$, $t \geq 0$, from Lemma 3.1 with its image $\mathcal{S}(t)$, $t \geq 0$, after the identification. Show that
$$\|\mathcal{S}(t)\| = e^t, \ t \geq 0 \ \text{and} \ \sigma(\mathcal{A}) = \{\pm i\lambda_n; n = 1, 2, \ldots\}.$$
where $\mathcal{A}$ is the generator of $\mathcal{S}(t)$, $t \geq 0$.

If a semigroup $S(t)$, $t \geq 0$, satisfies one of the conditions (3.2), (3.3), (3.4) then it is called *exponentially stable*, and its generator an *exponentially stable generator*.

§ 3.2. Liapunov's equation

As we know from § I.2.4, a matrix A is stable if and only if the Liapunov equation
$$A^*Q + QA = -I, \quad Q \geq 0, \tag{3.8}$$
has a nonnegative solution Q. A generalization of this result to the infinite dimensional case is the subject of the following theorem.

Theorem 3.2. *Assume that E is a real Hilbert space. The infinitesimal generator A of the semigroup is exponentially stable if and only if there exists a nonnegative, linear and continuous operator Q such that*
$$2\langle QAx, x\rangle = -|x|^2 \quad \text{for all} \ x \in D(A). \tag{3.9}$$

If the generator A is exponentially stable then equation (3.9) has exactly one nonnegative, linear and continuous solution Q.

Proof. Assume that $Q \geq 0$ is a linear and continuous operator solving (3.9). If $x \in D(A)$ then the function $v(t) = \langle Qz^x(t), z^x(t)\rangle$, $t \geq 0$, is differentiable, see Theorem 1.2. Moreover
$$\frac{d}{dt}v(t) = \langle Q\frac{dz^x}{dt}(t), z^x(t)\rangle + \langle Qz^x(t), \frac{dz^x}{dt}(t)\rangle$$
$$= 2\langle QAz^x(t), z^x(t)\rangle, \quad t \geq 0.$$

Therefore, for arbitrary $t \geq 0$,
$$\langle Qz^x(t), z^x(t)\rangle - \langle Qx, x\rangle = -\int_0^t |z^x(s)|^2 ds,$$
$$\int_0^t |z^x(s)|^2 ds + \langle Qz^x(t), z^x(t)\rangle = \langle Qx, x\rangle. \tag{3.10}$$

Letting t tend to $+\infty$ in (3.10) we obtain
$$\int_0^{+\infty} |S(s)x|^2 ds \leq \langle Qx, x\rangle, \quad x \in D(A). \tag{3.11}$$

Since the subspace $D(A)$ is dense in E, estimate (3.11) holds for $x \in E$. By Theorem 3.1, the semigroup $S(t)$, $t \geq 0$, is exponentially stable.

If the semigroup $S(t)$, $t \geq 0$, is exponentially stable, then $z^x(t) \longrightarrow 0$ as $t \to +\infty$, and, letting t tend to $+\infty$ in (3.10), we obtain
$$\int_0^{+\infty} \langle S^*(t)S(t)x, x\rangle \, dt = \langle Qx, x\rangle, \quad x \in E. \tag{3.12}$$

Therefore, there exists a solution Q to (3.9) that is given by

$$Qx = \int_0^{+\infty} S^*(t)S(t)x\, dt, \quad x \in E. \tag{3.13}$$

□

§3.3. Stabilizability and controllability

A linear control system

$$\dot{y} = Ay + Bu, \quad y(0) = x \tag{3.14}$$

is said to be *exponentially stabilizable* if there exists a linear, continuous operator $K\colon E \longrightarrow U$ such that the operator A_K,

$$A_K = A + BK \quad \text{with the domain} \quad D(A_K) = D(A), \tag{3.15}$$

generates an exponentially stable semigroup $S_K(t)$, $t \geq 0$. That the operator A_K generates a semigroup follows from the Phillips theorem, see §1.4. The operator K should be interpreted as the feedback law. If, in (3.14), $u(t) = Ky(t)$, $t \geq 0$, then

$$\dot{y} = (A + BK)y, \quad y(0) = x, \tag{3.16}$$

and

$$y(t) = S_K(t)x, \quad t \geq 0. \tag{3.17}$$

One of the main results of finite dimensional linear control theory was Theorem I.2.9 stating that controllable systems are stabilizable. In the infinite dimensional theory there are many concepts of controllability and stabilizability, and therefore relationships are much more complicated. Here is a typical result.

Theorem 3.3. (i) *Null controllable systems are exponentially stabilizable.*

(ii) *There are approximate controllable systems which are not exponentially stabilizable.*

Proof. (i). Let $u^x(\cdot)$ be a control transferring x to 0 in time $T^x > 0$ and $u^x(t) = 0$ for $t > T^x$. If $y^x(\cdot)$ is the output corresponding to $u^x(\cdot)$ then $y^x(t) = 0$ for $t \geq T^x$. Therefore

$$J(x, u^x(\cdot)) = \int_0^{+\infty} \left(|y^x(t)|^2 + |u^x(t)|^2\right) dt < +\infty, \tag{3.18}$$

and the assumptions of Theorem 4.3, from Chapter 4, are satisfied with $Q = I$ and $R = I$. Consequently, there exists a nonnegative continuous operator

$\tilde{P}$ on H such that for the feedback control $\tilde{u}(t) = \tilde{K}\tilde{y}(t) = -B^*\tilde{P}\tilde{y}(t)$ and the corresponding solution $\tilde{y}(t)$, $t \geq 0$, of

$$\dot{\tilde{y}}(t) = A\tilde{y}(t) + B\tilde{K}\tilde{y}(t) = (A + B\tilde{K})\tilde{y}(t), \quad t \geq 0,$$

one has

$$J(x, \tilde{u}(\,\cdot\,)) = \int_0^{+\infty} \left(|\tilde{y}(t)|^2 + |\tilde{u}(t)|^2\right) dt \leq J(x, u^x(\,\cdot\,)) < +\infty, \quad x \in E.$$

But $\tilde{y}(t) = S_{\tilde{K}}(t)x$, $t \geq 0$, where $S_{\tilde{K}}(\,\cdot\,)$ is the semigroup generated by $A_{\tilde{K}}$. Consequently,

$$\int_0^{+\infty} |S_{\tilde{K}}(t)x|^2 dt = \int_0^{+\infty} |\tilde{y}(t)|^2 dt < +\infty \text{ for all } x \in E,$$

and it follows from Theorem 3.1(i) that the semigroup $S_{\tilde{K}}(\,\cdot\,)$ is exponentially stable.

(ii) Let us consider a control system of the form

$$\dot{y} = Ay + hu, \quad y(0) = x, \tag{3.19}$$

in a separable Hilbert space E with an orthonormal and complete basis (e_m). Let A be a linear and bounded operator on E given by

$$Ax = \sum_{m=1}^{+\infty} \lambda_m \langle x, e_m \rangle e_m, \quad x \in E,$$

with (λ_m) a strictly increasing sequence converging to 0. Let $h \in E$ be an element such that $< h, e_m > \neq 0$ for all $m = 1, 2, \ldots$ and

$$\sum_{m=1}^{+\infty} \frac{|\langle h, e_m \rangle|^2}{|\lambda_m|} < +\infty.$$

Then (3.19) is approximately controllable, see Theorem 2.8.

To show that (3.19) is not exponentially stabilizable, consider an arbitrary bounded linear operator $K \colon E \longrightarrow U = \mathbf{R}^1$. It is of the form

$$Kx = \langle k, x \rangle, \quad x \in E,$$

where k is an element in E. We prove that $0 \in \sigma(A + BK)$. Let us fix $z \in E$ and consider the following linear equation

$$Ax + BKx = z, \quad x \in E. \tag{3.20}$$

§ 3.3. Stabilizability and controllability

Denote
$$\langle h, e_m \rangle = \gamma_m, \quad \langle x, e_m \rangle = \xi_m, \quad \langle z, e_m \rangle = \eta_m, \quad m = 1, 2, \ldots.$$

Then (3.20) is equivalent to an infinite system of equations
$$\lambda_m \xi_m + \gamma_m \langle k, x \rangle = \eta_m, \quad m = 1, 2, \ldots,$$

from which it follows that
$$\xi_m = \frac{\eta_m}{\lambda_m} - \frac{\gamma_m}{\lambda_m} \langle k, x \rangle, \quad m = 1, 2, \ldots.$$

Since
$$\sum_{m=1}^{+\infty} \left| \frac{\gamma_m}{\lambda_m} \right|^2 < +\infty \quad \text{and} \quad \sum_{m=1}^{+\infty} |\xi_m|^2 < +\infty,$$

therefore
$$\sum_{m=1}^{+\infty} \left| \frac{\eta_m}{\lambda_m} \right|^2 < +\infty,$$

and $z \in E$ cannot be an arbitrary element of E. Hence $0 \in \sigma(A + BK)$.

It follows from Theorem 3.1(ii) that (3.19) is not exponentially stabilizable. □

It follows from Theorem 3.3 that all exactly controllable systems are exponentially stabilizable. We will show that in a sense a converse statement is true.

We say that system (3.14) is *completely stabilizable* if for arbitrary $\omega \in \mathbf{R}$ there exist a linear, continuous operator $K \colon E \longrightarrow$ and a constant $M > 0$ such that
$$|S_K(t)| \leq M e^{\omega t} \quad \text{for } t \geq 0. \tag{3.21}$$

The following theorem together with Theorem 3.2(i) are an infinite dimensional version of Wonham's theorem (Theorem I.2.9).

Theorem 3.4. *If system (3.14) is completely stabilizable and the operator A generates a group of operators $S(t)$, $t \in \mathbf{R}$, then system (3.14) is exactly controllable in some time $T > 0$.*

Proof. By Theorem 1.1 there exist $N > 0$ and $\nu \in \mathbf{R}$ such that
$$|S(-t)| \leq N e^{\nu t}, \quad t \geq 0. \tag{3.22}$$

Assume that for a feedback $K \colon E \longrightarrow U$ and some constants $M > 0$ and $\omega \in \mathbf{R}^1$
$$|S_K(t)| \leq M e^{\omega t}, \quad t \geq 0.$$

Since

$$S(t)x = S_K(t)x - \int_0^t S(t-s)BKS_K(s)x\,ds, \quad x \in E,\ t \geq 0,$$

therefore, for arbitrary $f \in E^*$, $t \geq 0$, $x \in E$,

$$|S^*(t)f| \leq |S_K^*(t)f| + \int_0^t |S_K^*(s)K^*B^*S^*(t-s)f|\,ds$$

$$\leq Me^{\omega t}|f| + |K|M \int_0^t e^{\omega(t-s)}|B^*S^*(s)f|\,ds$$

$$\leq Me^{\omega t}|f| + |K|M \left(\int_0^t e^{2\omega s}\,ds\right)^{1/2} \left(\int_0^t |B^*S^*(s)f|^2\,ds\right)^{1/2}.$$

If $|f| = 1$, then $|S^*(-t)S^*(t)f| = 1$ and

$$|S^*(-t)|^{-1} = |S(-t)|^{-1} \leq |S^*(t)f|, \quad t \geq 0.$$

The above estimates imply that if $|f| = 1$ then

$$|S(-t)|^{-1} \leq Me^{\omega t} \tag{3.23}$$
$$+ |K|M \left(\int_0^t e^{2\omega s}\,ds\right)^{1/2} \left(\int_0^t |B^*S^*(s)f|^2\,ds\right)^{1/2}.$$

Assume that system (3.14) is not exactly controllable at a time $t > 0$. It follows from the results of § 2.1 that for arbitrary $c > 0$ there exists $f \in E^*$ such that

$$\int_0^t |B^*S^*(s)f|^2\,ds \leq c \text{ and } |f| = 1.$$

Since c is arbitrary and from (3.23)

$$|S(-t)|^{-1} \leq Me^{\omega t}. \tag{3.24}$$

If (3.14) is not exactly controllable at any time $t > 0$ then by (3.24) and (3.22)

$$M^{-1}e^{-\omega t} \leq |S(-t)| \leq Ne^{\nu t} \text{ for all } t > 0.$$

This way we have $\omega \geq -\nu$, and (3.14) cannot be completely stabilizable. □

Bibliographical notes

Theorem 3.1(i) and Theorem 3.1(iii) are due to R. Datko [19,20]. Lemma 3.1 is taken from [64]. Theorem 3.2 is due to R. Datko [20]. As noticed in [65], Theorem 3.3(i) is implicitly contained in R. Datko [20]. Theorem 3.4 is due to M. Megan [40]. For its generalization, see [66]. In condition (3.4) function $\|z^x(\cdot)\|^2$ can be replaced by the much more general functions $\mathcal{N}(\|z^x(\cdot)\|)$; see [62] and [50].

For a review of stability and stabilizability results we refer the reader to [47] and [52].

Chapter 4
Linear regulators in Hilbert spaces

In this chapter the existence of a solution to an operator Riccati equation related to the linear regulator problem in separable Hilbert spaces is shown first. Then a formula for the optimal solution on an arbitrary finite time interval is given. The existence of an optimal solution on the infinite time interval is dicussed as well. Some applications to the stabilizability problem are also included.

§4.1. Introduction

Our discussion of optimal control problems in infinite dimensions will be limited to the linear regulator problem. In what follows we will assume that the stated space denoted here by H as well as the space U of the control parameters are real, separable Hilbert spaces. The control system is given by
$$\dot{y} = Ay + Bu, \quad y(0) = x \in H, \tag{4.1}$$
where A is the infinitesimal generator of a semigroup $S(t)$, $t \geq 0$, and B is a linear and continuous operator from U into H. By a solution to (4.1) we mean the weak solution, see §1.7, given by
$$y(t) = S(t)x + \int_0^t S(t-s)Bu(s)\,ds, \quad r \geq 0. \tag{4.2}$$

Admissible controls are required to be Borel measurable and locally square integrable
$$\int_0^T |u(s)|^2 ds < +\infty \text{ for } T > 0.$$

We will also consider closed loop controls of the form
$$u(t) = K(t)y(t), \quad t \geq 0, \tag{4.3}$$
where $K(\cdot)$ is a function with values in the space $L(H,U)$ of all linear and continuous operators from H into U.

Let $Q \in L(H,H)$, $P_0 \in L(H,H)$, $R \in L(U,U)$ be fixed self-adjoint continuous operators, and let R be, in addition, an invertable operator with the continuous inverse R^{-1}.

§4.1. Introduction

The *linear regulator problem with finite horizon T* (compare §III.1.3), consists of finding a control $u(\,\cdot\,)$ minimizing the functional $J_T(x,\cdot)$:

$$J_T(x,u) = \int_0^T (\langle Qy(s), y(s)\rangle + \langle Ru(s), u(s)\rangle)\,ds + \langle P_0 y(T), y(T)\rangle. \quad (4.4)$$

The *linear regulator problem with infinite horizon* consists of minimizing the functional

$$J(x,u) = \int_0^{+\infty} (\langle Qy(s), y(s)\rangle + \langle Ru(s), u(s)\rangle)\,ds. \quad (4.5)$$

We present solutions to both problems.

We are particularly interested in optimal controls given in the closed loop form (4.3). Such controls define outputs $y(\,\cdot\,)$ only indirectly as solutions of the following integral equation:

$$y(t) = S(t)x + \int_0^t S(t-s)BK(s)y(s)\,ds, \quad t \geq 0, \quad (4.6)$$

and we start by giving conditions implying existence of a unique continuous solutions to (4.6).

We say that $K\colon [0,T] \longrightarrow L(H,U)$ is *strongly continuous* if for arbitrary $h \in H$ the function $K(t)h$, $t \in [0,T]$, is continuous. In the obvious way this definition generalizes to the function $K(\,\cdot\,)$ defined on $[0,+\infty)$. We check easily that if $K(\,\cdot\,)$ is strongly continuous on $[0,T]$ and an H-valued function $h(t)$, $t \in [0,T]$, is continuous, then also $K(t)h(t)$, $t \in [0,T]$, is continuous. There also exists a constant $M_1 > 0$ such that $|K(t)| \leq M_1$, $t \in [0,T]$.

Lemma 4.1. *If an operator valued function $K(\,\cdot\,)$ is strongly continuous on $[0,T]$ then equation (4.6) has exactly one continuous solution $y(t)$, $t \in [0,T]$.*

Proof. We apply the contraction mapping principle. Let us denote by $C_T = C(0,T;H)$ the space of all continuous, H-valued functions defined on $[0,T]$ with the norm $\|h\|_T = \sup\{|h(t)|; t \in [0,T]\}$. Let us remark that if $h \in C_T$ then the function

$$\mathcal{A}_T h(t) = \int_0^t S(t-s)K(s)h(s)\,ds, \quad t \in [0,T], \quad (4.7)$$

is well defined and continuous.

Let $M > 0$ and $\omega > 0$ be constants such that

$$|S(t)| \leq Me^{\omega t}, \quad t \geq 0.$$

Then

$$|\mathcal{A}_T h(t)| \leq \omega^{-1} M M_1 (e^{\omega T} - 1)\|h\|_T, \quad h \in C_T.$$

Therefore, if
$$MM_1(e^{\omega T_0} - 1) < \omega,$$
then $\|\mathcal{A}_{T_0}\| < 1$, and, by the contraction principle, see § A.1, equation
$$y(t) = S(t)x + \mathcal{A}_{T_0} y(t), \quad t \in [0, T_0],$$
has exactly one solution $y(\,\cdot\,)$ in C_{T_0}. Let k be a natural number such that $kT_0 \geq T$. By the above argument, equation (4.6) has exactly one solution on the arbitrary interval $[jT_0, (j+1)T_0]$, $j = 0, 1, \ldots, k-1$, therefore also on $[0, T]$. □

Let us assume in particular that $K(t) = K$ for all $t \geq 0$. By the Phillips theorem, see § 1.4, the operator $A_K = A + K$, $D(A_K) = D(A)$ defines a semigroup of operators $S_K(t)$, $t \geq 0$. Its relation with the solution $y(\,\cdot\,)$ of (4.6) is given by the following proposition, the proof of which is left as an exercise.

Proposition 4.1. *If $K(t) = K$, $t \geq 0$, and $S_K(t)$, $t \geq 0$, is the semigroup generated by $A_K = A + K$, $D(A_K) = D(A)$, then the solution $y(\,\cdot\,)$ to (4.5) is given by*
$$y(t) = S_K(t)x, \quad t \geq 0.$$

§ 4.2. The operator Riccati equation

To solve the linear regulator problem in Hilbert spaces we proceed similarly to the finite dimensional case, see § III.1.3, and start from an analysis of an infinite dimensional version of the Riccati equation
$$\dot{P} = A^*P + PA + Q - PBR^{-1}B^*P, \quad P(0) = P_0. \tag{4.8}$$

Taking into account that operators A and A^* are not everywhere defined, the concept of a solution of (4.8) requires special care.

We say that a function $P(t)$, $t \geq 0$, with values in $L(H, H)$, $P(0) = P_0$ is a *solution* to (4.8) if, for arbitrary $g, h \in D(A)$, the function $\langle P(t)h, g\rangle$, $t \geq 0$, is absolutely continuous and
$$\frac{d}{dt}\langle P(t)h, g\rangle = \langle P(t)h, Ag\rangle + \langle P(t)Ah, g\rangle \tag{4.9}$$
$$+ \langle Qh, g\rangle - \langle PBR^{-1}B^*Ph, g\rangle \text{ for almost all } t \geq 0.$$

Theorem 4.1. *A strongly continuous operator valued function $P(t)$, $t \geq 0$, is a solution to (4.8) if and only if, for arbitrary $h \in H$, $P(t)$, $t \geq 0$, is a solution to the following integral equation:*
$$P(t)h = S^*(t)P_0 S(t)h \tag{4.10}$$
$$+ \int_0^t S^*(t-s)(Q - P(s)BR^{-1}B^*P(s))S(t-s)h\,ds, \quad t \geq 0.$$

§ 4.2. The operator Riccati equation

The proof of the theorem follows immediately from Lemma 2.4 below concerned with linear operator equations

$$\dot{P}(t) = A^*P + PA + Q(t), \quad t \geq 0, \; P(0) = P_0. \tag{4.11}$$

In equation (4.11) values $Q(t)$, $t \geq 0$, are continuous, self-adjoint operators.

An operator valued function $P(t)$, $t \geq 0$, $P(0) = P_0$ is a *solution to* (4.11) if for arbitrary $g, h \in D(A)$, function $\langle P(t)h, g \rangle$, $t \geq 0$ is absolutely continuous and

$$\frac{d}{dt}\langle P(t)h, g \rangle = \langle P(t)h, Ag \rangle + \langle P(t)Ah, g \rangle + \langle Q(t)h, g \rangle \text{ for almost all } t \geq 0. \tag{4.12}$$

Lemma 4.2. *If $Q(t)$, $t \geq 0$, is strongly continuous then a solution to (4.11) exists and is given by*

$$P(t)h = S^*(t)P_0 S(t)h + \int_0^t S^*(t-s)Q(s)S(t-s)\,ds, \quad t \geq 0, \; h \in H. \tag{4.13}$$

Proof. It is not difficult to see that function $P(\,\cdot\,)$ given by (4.13) is well defined and strongly continuous. In addition,

$$\langle P(t)h, g \rangle = \langle P_0 S(t)h, S(t)g \rangle + \int_0^t \langle Q(s)S(t-s)h, S(t-s)g \rangle \, ds, \quad t \geq 0.$$

Since for $h, g \in D(A)$

$$\frac{d}{dt}\langle P_0 S(t)h, S(t)g \rangle = \langle P_0 S(t)Ah, S(t)g \rangle + \langle P_0 S(t)h, S(t)Ag \rangle, \quad t \geq 0,$$

and

$$\frac{d}{dt}\langle Q(s)S(t-s)h, S(t-s)g \rangle = \langle Q(s)S(t-s)Ah, S(t-s)g \rangle$$
$$+ \langle Q(s)S(t-s)h, S(t-s)Ag \rangle, \quad t \geq s \geq 0,$$

hence

$$\frac{d}{dt}\langle P(t)h, g \rangle = \langle S^*(t)P_0 S(t)Ah, g \rangle + \langle S^*(t)P_0 S(t)h, Ag \rangle$$
$$+ \langle Q(t)h, g \rangle + \int_0^t \big(\langle S^*(t-s)Q(s)S(t-s)Ah, g \rangle$$
$$+ \langle S^*(t-s)Q(s)S(t-s)h, Ag \rangle\big) \, ds.$$

Taking into account (4.13) we obtain (4.12).

Conversely, let us assume that (4.12) holds. For $s \in [0, t]$

$$\frac{d}{ds}\langle P(s)S(t-s)h, S(t-s)g\rangle$$
$$= \langle P(s)S(t-s)h, S(t-s)Ag\rangle + \langle P(s)S(t-s)Ah, S(t-s)g\rangle$$
$$+ \langle Q(s)S(t-s)h, S(t-s)g\rangle$$
$$- \langle P(s)S(t-s)Ah, S(t-s)g\rangle - \langle P(s)S(t-s)h, S(t-s)Ag\rangle$$
$$= \langle Q(s)S(t-s)h, S(t-s)g\rangle.$$

Integrating the above identity over $[0, t]$, we obtain

$$\langle P(t)h, g\rangle - \langle S^*(t)P_0 S(t)h, g\rangle = \int_0^t \langle S^*(t-s)Q(s)S(t-s)h, g\rangle\, ds. \quad (4.14)$$

Since $D(A)$ is dense in H, equality (4.14) holds for arbitrary elements $h, g \in H$. □

§4.3. The finite horizon case

The following theorem extends the results of §III.1.3 to Hilbert state spaces.

Theorem 4.2. (i) *Equation (4.8) has exactly one global solution $P(s)$, $s \geq 0$. For arbitrary $s \geq 0$ the operator $P(s)$ is self-adjoint and nonnegative definite.*

(ii) *The minimal value of functional (4.4) is equal to $\langle P(T)x, x\rangle$ and the optimal control $\hat{u}(\cdot)$ is given in the feedback form*

$$\hat{u}(t) = \hat{K}(t)\hat{y}(t),$$
$$\hat{K}(t) = -R^{-1}B^*P(T-t), \quad t \in [0, T].$$

Proof. We prove first the existence of a local solution to (4.10). We apply the following version of the contraction mapping theorem (compare Theorem A.1).

Lemma 4.3. *Let $\mathcal{A}$ be a transformation from a Banach space C into C, v an element of C and α a positive number. If $\mathcal{A}(0) = 0$, $\|v\| \leq \frac{1}{2}\alpha$ and*

$$\|\mathcal{A}(p_1) - \mathcal{A}(p_2)\| \leq \frac{1}{2}\|p_1 - p_2\|, \quad \text{if } \|p_1\| \leq \alpha, \|p_2\| \leq \alpha,$$

then equation

$$\mathcal{A}(p) + v = p$$

§4.3. The finite horizon case

has exactly one solution p satisfying $\|p\| \leq \alpha$.

Proof. Let $\tilde{\mathcal{A}}(p) = \mathcal{A}(p) + v$, $p \in C$. By an elementary induction argument

$$\|\tilde{\mathcal{A}}^k(0)\| \leq \alpha\left(1 - \left(\frac{1}{2}\right)^k\right), \quad k = 1, 2, \ldots.$$

Hence $\|\tilde{\mathcal{A}}^k(0)\| \leq \alpha$, $k = 1, 2, \ldots$. From the Lipschitz condition it follows that the sequence $(\tilde{\mathcal{A}}^k(0))$ converges to a fixed point of $\tilde{\mathcal{A}}$. □

Let us denote by C_T the Banach space of all strongly continuous functions $P(t)$, $t \in [0, T]$, having values being self-adjoint operators on H, with the norm

$$\|P(\cdot)\|_T = \sup\{|P(t)|;\ t \in [0, T]\}.$$

For $P(\cdot) \in C_T$ we define

$$\mathcal{A}_T(P)(t)h = -\int_0^t S^*(t-s)P(s)BR^{-1}B^*P(s)S(t-s)h\,ds$$

and

$$v(t)h = S^*(t)P_0 S(t)h + \int_0^t S^*(t-s)QS(t-s)h\,ds, \quad h \in H,\ t \in [0, T].$$

Equation (4.10) is equivalent to

$$P = v + \mathcal{A}_T(P). \tag{4.15}$$

It is easy to check that $\mathcal{A}_T$ maps C_T into C_T and $v(\cdot) \in C_T$.

If $\|P_1\|_T \leq \alpha$, $\|P_2\|_T \leq \alpha$, then

$$|P_1(s)BR^{-1}B^*P_1(s) - P_2(s)BR^{-1}B^*P_2(s)|$$
$$\leq |(P_1(s) - P_2(s))BR^{-1}B^*P_1(s)| + |P_2(s)BR^{-1}B^*(P_2(s) - P_1(s))|$$
$$\leq 2\alpha|BR^{-1}B^*|\,\|P_1 - P_2\|_T, \quad s \in [0, T].$$

Therefore

$$\|\mathcal{A}_T(P_1) - \mathcal{A}_T(P_2)\|_T \leq 2\alpha|BR^{-1}B^*|M\int_0^T e^{2\omega s}\,ds\,\|P_1 - P_2\|_T$$
$$\leq \beta(\alpha, T)\|P_1 - P_2\|_T,$$

where

$$\beta(\alpha, T) = 2\alpha\omega^{-1}M|BR^{-1}B^*|(e^{\omega T} - 1). \tag{4.16}$$

At the same time,

$$\|v\|_T \leq M^2 e^{2\omega T}|P_0| + M^2 \int_0^T e^{2\omega s}\,ds|Q| \leq M^2 e^{2\omega T}|P_0| + \frac{M^2 e^{2\omega T}}{2\omega}|Q|.$$

Consequently, if for numbers $\alpha > 0$ and $T > 0$,

$$2\alpha\omega^{-1}M|BR^{-1}B^*|(e^{\omega T} - 1) < \frac{1}{2}, \tag{4.17}$$

$$M^2 e^{2\omega T}|P_0| + \frac{M^2 e^{2\omega T}}{2\omega}|Q| < \frac{\alpha}{2}, \tag{4.18}$$

then, by Lemma 4.3, equation (4.10) has exactly one solution in the set

$$\left\{P(\cdot) \in C_T; \sup_{t \leq T}|P(t)| \leq \alpha\right\}.$$

For given operators P_0 and Q and for given numbers $\omega > 0$ and $M > 0$, one can find $\alpha > 0$ such that

$$M^2|P_0| + \frac{M^2}{2\omega}|Q| < \frac{\alpha}{2}.$$

One can therefore find $T = T(\alpha) > 0$ such that both conditions (4.17) and (4.18) are satisfied. By Lemma 4.3, equation (4.10) has a solution on the interval $[0, T(\alpha)]$. □

To proceed further we will need an interpretation of the local solution.

Lemma 4.4. *Let us assume that a function $P(t)$, $t \in [0, T_0]$, is a solution to (4.10). Then for arbitrary control $u(\cdot)$ and the corresponding output $y(\cdot)$,*

$$J_{T_0}(x, u) = \langle P(T_0)x, x\rangle \tag{4.19}$$

$$+ \int_0^{T_0}|R^{1/2}u(s) + R^{-1/2}B^*P(T_0 - s)y(s)|^2\,ds.$$

Proof. Assume first that $x \in D(A)$ and that $u(\cdot)$ is of class C^1. Then $y(\cdot)$ is the strong solution of the equation

$$\frac{d}{dt}y(t) = Ay(t) + Bu(t), \quad y(0) = x.$$

It follows from (4.9) that for arbitrary $z \in D(A)$ function $\langle P(T_0 - t)z, z\rangle$, $t \in [0, T_0]$, is of class C^1, and its derivative is given by

$$\frac{d}{dt}\langle P(T_0 - t)z, z\rangle = -\langle P(T_0 - t)z, Az\rangle - \langle P(T_0 - t)Az, z\rangle - \langle Qz, z\rangle$$

$$+ \langle P(T_0 - t)BR^{-1}B^*P(T_0 - t)z, z\rangle, \quad t \in [0, T_0].$$

§ 4.3. The finite horizon case

Hence

$$\frac{d}{dt}\langle P(T_0 - t)y(t), y(t)\rangle$$
$$= -\langle P(T_0 - t)y(t), Ay(t)\rangle - \langle P(T_0 - t)Ay(t), y(t)\rangle - \langle Qy(t), y(t)\rangle$$
$$+ \langle P(T_0 - t)BR^{-1}B^*P(T_0 - t)y(t), y(t)\rangle$$
$$+ \langle P(T_0 - t)\dot{y}(t), y(t)\rangle + \langle P(T_0 - t)y(t), \dot{y}(t)\rangle$$
$$= -\langle Qy(t), y(t)\rangle + \langle P(T_0 - t)BR^{-1}B^*P(T_0 - t)y(t), y(t)\rangle$$
$$+ \langle P(T_0 - t)Bu(t), y(t)\rangle + \langle P(T_0 - t)y(t), Bu(t)\rangle$$
$$= -\langle Qy(t), y(t)\rangle + |R^{1/2}u(t) + R^{-1/2}B^*P(T_0 - t)y(t)|^2$$
$$- \langle Ru(t), u(t)\rangle, \quad t \in [0, T_0].$$

Integrating this equality over $[0, T_0]$ we obtain

$$-\langle P(T_0)x, x\rangle = -J_{T_0}(x, u) + \int_0^{T_0} |R^{1/2}u(t) + R^{-1/2}B^*P(T_0 - t)y(t)|^2 dt,$$

and therefore (4.19) holds. Since the solutions $y(\cdot)$ depend continuously on the initial state x and on the control $u(\cdot)$, equality (4.19) holds for all $x \in H$ and all $u(\cdot) \in L^2(0, T_0; U)$. □

It follows from Lemma 4.4 that for arbitrary $u(\cdot) \in L^2(0, T_0; U)$,

$$\langle P(T_0)x, x\rangle \leq J_{T_0}(x, u).$$

By Lemma 4.1, equation

$$\hat{y}(t) = S(t)x - \int_0^t S(t - s)BR^{-1}B^*P(T_0 - s)\hat{y}(s)\, ds, \quad t \in [0, T_0],$$

has exactly one solution $\hat{y}(t)$, $t \in [0, T]$. Define control $\hat{u}(\cdot)$:

$$\hat{u}(t) = -R^{-1}B^*P(T_0 - t)\hat{y}(t), \quad t \in [0, T_0].$$

Then $\hat{y}(\cdot)$ is the output corresponding to $\hat{u}(\cdot)$ and by (4.19),

$$J_{T_0}(x, \hat{u}(\cdot)) = \langle P(T_0)x, x\rangle. \tag{4.20}$$

Therefore $\hat{u}(\cdot)$ is the optimal control.

It follows from (4.20) that $P(T_0) \geq 0$. Setting $u(\cdot) = 0$ in (4.19) we obtain

$$0 \leq \langle P(T_0)x, x\rangle \leq \int_0^{T_0} \langle QS(t)x, S(t)x\rangle\, dt,$$

$$0 \leq \langle P(T_0)x, x\rangle \leq \langle \int_0^{T_0} S^*(t)QS(t)\, dt\, x, x\rangle.$$

Consequently

$$|P(T)| \leq \int_0^{T_0} |S^*(t)QS(t)|\,dt \quad \text{for } T \leq T_0.$$

Let now T_1 and α be numbers such that

$$M^2 \int_0^{T_1} |S^*(t)QS(t)|\,dt + \frac{M^2}{2\omega}|Q| < \frac{\alpha}{2}.$$

We can assume that $T_1 < T(\alpha)$. It follows from the proof of the local existence of a solution to (4.10) that the solution to (4.10) exists on $[0, T_1]$. Repeating the proof on consecutive intervals $[T_1, 2T_1]$, $[2T_1, 3T_1], \ldots$, we obtain that there exists a unique global solution (4.8).

Nonnegativeness of $P(\,\cdot\,)$ as well as the latter part of the theorem follow from Lemma 4.4. □

§4.4. The infinite horizon case: Stabilizability and detectability

We proceed to the regulator problem on the interval $[0, +\infty)$ (compare to (4.5)).

An *operator algebraic Riccati equation* is of the form

$$2\langle PAx, x\rangle - \langle PBR^{-1}B^*Px, x\rangle + \langle Qx, x\rangle = 0, \quad x \in D(A), \qquad (4.21)$$

where a nonnegative operator $P \in L(H, H)$ is unknown.

Theorem 4.3. *Assume that for arbitrary $x \in H$ there exists a control $u^x(t)$, $t \geq 0$, such that*

$$J(x, u^x(\,\cdot\,)) < +\infty. \qquad (4.22)$$

Then there exists a nonnegative operator $\widetilde{P} \in L(H, H)$ satisfying (4.21) such that $\widetilde{P} \leq P$ for an arbitrary nonnegative solution P of (4.21). Moreover, the control $\tilde{u}(\,\cdot\,)$ given in the feedback form

$$\tilde{u}(t) = -R^{-1}B^*\widetilde{P}y(t), \quad t \geq 0,$$

minimizes functional (1.5). The minimal value of this functional is equal to

$$\langle \widetilde{P}x, x\rangle.$$

Proof. Let $P(t)$, $t \geq 0$, be the solution of the Riccati equation (4.9) with the initial condition $P_0 = 0$. It follows from Theorem 4.2(ii) that, for arbitrary $x \in H$, function $\langle P(t)x, x\rangle$, $t \geq 0$, is nondecreasing. Moreover,

$$\langle P(t)x, x\rangle \leq J(x, u^x(\,\cdot\,)) < +\infty, \quad x \in H.$$

§ 4.4. The infinite horizon case: Stabilizability and detectability

Therefore there exists a finite $\lim_{t\uparrow+\infty}\langle P(t)x,x\rangle$, $x\in H$. On the other hand,

$$\langle P(t)x,y\rangle = \frac{1}{2}(\langle P(t)(x+y),x+y\rangle - \langle P(t)(x-y),x-y\rangle) \quad (4.23)$$

for arbitrary $x,y\in A$.

Applying Theorem A.5 (the Banach-Steinhaus theorem), first to the family of functionals $\langle P(t)x,\cdot\rangle$, $t\geq 0$, for arbitrary $x\in H$ and then to the family of operators $P(t)$, $t\geq 0$, we obtain that $\sup_{t\geq 0}|P(t)|=c<+\infty$. Hence

$$|a(x,y)| \leq (\sup_{t\geq 0}|P(t)|)|x|\,|y| \leq c|x|\,|y|, \quad x,y\in H.$$

Therefore there exists $\widetilde{P}\in L(H,H)$ such that

$$a(x,y)=\langle \widetilde{P}x,y\rangle, \quad x,y\in H.$$

The operator $\widetilde{P}$ is self-adjoint, nonnegative definite, because $a(x,y)=a(y,x)$, $a(x,x)\geq 0$, $x,y\in H$.

To show that $\widetilde{P}$ satisfies (4.21) let us fix $x\in D(A)$ and consider (4.9) with $h=y=x$. Then

$$\frac{d}{dt}\langle P(t)x,x\rangle = \langle P(t)x,Ax\rangle + \langle P(t)Ax,x\rangle \quad (4.24)$$
$$+ \langle Qx,x\rangle - \langle PBR^{-1}B^*Px,x\rangle.$$

Letting t tend to $+\infty$ in (4.24) and arguing as in the proof of Theorem III.1.4 we easily show that $\widetilde{P}$ is a solution to (4.21).

The proof of the final part of the theorem is completely analogous to that of Theorem III.1.4. □

Let A be the infinitesimal generator of a C_0-semigroup $S(t)$, $t\geq 0$, and let $B\in L(U,H)$. The pair (A,B) is said to be *exponentially stabilizable* if there exists $K\in L(H,U)$ such that the operator $A_K=A+BK$, $D(A_K)=D(A)$ generates an exponentially stable semigroup (compare to § 3.3). Note that in particular if the pair (A,B) is null controllable then it is exponentially stabilizable.

Let $C\in L(H,V)$ where V is another separable Hilbert space. The pair (A,C) is said to be *exponentially detectable* if the pair (A^*,C^*) is *exponentially stabilizable*.

The following theorem is a generalization of Theorem III.1.5.

Theorem 4.4. (i) *If the pair (A,B) is exponentially stabilizable then the equation (4.21) has at least one nonnegative solution $P\in L(H,H)$.*

(ii) *If $Q = C^*C$ and the pair (A,C) is exponentially detectable then equation (4.21) has at most one solution and if P is the solution of (4.21) then the operator $A - BR^{-1}B^*P$ is exponentially stable and the feedback $K = -R^{-1}B^*P$ exponentially stabilizes (4.1).*

Proof. (i) If the pair (A, B) is exponentially stabilizable then the assumptions of Theorem 4.3 are satisfied and therefore (4.21) has at least one solution.

(ii) We prove first a generalization of Lemma III.1.3.

Lemma 4.5. *Assume that for a nonnegative operator $M \in L(H, H)$ and $K \in L(H, U)$*

$$2\langle M(A + BK)x, x\rangle + \langle C^*Cx, x\rangle + \langle K^*RKx, x\rangle = 0 \quad x \in D(A). \quad (4.25)$$

(i) *If the pair (A, C) is exponentially detectable then the operator $A_K = A + BK$, $D(A_K) = D(A)$ is exponentially stable.*
(ii) *If in addition $P \in L(H, H)$, $P \geq 0$ is a solution to (4.21) with $Q = C^*C$, then*

$$P \geq M. \quad (4.26)$$

Proof. (i) Let $S_1(\cdot)$ be the semigroup generated by A_K. Since the pair (A, C) is exponentially detectable, there exists an operator $L \in L(V, H)$ such that the operator $A_{L^*}^* = A^* + C^*L^*$, $D(A_{L^*}^*) = D(A^*)$, is exponentially stable. Hence the operator $\widetilde{A} = (A^* + C^*L^*)^* = A + LC$, $D(\widetilde{A}) = D(A)$ generates an exponentially stable semigroup $S_2(\cdot)$. Let $y(t) = S_1(t)x$, $t \geq 0$. Since

$$A + BK = (A + LC) + (LC + BK),$$

therefore, by Proposition 4.1,

$$y(t) = S_2(t)x + \int_0^t S_2(t-s)(LC + BK)y(s)\,ds. \quad (4.27)$$

We show that

$$\int_0^{+\infty} |Cy(s)|^2 ds < +\infty \quad \text{and} \quad \int_0^{+\infty} |Ky(s)|^2 ds < +\infty. \quad (4.28)$$

Assume first that $x \in D(A)$. Then

$$\dot{y}(t) = (A + BK)y \quad \text{and} \quad \frac{d}{dt}\langle My(t), y(t)\rangle = 2\langle M\dot{y}(t), y(t)\rangle, \quad t \geq 0.$$

It follows from (4.25) that

$$\frac{d}{dt}\langle My(t), y(t)\rangle + \langle C^*y(t), Cy(t)\rangle + \langle RKy(t), Ky(t)\rangle = 0.$$

Hence

$$\langle My(t), y(t)\rangle + \int_0^t |Cy(s)|^2 ds + \int_0^t \langle RKy(s), Ky(s)\rangle ds = \langle Mx, x\rangle. \quad (4.29)$$

By a standard limit argument, (4.29) holds for all $x \in H$.

Applying Theorem A.8 (Young's inequality) in the same way as in the proof of Lemma III.1.3, we obtain that

$$\int_0^{+\infty} |S_1(t)x|^2 dt = \int_0^{+\infty} |y(t)|^2 dt < +\infty.$$

By Theorem 3.1(i), the semigroup $S_1(\cdot)$ is exponentially stable. This proves (i).

(ii) Denote $M - P = W$. Then for $x \in D(A)$

$$2\langle W(A+BK)x, x\rangle = -\langle C^*Cx, x\rangle - \langle K^*RKx, x\rangle$$
$$- 2\langle PAx, x\rangle - 2\langle PBKx, x\rangle.$$

Since P satisfies (4.21) with $Q = C^*C$,

$$2 < W(A+BK)x, x\rangle = -\langle K^*RKx, x\rangle - \langle PBR^{-1}B^*Px.x\rangle \quad (4.30)$$
$$- 2\langle PBKx, x\rangle.$$

Let $K_0 = -R^{-1}B^*P$, then $RK_0 = -B^*P$, $PB = -K_0^*R$, and from (4.30) we easily obtain that

$$2 < W(A+BK)x, x\rangle = -\langle (K-K_0)^*R(K-K_0)x, x\rangle, \quad x \in D(A).$$

Since $A+BK$ is the exponentially stable generator, operator W is nonnegative by Theorem 3.2. □

We return now to the proof of part (ii) of Theorem 4.4 and assume that operators $P \geq 0$ and $P_1 \geq 0$ are solutions of (4.21). Let $K = -R^{-1}B^*P$, then

$$2\langle P(A+BK)x, x\rangle + \langle K^*RKx, x\rangle + \langle C^*Cx, x\rangle$$
$$= 2\langle PAx, x\rangle - \langle PBR^{-1}B^*Px, x\rangle + \langle C^*Cx, x\rangle \quad (4.31)$$
$$= 0.$$

Consequently, by Lemma 4.5(ii), $P_1 \leq P$. In the same way $P \leq P_1$. Hence $P = P_1$. Moreover, by Lemma 4.5(i) and (4.31) the operator $A+BK$ is exponentially stable. □

Bibliographical notes

The direct proof of the existence of the solution $P(\cdot)$ to the Riccati equation (4.9) based on Lemma 4.3 is due to Da Prato [18]. Theorem 4.4 is taken from the author's paper [65]. More information on the subject can be found in the monograph by Curtain and Pritchard [17].

APPENDIX

§ A.1. Metric spaces

A *metric space* is a pair composed of a set E and a function $\varrho\colon E\times E \longrightarrow \mathbf{R}_+$ such that

$$\varrho(x,y) = 0 \quad \text{if and only if} \quad x = y$$
$$\varrho(x,y) = \varrho(y,x) \quad \text{for all} \quad x,y \in E$$
$$\varrho(x,y) \leq \varrho(x,z) + \varrho(z,y), \quad \text{for arbitrary} \quad x,y,z \in E.$$

The function ϱ is called a *metric*.

For any $r \geq 0$ and $a \in E$ the set

$$B(a,r) = \{x \in E;\ \varrho(a,x) < r\} \tag{A.1}$$

is called a *ball* of radius r and centre a. A union of any family of balls is an open subset of E and the complement A^c of any open set $A \subset E$ is called a *closed* subset of E. The smallest closed set containing a set $A \subset E$ is denoted by $\overline{A}$ or $\operatorname{cl} A$ and called the *closure* of A.

If there exists a countable set $E_0 \subset E$ such that $\overline{E_0} = E$ then E is a *separable* metric space.

A metric space (E, ϱ) is *complete* if for arbitrary sequence (x_n) of elements of E such that $\lim_{n,m\to\infty} \varrho(x_n, x_m) = 0$ there exists $x \in E$ for which $\lim_{n\to\infty} \varrho(x_n, x) = 0$.

If E_1, E_2 are two metric spaces and $F\colon E_1 \longrightarrow E_2$ a transformation such that for any open set $\Gamma_2 \subset E_2$ the set

$$F^{-1}(\Gamma_2) = \{x \in E_1;\ F(x) \in \Gamma_2\}$$

is open in E_1, then F is called a *continuous* mapping from E_1 onto E_2.

A transformation $F\colon E \longrightarrow E$ is called a *contraction* if there exists $c \in (0,1)$ such that

$$\varrho(F(x), F(y)) \leq c\varrho(x,y), \quad \text{for all} \quad x,y \in E. \tag{A.2}$$

Theorem A.1. (The contraction mapping principle.) *Let E be a complete metric space and $F\colon E \longrightarrow E$ a contraction. Then the equation*

$$F(x) = x, \quad x \in E \tag{A.3}$$

has a unique solution $\hat{x}$. Moreover, for arbitrary $x_0 \in E$,

$$\lim_{n\to\infty} \varrho(x_n, \hat{x}) = 0,$$

where $x_{n+1} = F(x_n)$, $n = 0, 1, \ldots$.

§ A.2. Banach spaces

A complete, metric space (E, ϱ) which is also a linear space over **R** (or **C**) and such that

$$\varrho(x, y) = \varrho(x - y, 0), \quad x, y \in E$$
$$\varrho(\alpha x, 0) = |\alpha|\varrho(x, 0), \quad x \in E, \alpha \in \mathbf{R} \ (\alpha \in \mathbf{C})$$

is called a *Banach space*. The function

$$\|x\| = \varrho(x, 0), \quad x \in E$$

is called the norm of E.

An operator A acting from a linear subspace $D(A)$ of a Banach space E onto a Banach space E_2 such that

$$A(\alpha x + \beta y) = \alpha Ax + \beta Ay, \quad x, y \in D(A), \ \alpha, \beta \in \mathbf{R} \ (\alpha, \beta \in \mathbf{C})$$

is called *linear*.

If $D(A) = E_1$ and the linear operator A is continuous then it is a *bounded operator* and the number

$$\|A\| = \sup\{\|Ax\|_2; \ \|x\|_1 \leq 1\} \tag{A.4}$$

is its *norm*. Here $\|\cdot\|_1, \|\cdot\|_2$ stand for the norms on E_1 and E_2 respectively.

A linear operator A is *closed* if the set $\{(x, Ax); \ x \in D(A)\}$ is a closed subset of $E_1 \times E_2$. Equivalently, an operator A is closed if, for an arbitrary sequence of elements (x_n) from $D(A)$ such that for some $x \in E_1$, $y \in E_2$, $\lim_{n\to\infty} \|x_n - x\|_1 = 0$, $\lim_n \|Ax_n - y\|_2 = 0$, one has $x \in D(A)$ and $Ax = y$.

Theorem A.2. (The closed graph theorem.) *If a closed operator* $A: E_1 \longrightarrow E_2$ *is defined everywhere on* E_1, $(D(A) = E_1)$, *then A is bounded.*

Theorem A.3. (The inverse mapping theorem.) *If a bounded operator* $A: E_1 \longrightarrow E_2$ *is one-to-one and onto E_2, then the inverse operator F^{-1} is bounded.*

Theorem A.4. (The Hahn-Banach theorem.) *Assume that $E_0 \subset E$ is a linear subspace of E and $p: E \longrightarrow \mathbf{R}_+$ a functional such that*

$$p(a + b) \leq p(a) + p(b), \quad a, b \in E$$
$$p(\alpha a) = \alpha p(a), \quad \alpha \geq 0, \ a \in E.$$

If $\varphi\colon E_0 \longrightarrow \mathbf{R}$ is a linear functional such that $\varphi_0(a) \leq p(a)$, $a \in E_0$, then there exists a linear functional $\varphi\colon E \longrightarrow \mathbf{R}$ such that

$$\varphi(x) = \varphi_0(x), \quad x \in E_0$$

and

$$\varphi(x) \leq p(x), \quad x \in E.$$

The space of all linear, continuous functionals on a Banach space E, with values in $\mathbf{R}$ (in $\mathbf{C}$) equipped with the norm $\|\cdot\|_*$,

$$\|\varphi\|_* = \sup\{|\varphi(x)|;\ \|x\| \leq 1\},$$

is a Banach space over $\mathbf{R}$ (or $\mathbf{C}$) denoted by E^* and called the *dual* or *adjoint* space to E.

Assume that A is a linear mapping from a Banach space E_1 onto a Banach space E_2 with the dense domain $D(A)$. Denote by $D(A^*)$ the set of all elements $\varphi \in E_2^*$ such that the linear functional $x \in D(A) \longrightarrow \varphi(Ax)$ has a continuous extension Ψ to E_1. Then $D(A^*)$ is a linear subspace of E_2^* and the extension Ψ is unique and denoted by $A^*\varphi$. The operator A^* with the domain $D(A^*)$ is called the *adjoint operator* of A.

If a bounded operator A from a Hilbert space E onto E is such that $\|Fx\| = \|x\|$ for all $x \in E$, then A is called a *unitary operator*. A bounded operator A is unitary if and only if $A^* = F^{-1}$.

Theorem A.5. (The Banach-Steinhaus theorem.)

(i) *If (A_n) is a sequence of bounded operators from a Banach space E_1 onto E_2 such that*

$$\sup_n \|A_n x\|_2 < +\infty \quad \text{for all } x \in E_1$$

then $\sup_n \|A_n\| < +\infty$.

(ii) *If, in addition, sequence $(A_n x)$ is convergent for all $x \in E_0$, where E_0 is a dense subset of E, then there exists the limit*

$$\lim_{n \to \infty} A_n x = Ax,$$

for all $x \in E$ and A is a bounded operator.

Let X and Z be Banach spaces and $F\colon U \longrightarrow Z$ a transformation from an open set $U \subset X$ into Z. The *Gateaux derivative* of F at $x_0 \in U$ is a linear bounded operator A from X into Z such that for arbitrary $h \in X$

$$\lim_{t \to 0} \frac{1}{t}\left(F(x_0 + th) - F(x_0)\right) = Ah.$$

If, in addition,
$$\lim_{\|h\|\to 0} \frac{\|F(x_0+h) - F(x_0) - Ah\|}{\|h\|} = 0,$$
then F is called *Fréchet differentiable* at x_0 and A the *Fréchet derivative* of F at x_0 denoted by $F_x(x_0)$ or $dF(x_0)$. The vector $Ah = F_x(x_0)h = dF(x_0; h)$ is the *directional derivative* of F at x_0 and at the direction $h \in X$.

Theorem A.6. (The mean value theorem.) *If a continuous mapping $F: U \longrightarrow Z$ has the Gateaux derivative $F_x(z)$ at an arbitrary point z in the open set U, then*
$$\|F(b) - F(a) - F_x(a)(b-a)\| \qquad (A.5)$$
$$\leq \sup_{\eta \in I(a,b)} \|F_x(\eta) - F_x(a)\| \|b-a\|,$$
where $I(a,b) = \{a + s(b-a); s \in [0,1]\}$ is the interval with the endpoints a, b.

§A.3. Hilbert spaces

If E is a Banach space over $\mathbf{R}$ (or $\mathbf{C}$) equipped with the transformation $\langle \cdot, \cdot \rangle : E \times E \longrightarrow \mathbf{R}$ (or $\mathbf{C}$) such that
$$\langle a, b \rangle = \langle b, a \rangle, \quad (= \overline{\langle b, a \rangle})$$
$$\langle a, a \rangle \geq 0 \text{ and } \langle a, a \rangle = 0 \text{ if and only if } a = 0$$
$$\langle \alpha a + \beta b, c \rangle = \alpha \langle a, c \rangle + \beta \langle b, c \rangle$$
$$\text{for } a, b \in E \text{ and scalars } \alpha, \beta$$
$$\|a\| = \langle a, a \rangle^{1/2}, \quad a \in E,$$
then E is called a Hilbert space and the form $\langle \cdot, \cdot \rangle$ is the *scalar product* on E.

Theorem A.7. (The Riesz representation theorem.) *If E is a Hilbert space then for arbitrary $\varphi \in E^*$ there exists exactly one element $a \in E$ such that*
$$\varphi(x) = \langle x, a \rangle, \quad x \in E. \qquad (A.6)$$
Moreover,
$$\|\varphi\|_* = \|a\|.$$
This way E^* can be identified with E.

Norms on Hilbert spaces are also denoted by $|\cdot|$.

Typical examples of Hilbert spaces are l^2 and $l^2_{\mathbf{C}}$. The space l^2 consists of all real sequences (ξ_n) such that $\sum_{n=1}^{+\infty} |\xi_n|^2 < +\infty$ and the scalar product is defined by
$$\langle (\xi_n), (\eta_n) \rangle = \sum_n \xi_n \eta_n, \quad (\xi_n), (\eta_n) \in l^2.$$

The space $l_\mathbb{C}^2$ consists of all complex sequences (ξ_n) such that $\sum_{n=1}^{+\infty} |\xi_n|^2 < +\infty$ and the scalar product is given by

$$\langle (\xi_n), (\eta_n) \rangle = \sum_{n=1}^{+\infty} \xi_n \bar{\eta}_n, \quad (\xi_n), (\eta_n) \in l_\mathbb{C}^2.$$

§A.4. Bochner's integral

Let E be a metric space. The smallest family $\mathcal{E}$ of subsets of a metric space E such that

(i) all open sets are in $\mathcal{E}$
(ii) if $\Gamma \in \mathcal{E}$ then $\Gamma^c \in \mathcal{E}$
(iii) if $\Gamma_1, \Gamma_2, \ldots \in \mathcal{E}$ then $\bigcup_{n=1}^{+\infty} \Gamma_n \in \mathcal{E}$

is called the *Borel σ-field* of E and is denoted by $\mathbf{B}(E)$. Elements of $\mathbf{B}(E)$ are called Borel sets of E.

A transformation $F: E_1 \longrightarrow E_2$ is called *Borel* if for arbitrary $\Gamma_2 \mathbf{B}(E_2)$, $F^{-1}(\Gamma_2) \in \mathbf{B}(F_1)$. If $E_2 = \mathbf{R}$ then F is called a *Borel function*. Continuous mappings are Borel.

Assume that E is a separable Banach space and let $E_1 = (\alpha, \beta) \subset (-\infty, +\infty)$ and $E_2 = E$. If $f: (\alpha, \beta) \longrightarrow E$ is a Borel transformation then also function $\|F(t)\|$, $t \in (\alpha, \beta)$ is Borel measurable. The transformation F is called *Bochner integrable* if

$$\int_\alpha^\beta \|F(t)\| \, dt < +\infty,$$

where the integral is the usual Lebesgue integral of the real-valued function $\|F(\cdot)\|$. Assume, in addition, that F is a *simple transformation*, i.e., there exist Borel, disjoint subsets $\Gamma_1, \ldots, \Gamma_m$ of (α, β), $\bigcup_{j=1}^m \Gamma_j = (\alpha, \beta)$ and vectors $a_1, \ldots, a_m$ such that

$$F(t) = a_i, \quad t \in \Gamma_i.$$

Then the sum

$$\sum_{j=1}^m a_j \int_\alpha^\beta \chi_{\Gamma_j}(t) \, dt$$

is well defined and by definition equal to the Bochner integral of f over (α, β). It is denoted by

$$\int_\alpha^\beta F(t) \, dt.$$

Note that for simple transformations F

$$\left\| \int_\alpha^\beta F(t) \, dt \right\| \leq \int_\alpha^\beta \|F(t)\| \, dt. \tag{A.7}$$

Let $\{e_n\}$ be a dense and a countable subset of E and F a Borel mapping from (α, β) into E. Define

$$\gamma_m(t) = \min\{\|f(t) - e_k\|;\ k = 1, \ldots, m\},$$
$$k_m(t) = \min\{k \leq m;\ \gamma_m(t) = \|f(t) - e_k\|\}$$

and $F_m(t) = e_{k_m(t)}$, $t \in (\alpha, \beta)$, $m = 1, 2, \ldots$. Then F_m is a Borel, simple transformation and $\|F_m(t) - F(t)\|$, $t \in (\alpha, \beta)$, decreases monotonically to 0 as $m \to +\infty$. In addition,

$$\|F_m(t) - F(t)\| \leq \|e_1 - F(t)\|$$
$$\leq \|e_1\| + \|F(t)\|, \quad t \in (\alpha, \beta),\ m = 1, 2 \ldots.$$

Consequently, if F is Bochner's integrable

$$\lim_{n,m \to \infty} \|\int_\alpha^\beta F_m(t)\,dt - \int_\alpha^\beta F_n(t)\,dt\|$$
$$\leq \lim_{m \to \infty} \int_\alpha^\beta \|F_m(t) - F(t)\|\,dt + \lim_{n \to \infty} \int_\alpha^\beta \|F_n(t) - F(t)\|\,dt = 0,$$

and therefore there exists the limit

$$\lim_{n \to \infty} \int_\alpha^\beta F_n(t)\,dt,$$

which is also denoted by $\int_\alpha^\beta F(t)\,dt$ and called the Bochner integral of F. The estimate $(A.7)$ holds for arbitrary Bochner integrable functions. Moreover, if (G_m) is any sequence of Borel transformations such that

$$\lim_{m \to \infty} \int_\alpha^\beta \|G_m(t) - F(t)\|\,dt = 0$$

then

$$\lim_{m \to \infty} \|\int_\alpha^\beta G_m(t)\,dt - \int_\alpha^\beta F(t)\,dt\| = 0.$$

One can asily show, using Theorem A.9, that for an arbitrary Bochner integrable transformation $F(\,\cdot\,)$ there exists a sequence of continuous transformations (G_m) such that

$$\lim_{m \to \infty} \int_\alpha^\beta \|G_m(t) - F(t)\|\,dt = 0.$$

§A.5. Spaces of continuous functions

Let $-\infty < \alpha < \beta < +\infty$ and E be a Banach space. The space of all continuous functions from $[\alpha,\beta]$ into E with the norm

$$\|f\| = \sup\{\|f(t)\|;\ t \in [\alpha,\beta]\}$$

is a Banach space, denoted by $C(\alpha,\beta;E)$. If the space E is separable, the space $C(\alpha,\beta;E)$ is separable as well. We have also the following result:

Theorem A.8. (The Ascoli theorem.) *Let K be a closed subset of $C(\alpha,\beta;\mathbf{R}^n)$. Every sequence of elements from K contains a convergent subsequence if and only if*

(i) $\sup\{\|f(t)\|;\ t \in [\alpha,\beta],\ f \in K\} < +\infty$

(ii) *for arbitrary $\varepsilon > 0$ there exists $\delta > 0$ such that*

$$\|f(t) - f(s)\| \leq \varepsilon \quad \text{for all}\ f \in K$$

and all $t,s \in [\alpha,\beta]$ such that $|t-s| < \delta$.

§A.6. Spaces of measuarable functions

Let $p \geq 1$. The space of all equivalence classes of Bochner's integrable functions f from $(\alpha,\beta) < (-\infty,+\infty)$ into E such that

$$\int_\alpha^\beta \|f(t)\|^p\,dt < +\infty$$

is denoted by $L^p(\alpha,\beta;E)$. It is a Banach space equipped with the norm

$$\|f\| = \left(\int_\alpha^\beta \|f(t)\|^p\,dt\right)^{1/p}.$$

If E is a Hilbert space and $p=2$ then $L^2(\alpha,\beta;E)$ is a Hilbert space as well and the scalar product $\langle\langle\cdot,\cdot\rangle\rangle$ is given by

$$\langle\langle f,g\rangle\rangle = \int_\alpha^\beta \langle f(t),g(t)\rangle_E\,dt.$$

If $E = R$ or $E = \mathbf{C}$ we write $L^p(\alpha,\beta)$ for short.

Theorem A.9. (Young's inequality.) *If $1 \leq p,q < +\infty$, $\frac{1}{r} = \frac{1}{p} + \frac{1}{q} - 1$ and $f \in L^p(0,+\infty)$, $g \in L^q(0,+\infty)$, then $f*g \in L^r(0,+\infty)$ ($f*g(t) = \int_0^t f(t-s)g(s)\,ds,\ t \geq 0$) and*

$$\left(\int_0^{+\infty} |f*g(t)|^r\,dt\right)^{1/r} \leq \left(\int_0^{+\infty} |f(t)|^p\,dt\right)^{1/p}\left(\int_0^{+\infty} |g(t)|^q\,dt\right)^{1/q}.$$

Theorem A.10. *If $f \in L^p(-\infty, +\infty)$ then*

$$\lim_{\delta \to 0} \int_{-\infty}^{+\infty} |f(t+\delta) - f(t)|\, dt = 0.$$

Theorem A.11. (Luzin's theorem.) *If $f: [\alpha, \beta] \longrightarrow \mathbf{R}$ is a Borel function then for arbitrary $\varepsilon > 0$ there exists a closed set $\Gamma \subset [\alpha, \beta]$ such that the Lebesgue measure of $[\alpha, \beta] \setminus \Gamma$ is smaller then ε and f restricted to Γ is a continuous function.*

A function $f: [\alpha, \beta] \longrightarrow \mathbf{R}^n$ is called *absolutely continuous* if for arbitrary $\varepsilon > 0$ one can find $\delta > 0$ such that if $\alpha \leq t_0 < t_1 < t_2 < \ldots < t_{2k} < t_{2k+1} \leq \beta$ and

$$\sum_{j=0}^{k} (t_{2j+1} - t_j) < \delta$$

then

$$\sum_{j=0}^{k} |f(t_{2j+1}) - f(t_j)| < \varepsilon.$$

A function $f: [\alpha, \beta] \longrightarrow \mathbf{R}^n$ is absolutely continuous if and only if there exists $\Psi \in L^1(\alpha, \beta; \mathbf{R}^n)$ such that

$$f(t) = f(\alpha) + \int_{\alpha}^{t} \Psi(s)\, ds, \quad t \in [\alpha, \beta].$$

Moreover, f is differentiable at almost all $t \in (\alpha, \beta]$, and

$$\frac{df}{dt}(t) = \Psi(t).$$

References

[1] D. AEYELS and M. SZAFRANSKI, *Comments on the stability of the angular velocity of a rigid body*, Systems and Control Letters, 10(1988), 35-39.
[2] W.I. ARNOLD, *Mathematical Methods of Classical Mechanics*, Springer-Verlag, New York 1978.
[3] A.V. BALAKRISHNAN, *Semigroup theory and control theory*, Proc. IFIP Congress, Tokyo 1965.
[4] S. BARNETT, *Introduction to Mathematical Control Theory*, Clarendon Press, Oxford 1975.
[5] R. BELLMAN, *Dynamic Programming*, Princeton University Press, 1977.
[6] A. BENSOUSSAN and J.L. LIONS, *Contrôle impulsionnel et inéquations quasivariationnelles*, Dunod, Paris 1982.
[7] A. BENSOUSSAN, G.Da PRATO, M. DELFOUR and S.K. MITTER, *Representation and Control of Infinite Dimensional Systems*, Birkhäser, (to appear).
[8] N.P. BHATIA and G.P. SZEGÖ, *Stability Theory of Dynamical Systems*, Springer-Verlag, New York 1970.
[9] A. BLAQUIERE, *Impulsive optimal control with finite or infinite time horizon*, Journal of Optimization Theory and Applications, 46(4) (1985), 431-439.
[10] R.W. BROCKETT, *Finite Dimensional Linear Systems*, Wiley, New York 1970.
[11] R.W. BROCKETT, Asymptotic stability and feedback stabilization. In: *Differential Geometric Control Theory*, R.W. Brockett, R.S. Millman, H.J. Sussmann (eds.), Birkhäuser, Basel 1983, 181–191.
[12] A.G. BUTKOVSKI, *Teorija upravlenija sistemami s raspredelennymi parametrami*, Nauka, Moscow 1975.
[13] L. CESARI, *Optimization Theory and Applications*, Springer–Verlag, New York 1963.
[14] F.H. CLARKE, *Optimization and Nonsmooth Analysis*, Wiley Interscience, New York 1983.
[15] E.A. CODDINGTON and N. LEVINSON, *Theory of Ordinary Differential Equations*, McGraw-Hill, New York 1955.
[16] Y. COHEN (ed.), *Applications of control theory in ecology*, Lecture Notes in Biomathematics, Springer–Verlag, New York 1988.
[17] R.F. CURTAIN and A.J. PRITCHARD, *Infinite Dimensional Linear Systems Theory*, Springer-Verlag, Lecture Notes in Control and Information Sciences, New York 1978.

[18] G. DA PRATO, *Quelques résultats d'existence et régularité pour un problème non linéaire de la théorie du controle*, J. Math. Pures et Appl., 52(1973), 353–375.

[19] R. DATKO, *Extending a theorem of A.M. Lapunov to Hilbert spaces*, J. Math. Anal. Appl. 32(1970), 610–616.

[20] R. DATKO, *Uniform asymptotic stability of evolutionary processes in a Banach space*, SIAM J. Math. Anal., 3(1972), 428–445.

[21] S. DOLECKI and D.L. RUSSELL, *A general theory of observation and control*, SIAM J. Control, 15(1977), 185–221.

[22] R. DOUGLAS, *On majorization, factorization and range inclusion of operators in Hilbert spaces*, Proc. Amer. Math. Soc. 17, 413–415.

[23] J. DUGUNDJI, *Topology*, Allyn and Bacon, Boston 1966.

[24] N. DUNFORD and J. SCHWARTZ, *Linear operators*, Part I, Interscience Publishers, New York, London 1958.

[25] N. DUNFORD and J. SCHWARTZ, *Linear operators*, Part II, Interscience Publishers, New York, London 1963.

[26] H.O. FATTORINI, *Some remarks on complete controllability*, SIAM J. Control, 4(1966), 686–694.

[27] W.H. FLEMING and R.W. RISHEL, *Deterministic and Stochastic Optimal Control*, Springer-Verlag, Berlin, Heidelberg, New York 1975.

[28] F.R. GANTMACHER, *Applications of the Theory of Matrices*, Interscience Publishers Inc., New York 1959.

[29] I.V.GIRSANOV, *Lectures on mathematical theory of extremum problems*, Springer-Verlag, New York 1972.

[30] R. HERMAN and A. KRENER, *Nonlinear controllability and observability*, IEEE Trans. Automatic Control, AC-22(1977), 728–740.

[31] A. ISIDORI, *Nonlinear Control Systems: An Introduction*, Lecture Notes in Control and Information Sciences, vol. 72, Springer-Verlag, New York 1985.

[32] B. JAKUBCZYK, *Local realization theory for nonlinear systems*, in: *Geometric Theory of nonlinear control systems*, Polytechnic of Wroclaw, Wroclaw 1985, 83–104.

[33] R.E. KALMAN, *On the general theory of control systems*, Automatic and Remote Control, Proc. First Int. Congress of IFAC, Moskow, 1960, vol. 1, 481–492.

[34] J. KISYNSKI, Semigroups of operators and some of their applications to partial differential equations. In: *Control Theory and Topics in Functional Analysis*, v.3, IAEA, Vienna 1976, 305–405.

[35] M.A. KRASNOSELSKI and P.P. ZABREIKO, *Geometrical Methods of Nonlinear Analysis*, Springer-Verlag, Berlin, Heidelberg, New York 1984.

[36] E.B. LEE and L. MARKUS, *Foundations of Optimal Control Theory*, Wiley, New York 1967.

[37] G. LEITMAN, *An introduction to optimal control*, Mc Graw-Hill, New York 1966.
[38] J.L. LIONS, *Contrôle optimale de systèmes gouvernés par des équations aux dérivées partielles*. Dunod, Paris 1968.
[39] J.C. MAXWELL, *On governers*, Proc. Royal Society, 1868.
[40] M. MEGAN, *On the stabilizability and controllability of linear dissipative systems in Hilbert space*, S. E. F., Universitatea din Timisoara, 32(1975).
[41] C. OLECH, Existence theory in optimal control. In: *Control Theory and Topics in Functional Analysis*, v. 1, IAEA, Vienna 1976, 291-328.
[42] R. PALLU de la BARRIERE, *Cours d'automatique théorique*, Dunod, Paris 1966.
[43] P.C. PARKS, *A new proof of the Routh-Hurwitz stability criterion using the second method of Liapunov*, Proc. of the Cambridge Philos. Soc. Math. and Phys. Sciences, Oct. 1962, v. 58, part 4, 694-702.
[44] A. PAZY, *Semigroups of Linear Operators and Applications to Partial Differential Equations*, Springer-Verlag, New York 1983.
[45] L.S. PONTRIAGIN, V.G. BOLTIANSKI, R.V. GAMKRELIDZE and E.F. MIŠČENKO, *Matematičeskaja teorija optymal'nych processov*, Nauka, Moscow 1969.
[46] L.S. PONTRIAGIN, *Ordinary differential equation*, Addison-Wesley, Mass. 1962.
[47] A. PRITCHARD and J. ZABCZYK, *Stability and stabilizability of infinite dimensional systems*, SIAM Review, vol. 23, No. 1, 1981, 25-52.
[48] R. REMPALA, *Impulsive control problems*, Proceedings of the 14th IFIP Conference on System Modeling and Optimization, Leipzig, Springer-Verlag, New York 1989.
[49] R. REMPALA and J. ZABCZYK, *On the maximum principle for determistic impulse control problems*, Journal of Optimization Theory and Applications, 59(1988), 281-288.
[50] S. ROLEWICZ, *Functional analysis and control theory*, Polish Scientific Publisher, Warszawa, and D. Reidel Publishing Company, Dordrecht, Boston, Lancaster, Tokyo 1987.
[51] E.J. ROUTH, *Treatise on the Stability of a Given State of Motion*, Macmillan and Co., London 1877.
[52] D.L. RUSSEL, *Controllability and stabilizability theory for linear partial differential equations: recent progress and open questions*, SIAM Review, vol. 20, No 4(1978), 639-739.
[53] T. SCHANBACHER, *Aspects of positivity in control theory*, SIAM J. Control and Optimization, vol. 27, No 3(1989), 457-475.
[54] H.J. SUSSMANN and V. JUDRJEVIC, *Controllability of nonlinear systems*, J. Differential Equations, 12(1972), 95-116.

[55] M. SZAFRANSKI, *Stabilization of Euler's equation*, University of Warsaw, Dept. of Mathematics, Informatics and Mechanics, Master Thesis, 1987, in Polish.
[56] R. TRIGGIANI, *Constructive steering control functions for linear systems and abstract rank condition*, to appear.
[57] T. WAZEWSKI, *Systèmes de commande et équations au contigent*, Bull. Acad. Pol. Sc. 9(1961), 151–155.
[58] N. WIENER, *Cybernetics or control and communication in the animal and the machine*, The MIT Press, Cambridge, Massachusetts 1948.
[59] W.A. WOLVICH, *Linear Multivariable Systems*, Springer-Verlag, New York 1974.
[60] W.M. WONHAM, *On pole assignment in multi-input controllable linear systems*, IEEE Trans. Automat. Control, AC-12(1967), 660–665.
[61] W.M. WONHAM, *Linear multivariable control: A geometric approach*, Springer-Verlag, New York 1979.
[62] J. ZABCZYK, *Remarks on the control of discrete-time distributed parameter systems*, SIAM, J. Control, 12(1974), 161–176.
[63] J. ZABCZYK, *On optimal stochastic control of discrete-time parameter systems in Hilbert spaces*, SIAM, J. Control 13(1975), 1217–1234.
[64] J. ZABCZYK, *A note on C_0-semigroups*, Bull. Acad. Pol. Sc. Serie Math. 23(1975), 895–898.
[65] J. ZABCZYK, *Remarks on the algebraic Riccati equation in Hilbert space*, Appl. Math. Optimization 3(1976), 383–403.
[66] J. ZABCZYK, *Complete stabilizability implies exact controllability*, Seminarul Ecuati Functionale 38(1976), Timisoara, Romania, 1–7.
[67] J. ZABCZYK, *Stability properties of the discrete Riccati operator equation*, Kybernetika 13(1977), 1–10.
[68] J. ZABCZYK, *Infinite dimensional systems in optimal control*, Bulletin of the International Statistical Institute, vol. XLII, Book II, Invited papers, 1977, 286–310.
[69] J. ZABCZYK, *Stopping problems in stochastic control*, Proceedings of the International Congress of Mathematicians, Warsaw 1984, 1425–1437.
[70] J. ZABCZYK, *Some comments on stabilizability*, Appl. Math. and Optimization 19(1988), 1–9.

Notations

- **R**: The set of all real numbers.
- **R$_+$**: The set of all nonnegative real numbers.
- **C**: The set of all complex numbers.
- E^n: The Cartesian product of n-copies of E.
- $\mathsf{M}(n,m)$: Set of all $n \times m$ matrices with real elements.
- $\mathsf{M}(n,m;\mathsf{C})$: Set of all $n \times m$ matrices with complex elements.
- l^2: Hilbert space of all real sequences (ξ_n) satisfying $\sum_n |\xi_n|^2 < +\infty$ with the scalar product

$$\langle (\xi_n),(\eta_n)\rangle = \sum_n \xi_n \eta_n.$$

- l^2_{C}: Hilbert space of all complex sequences (ξ_n) satisfying $\sum_n |\xi_n|^2 < +\infty$ with the scalar product

$$\langle (\xi_n),(\eta_n)\rangle = \sum_n \xi_n \bar{\eta}_n.$$

- χ_Γ: Indicator function of the set Γ: $\chi_\Gamma(x) = 1$ if $x \in \Gamma$, $\chi_\Gamma(x) = 0$ if $x \in \Gamma^c$.
- $L^p(\alpha,\beta;E)$: The space of all E-valued, Borel functions integrable in power p on the interval (α,β).
- $L(E,F)$: The space of all linear bounded operators from E into F.
- $C_0^\infty(\alpha,\beta)$: The space of all infinite differentiable real functions with compact supports included in (α,β).

INDEX

Adjoint operator 190, 246
Adjoint semigroup 183
Adjoint space 162, 246
Algebraic Riccati equation 134, 240
Antipodal mapping 108
Asymptotically stable point 104
Attainable state 14, 16, 78, 212

Bochner's integral 180, 248

Canonical representation 28
Characteristic polynomial 17
– stable 32, 34-5, 40
Closed operator 181, 245
Coincidence set 145
Contraction mapping principle 244
Control 14, 73
– admissible 171
– closed loop 1
– elementary 84
– impulse 142
– optimal 3, 127-8, 143
– open loop 1
Control system 1, 10, 73, 202
– infinite dimensional 3, 176, 202
Controllability 2, 14, 77, 212
– approximate 212
– positive 65
– exact 212
– local 77
– wave equation 218
Controllability matrix 14
Controllable part 24

Controllable system 16
Convex cone 145
Concave mapping 146

Degree of a mapping 108
Detectability 49, 241
Detectable pair 49, 241
Differential inequalities 92-3
Dimension of observer 47
Dissipative mapping 100

Eigenvalue 28
Eigenvector 28
Electrical filter 6
Electrically heated oven 3
Embedding method 128
Equation adjoint 13
– algebraic Riccati 136, 240
– Bellman's 128
– for impulse control 144
– differential 73
– – linear 10
– – nonlinear 73
– Liapunov's 40, 141, 226
– Liénard's 108
– of observation 2
– Riccati 133, 234
Equilibrium state 100
Equivalent systems 22
Exact controllability 212
Exponentially stable generator 225
– semigroup 225

Function strongly continuous 233

Feedback stabilizing 112, 141, 242
– in Watt's regulator 6

First Liapunov's method 102, 120
Frechet's derivative 246

Gateaux' derivative 246
Generator of a semigroup 178, 181
Gronwall's lemma 92
Group of unitary transformations 180, 246

Hamiltonian 154, 162
Harmonic input 52
Heat equation 8, 176
Heating of a rod 8
Homotopic mapping 108
Homotopy 108

Image of an operator 207
Impulse response function 50, 54
– of linearization 122
Index of a transformation 109
Input 1
– reference 124
– harmonic 52
– impulsive 50
– periodic 52

Input-output mapping 50, 121
– map of a system 50, 121
– – regular 124
Interval of maximal existence 74

Jordan block 29, 96

Kalman's condition 17
– rank condition 17
Kernel of an operator 207

Lemma on measurable selector 173
Liapunov's function 103
Lie bracket 81
Limit set 106

Linear regulator problem 133, 232
Linearization of a system 78
– method 100
Lipschitz' condition 106
–local 106
Local solution 73

Mapping concave 146
– monotonic 146
Matrix 10
– controllability 14
– nonnegative 11
– observability 25
– stable 28, 32
Maximal interval of existence 74
– solution 74
Maximum principle 128, 152
– for linear regulator 155
– for impulse problems 158
– for time-optimal problems 164
Measurable selector 173
Method of linearization 92, 102
Model of electrically heated oven 3
Monotonic mapping 146

Needle variation 153

Observability 2, 25, 88
Observable pair 25
Observer 47
Operator adjoint 246
– bounded 245
– closed 181, 245
– controllability 209
– infinitesimal 178, 181
– linear 245
– pseudoinverse 209
– self-adjoint 191
Optimal consumption 7
– stopping 144
Optimality criteria 3, 127
Orbit 106
Output 1
Oven electrically heated 3

Pair controllable 16, 21
- detectable 46
- observable 25-6
- stabilizable 43
Parametrization of a surface 81
Periodic component 52
Positive controllability 65
- system 64
Principle contraction mapping 244
Problem of optimal consumption 7

Reachable point 2, 14, 16
Real Jordan block 29
Realizable mapping 3
Realization controllable-observable 58
- impulse response function 50, 54
- partial 124
- transfer function 51-2
Reference inputs 124
- sequence 124
Reflexive space 209
Regulator linear 133, 232
- of Watt 6, 102
Response 1
Rigid body 4
Robustness index 115
- of a system 115
Routh sequence 40

Second Liapunov's method 103, 120
Selector 130, 173
Semigroup of operators 177
Set attainable 212
- invariant 103
- nonattainable 105
- of control parameters 1
- of eigenvalues 28
Soft landing 6
Solution fundamental 13
- local 73
- maximal 74
- minimal 136

- strong 160
- weak 204
Stability criteria 33, 40, 96
- asymptotic 105
- exponential 101
- in Liapunov's sense 103
Stabilizability asymptotic 112, 116
- complete 44
- exponential 112-3
- in Liapunov's sense 112
Stabilization dynamic 47
- of Euler's equation 117
Stable matrix 28
- polynomial 32
State assymptotically stable 105
- exponentially stable 100
- equilibrium 100
- stable in Liapunov's sense 103
- reachable 14
Stopping optimal 144
Strategy impulse 142
- optimal 134
- stationary 143
Strongly continuous function 233
Surface 81
Symmetric matrix 11
System approximately controllable 212
- closed-loop 1
- completely stabilizable 44
- controllable 16
- controllable approximately 212
- - to zero 212
- detectable 49, 241
- exactly controllable 212
- exponentially stable 225
- locally controllable 77
- observable 25, 88
- on manifold 3
- stabilizable 43
- symmetric 87
- positive 64
- positively controllable 65
- with self-adjoint generator 213

Theorem Ascoli 250
– Banach–Steinhaus 246
– Cayley–Hamilton 17
– closed graph 245
– Datko 231
– Fillipov 171
– Hahn–Banach 245
– Hille–Yosida 185
– Jordan 29, 96
– La Salle's 105
– Lax–Milgram 192
– Lions 192
– Luzin 251
– on regular dependence 75
– on separation 162
– mean-value 247

– Philips 188
– Riesz 2
– Routh 34

Time-optimal problem 3, 164
Transfer function 51

Uncontrollable part 24

Value function 128, 130
Vector field of class C^k 81

Weak convergence 165
– limit 165

Systems & Control: Foundations & Applications

Series Editor
Christopher I. Byrnes
School of Engineering and Applied Science
Washington University
Campus P.O. 1040
One Brookings Drive
St. Louis, MO 63130-4899
U.S.A.

Systems & Control: Foundations & Applications publishes research monographs and advanced graduate texts dealing with areas of current research in all areas of systems and control theory and its applications to a wide variety of scientific disciplines.

We encourage the preparation of manuscripts in TeX, preferably in Plain or AMS TeX— LaTeX is also acceptable—for delivery as camera-ready hard copy which leads to rapid publication, or on a diskette that can interface with laser printers or typesetters.

Proposals should be sent directly to the editor or to: Birkhäuser Boston, 675 Massachusetts Avenue, Cambridge, MA 02139, U.S.A.

Estimation Techniques for Distributed Parameter Systems
H.T. Banks and K. Kunisch

Set-Valued Analysis
Jean-Pierre Aubin and Hélène Frankowska

Weak Convergence Methods and Singularly Perturbed
Stochastic Control and Filtering Problems
Harold J. Kushner

Methods of Algebraic Geometry in Control Theory: Part I
Scalar Linear Systems and Affine Algebraic Geometry
Peter Falb

H^∞ - Optimal Control and Related Minimax Design Problems
Tamer Başar and Pierre Bernhard

Identification and Stochastic Adaptive Control
Han-Fu Chen and Lei Guo

Viability Theory
Jean-Pierre Aubin

Representation and Control of Infinite Dimensional Systems, Vol. I
A. Bensoussan, G. Da Prato, M.C. Delfour and S.K. Mitter